Lecture Notes in Computer Science 15904

Founding Editors

Gerhard Goos
Juris Hartmanis

AF147710

The series Lecture Notes in Computer Science (LNCS), including its subseries Lecture Notes in Artificial Intelligence (LNAI) and Lecture Notes in Bioinformatics (LNBI), has established itself as a medium for the publication of new developments in computer science and information technology research, teaching, and education.

LNCS enjoys close cooperation with the computer science R & D community, the series counts many renowned academics among its volume editors and paper authors, and collaborates with prestigious societies. Its mission is to serve this international community by providing an invaluable service, mainly focused on the publication of conference and workshop proceedings and postproceedings. LNCS commenced publication in 1973.

Michael H. Lees · Wentong Cai ·
Siew Ann Cheong · Yi Su · David Abramson ·
Jack J. Dongarra · Peter M. A. Sloot
Editors

Computational Science – ICCS 2025

25th International Conference
Singapore, Singapore, July 7–9, 2025
Proceedings, Part II

 Springer

Editors
Michael H. Lees [ID]
University of Amsterdam
Amsterdam, The Netherlands

Siew Ann Cheong [ID]
Nanyang Technological University
Singapore, Singapore

David Abramson [ID]
The University of Queensland
Brisbane, QLD, Australia

Peter M. A. Sloot [ID]
University of Amsterdam
Amsterdam, The Netherlands

Wentong Cai [ID]
Nanyang Technological University
Singapore, Singapore

Yi Su
Institute for High Performance Computing
A*STAR
Singapore, Singapore

Jack J. Dongarra [ID]
The University of Tennessee
Knoxville, TN, USA

ISSN 0302-9743   ISSN 1611-3349 (electronic)
Lecture Notes in Computer Science
ISBN 978-3-031-97628-5   ISBN 978-3-031-97629-2 (eBook)
https://doi.org/10.1007/978-3-031-97629-2

Preface

Welcome to the 25th International Conference on Computational Science (ICCS - https://www.iccs-meeting.org/iccs2025/), held on July 7–9, 2025 at Nanyang Technological University (NTU), Singapore.

This 25th edition in Singapore marked our return to a fully in-person event. Although the challenges of our present times are manifold, we have always tried our best to keep the ICCS community as dynamic, creative, and productive as possible. We are proud to present the proceedings you are reading as a result.

ICCS 2025 was jointly organized by Nanyang Technological University, the A*STAR Institute of High Performance Computing, the University of Amsterdam, and the University of Tennessee.

Considered one of the most developed countries in the world, the island country of Singapore is a major aviation, financial, and maritime shipping hub in Asia. Singapore is multilingual, multiethnic, and multicultural, and as such a very popular, safe tourism destination.

NTU Singapore is a public university ranked among the world's best, with 35,000 students, and home to the world-renowned autonomous National Institute of Education and S. Rajaratnam School of International Studies. In addition to many research institutes and centers at the university, college, and school levels, NTU also hosts two National Research Foundation (NRF) and Ministry of Education (MOE) Research Centers of Excellence, namely the Singapore Center for Environmental Life Sciences Engineering (SCELSE) and the Institute for Digital Molecular Analytics & Science (IDMxS), and 11 Corporate Labs in partnership with various industries. ICCS 2025 took place on the One-north campus.

The Institute of High Performance Computing (IHPC) is a national research institute under the Agency for Science, Technology and Research (A*STAR), dedicated to advancing science and technology through computational modeling, simulation, AI, and high-performance computing. With a multidisciplinary team of scientists and engineers, IHPC drives innovation across sectors such as advanced manufacturing, microelectronics, sustainability, maritime, and biomedical sciences. It leads Singapore's national efforts in hybrid quantum-classical computing and digital twin platforms, and partners extensively with industry and government agencies to translate deep tech into real-world impact.

The International Conference on Computational Science is an annual conference that brings together researchers and scientists from mathematics and computer science as basic computing disciplines, as well as researchers from various application areas who are pioneering computational methods in sciences such as physics, chemistry, life sciences, engineering, arts, and humanitarian fields, to discuss problems and solutions in the area, identify new issues, and shape future directions for research.

The ICCS proceedings series has become a primary intellectual resource for computational science researchers, defining and advancing the state of the art in this field.

We are proud to note that this 25th edition, with 23 workshops (the Workshops on Computational Science), one co-located event (the Asian Network of Complexity Scientists Workshop), and over 300 participants, kept to the tradition and high standards of previous editions.

The theme for 2025, "**Making Complex Systems tractable through Computational Science**", highlighted the role of Computational Science in tackling the complex problems of today and tomorrow. This conference was a unique event, focusing on recent developments in scalable scientific algorithms; advanced software tools; computational grids; advanced numerical methods; and novel application areas. These innovative novel models, algorithms, and tools drive new science through efficient application in physical systems, computational and systems biology, environmental systems, finance, and others.

ICCS is well known for its lineup of keynote speakers. The keynotes for 2025 were:

- **Johan Bollen**, Indiana University Bloomington, USA
- **Jack Dongarra**, University of Tennessee, USA
- **Mile Gu**, Nanyang Technological University, Singapore
- **Erika Fille Legara**, Center for AI Research|Asian Institute of Management, Philippines
- **Yong-Wei Zhang**, Institute of High Performance Computing, A*STAR, Singapore

This year, the main track of ICCS registered 162 submissions, of which 64 were accepted as full papers, and 52 as short papers. There were on average 2.4 single-blind reviews per submission.

We would like to thank all committee members from the main track and workshops for their contribution to ensuring a high standard for the accepted papers. We would also like to thank *Springer, Elsevier,* and *Intellegibilis* for their support. Finally, we appreciate all the local organizing committee members for their hard work in preparing this conference.

We hope you enjoyed the conference and the beautiful country of Singapore.

July 2025

Michael H. Lees
Wentong Cai
Siew Ann Cheong
Yi Su
David Abramson
Jack J. Dongarra
Peter M. A. Sloot

Organization

Program Committee Chairs

Peter M. A. Sloot   University of Amsterdam, The Netherlands
Jack J. Dongarra   University of Tennessee, USA
Michael H. Lees   University of Amsterdam, The Netherlands
David Abramson   University of Queensland, Australia
Wentong Cai   Nanyang Technological University, Singapore
Cheong Siew Ann   Nanyang Technological University, Singapore
Su Yi   Institute for High Performance Computing,
A*Star, Singapore

Local Program Committee at NTU Singapore

Ee Hou Yong   Nanyang Technological University, Singapore
Kang Hao   Nanyang Technological University, Singapore

Publicity Chairs

Leonardo Franco   University of Málaga, Spain
Muhamad Azfar Ramli   Institute for High Performance Computing,
A*Star, Singapore

Impact Chair

Valeria Krzhizhanovskaya   University of Amsterdam, The Netherlands

Outreach Chair

Alfons Hoekstra   University of Amsterdam, The Netherlands

Program Committee Chair – Workshops on Computational Science

Maciej Paszynski   AGH University of Krakow, Poland

Program Committee – Workshops on Computational Science

Amanda S. Barnard   Australian National University, Australia
Yongjie Jessica Zhang   Carnegie Mellon University, USA

Reviewers

Julen Alvarez-Aramberri   University of the Basque Country, Spain
Philipp Andelfinger   Nanyang Technological University, Singapore
Adrian Bekasiewicz   Gdańsk University of Technology, Poland
Nik Brouw   University of Amsterdam, Netherlands
Roland V. Bumbuc   University of Amsterdam, Netherlands
Wentong Cai   Nanyang Technological University, Singapore
Pedro J. S. Cardoso   Universidade do Algarve, Portugal
Eddy Caron   ENS-Lyon/Inria/LIP, France
Lock-Yue Chew   Nanyang Technological University, Singapore
Ana Cortes   Universitat Autònoma de Barcelona, Spain
Daan Crommelin   CWI Amsterdam, Netherlands
Carlo Cunha   Northern Arizona University, USA
Bartosz Czaplewski   Gdańsk University of Technology, Poland
Venkata Rupesh Kumar Dabbir   Google LLC, USA
Eric Dignum   University of Amsterdam, Netherlands
Vitor Duarte   Universidade NOVA de Lisboa, Portugal
Mariusz Dzwonkowski   Medical University of Gdańsk, Poland
Nahid Emad   Paris-Saclay University, France
Roberto R. Expósito   Universidade da Coruña, CITIC, Spain
Ruy Freitas Reis   Universidade Federal de Juiz de Fora, Brazil
Wlodzimierz Funika   AGH University of Krakow, Poland
Victoria Garibay   University of Amsterdam, Netherlands
Paweł Gepner   Warsaw University of Technology, Poland
Alex Gerbessiotis   New Jersey Institute of Technology, USA
Maziar Ghorbani   Brunel University London, UK
Konstantinos Giannoutakis   University of Macedonia, Greece
Jorge González-Domínguez   Universidade da Coruña, Spain
Yuriy Gorbachev   Soft-Impact LLC, Russia
Michael Gowanlock   Northern Arizona University, USA

George Gravvanis	Democritus University of Thrace, Greece
Derek Groen	Brunel University London, UK
Loïc Guégan	UiT the Arctic University of Norway, France
Rafiazka Hilman	University of Amsterdam, Netherlands
Cillian Hourican	University of Amsterdam, Netherlands
Neil Huynh	Institute of High Performance Computing, A*STAR, Singapore
Alireza Jahani	Brunel University London, UK
Song Jie	Institute of High Performance Computing, A*STAR, Singapore
Zhong Jin	Computer Network Information Center, Chinese Academy of Sciences, China
David Johnson	Uppsala University, Sweden
Takahiro Katagiri	Nagoya University, Japan
Sotiris Kotsiantis	University of Patras, Greece
Sergey Kovalchuk	Huawei, Russia
Valeria Krzhizhanovskaya	University of Amsterdam, Netherlands
Michael Kuhn	Otto von Guericke University Magdeburg, Germany
Jaeyoung Kwak	Nanyang Technological University, Singapore
Michael Lees	University of Amsterdam, Netherlands
Malcolm Low	Singapore Institute of Technology, Singapore
Lukasz Madej	AGH University of Science and Technology, Poland
Tomas Margalef	Universitat Autònoma de Barcelona, Spain
Paula Martins	University of Algarve, Portugal
Pedro Medeiros	Universidade Nova de Lisboa, Portugal
Isaak Mengesha	University of Amsterdam, Netherlands
Marianna Milano	Università Magna Græcia di Catanzaro, Italy
Dhruv Mittal	University of Amsterdam, Netherlands
Francisco J. Moreno-Barea	Universidad de Málaga, Spain
Marcin Paprzycki	IBS PAN and WSM, Poland
Giulia Pederzani	Universiteit van Amsterdam, Netherlands
Alberto Perez de Alba Ortiz	University of Amsterdam, Netherlands
Dana Petcu	West University of Timisoara, Romania
Jolan Philippe	IMT Atlantique, France
Dirk Pleiter	University of Groningen, Netherlands
Alexander Pyayt	EPAM Systems, Russia
Rick Quax	University of Amsterdam, Netherlands
Muhamad Azfar Ramli	Institute of High Performance Computing, A*STAR, Singapore
Amir Raoofy	Technical University of Munich, Germany

Contents – Part II

ICCS 2025 Main Track Full Papers

CommPlex: <u>Comm</u>unity in Multi<u>Plex</u>es - Definition and a Suite of Algorithms for Analysis

Abhishek Santra[1(✉)], Sanjukta Bhowmick[2], and Sharma Chakravarthy[1]

[1] CSE Department and IT Lab, University of Texas at Arlington, Arlington, TX, USA
abhishek.santra@uta.edu, sharmac@cse.uta.edu
[2] Department of Computer Science, University of North Texas, Denton, TX, USA
sanjukta.bhowmick@unt.edu

Abstract. Multiplexes (also termed Multilayer Networks or networks of networks) are useful for modeling data sets with *multiple entity types, and relationships among them.* The notion of a community is well-defined for simple graphs (or a monoplex/network) and is widely used for aggregate analysis on graphs. Several simple graph algorithms (e.g., Infomap, Louvain) for computing a community and algorithms for computing other metrics (e.g., centrality, substructure, etc.) exist as well. Although multilayer networks (MLNs) are used for modeling, the concept of a community and algorithms for its computation are lacking. Ideally, an MLN community definition should be comparable to the simple graph definition and be a generalization. As MLNs have structure in terms of layers, including inter-layer edges, it is important to define a community that includes its structure and semantics. The resulting community should also be an MLN. The focus of this paper is on heterogeneous MLN (or HeMLN), which is a type of MLN with explicitly defined inter-layer edges.

In this paper, we introduce a community definition for HeMLNs that is structure-preserving and is also consistent with the traditional definition. Layer semantics are also preserved for drill-down and visualization. First, we define a community for any k connected layers of a HeMLN (termed *k-community* (1-community is the same as the traditional community on a simple graph or a layer of HeMLN.)) using binary composition. Then, we propose an algorithm for its computation using the concept of bipartite graphs. Further, we show how weight metrics can be customized to include the semantics of participating community characteristics. Our definition: i) leverages extant simple graph community computation algorithms, ii) composes partial results from different layers for computing HeMLN communities (i.e., uses the decoupling approach), iii) is customizable using weight metrics based on participating communities, and iv) is computationally efficient. We have experimentally validated the community concept (definition and computation) on several real-world and synthetic data sets.

Keywords: Community Definition and Detection · Heterogeneous Multilayer Networks · Decoupling Approach · Structure and Semantic Preservation

M. H. Lees et al. (Eds.): ICCS 2025, LNCS 15904, pp. 3–18, 2025.
https://doi.org/10.1007/978-3-031-97629-2_1

1   Motivation

As data sets become more complex in terms of the number and types of entities and relationships, approaches for their modeling and *analysis* also warrant extensions or new alternatives to match the data set complexity. With the advent of social networks and large data sets, we have already seen a surge in the use of graphs for modeling, along with a renewed interest in concepts, such as community, substructures, and centrality (e.g., hubs) being used for analysis.

Informally, Multilayer Networks (or MLNs) are *layers of networks* where each layer is a simple graph capturing the relationship semantics between two entity instances (either of the same or different type) using an edge. Entities from different layers can also be connected. If each MLN layer has a *common subset of entities of a single type*, it is termed a homogeneous MLN (or HoMLN.) For HoMLN, intra-layer edges are shown explicitly and inter-layer edges are considered implicit (and

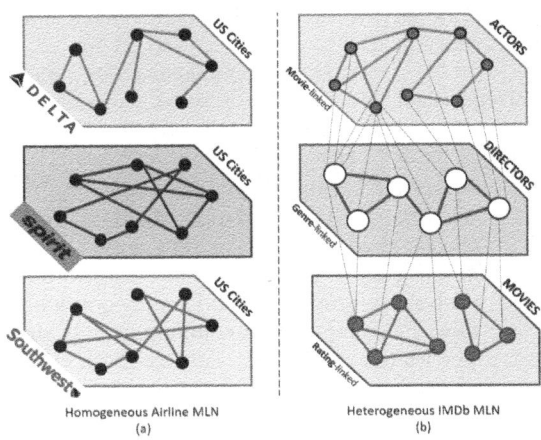

Fig. 1. Homogeneous and Heterogeneous MLNs

hence not shown.) For example, US cities are linked based on a direct flight between them operated by a *specific airline* (Fig. 1 (a)). On the other hand, if the *entity types are different for each layer*, then relationships between entities across layers are shown using explicit inter-layer edges. This distinguishes a heterogeneous MLN (or HeMLN) from the previous one. For example, relationships among actors (connected based on co-acting), directors (connected if they direct movies of similar genres), and movies (related by pre-defined average rating ranges) are modeled through a heterogeneous MLN (Fig. 1 (b)). The inter-layer edges represent the relationship across layers, such

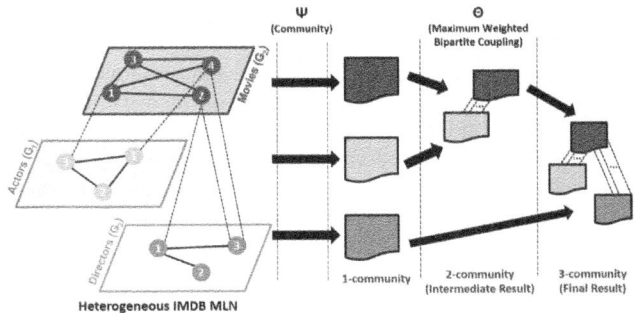

Fig. 2. Decoupling Approach: Compute 3-community $((G_2$ $\Theta_{2,1}$ $G_1)$ $\Theta_{2,3}$ $G_3)$ ω_e

as directs-movie, directs-actor, and acts-in-movie (not illustrated). Our focus, in this paper, is on HeMLNs.

For aggregate computations on MLNs, a novel decoupling approach has been proposed in [18, 22, 23]. Figure 2 shows the **decoupling approach**. Three layers and their inter-layer connections are shown. HeMLN community computation is accomplished by combining communities from two layers of a HeMLN using a binary *composition function* (Θ) and is extended to k layers by composing the result with additional layers one at a time. Figure 2 also shows how a 3-layer HeMLN community is expressed for computation. Composing partial results from individual (or previously computed) layers is central to the efficiency of the approach as elaborated in Sect. 7. This approach also preserves the structure of the MLN and its semantics for drill-down and visualization.

The contributions of this paper are shown below with related work in Sect. 2 and conclusions in Sect. 8.

– Definition and some properties of k-community for a HeMLN (Sect. 3),
– Composition function for k-community computation (Sect. 4),
– A new bipartite match algorithm for composition (Sect. 5),
– Experimental analysis to establish the validity of the proposed approach along with performance analysis (Sects. 6 and 7)

2   Related Work

As this paper focuses on the HeMLN community definition and its efficient detection, we present relevant work on simple graphs and HeMLNs. The advantages of modeling using MLNs are discussed in [4, 14, 15, 20, 23].

Community detection on a simple graph involves identifying groups of vertices that are more connected to each other than to other vertices in the network. Most of the work in the literature considers **single networks or simple graphs** where this objective is translated to optimizing network measures such as modularity [3], conductance [16] or map equation [6]. As the combinatorial optimization of community detection is NP-complete [7], many competitive approximation algorithms and deep learning based methods have been developed (see reviews in [11, 13, 25].) Algorithms for community detection have been developed for different types of input graphs, including directed, edge-weighted, and dynamic networks. However, to the best of our knowledge, there is no community definition for HeMLNs, let alone its computation that preserves structure along with node and edge labels for drill-down (semantics).

The majority of the work on analyzing HeMLN (reviewed in [5, 24, 26]) focuses on developing meta-path based techniques for determining clustering, similarity of objects, classification of objects, predicting the missing links, ranking/co-ranking, and recommendations.

The type-independent [8] and projection-based [2] approaches used for ground truth (GT) for HeMLNs use the existing community definition and *do not preserve the structure or semantics of the community*. Both approaches, in slightly different ways, conflate layers into a simple graph keeping *all* nodes and edges (including inter-layer edges) sans their types and labels. This has been shown to result in information loss [15]. Most of the community detection work in MLNs has focused on homogeneous MLNs, where the common set of nodes is present in each layer ([12,14,21]). However, the presence of different sets of entities in each layer and the presence of intra- and inter-layer edges make the structure-preserving definition more challenging for HeMLNs and also warrants an alternate technique (the decoupling approach.) A few existing works have proposed techniques for generating clusters of entities [10,17], but they have only considered the *inter-layer* links and *not* the networks themselves.

This paper fills the gap by providing a clear **new** formal definition of community for HeMLNs that is *structure- and semantics-preserving*. This definition can be shown to be similar to the traditional modularity definition for communities. A distinct advantage of the definition and the use of the decoupling approach is that it leverages existing community detection algorithms (and several of them are currently available.) Infomap and Louvain are more popular than others. This paper also established the efficiency of the decoupling approach for HeMLN community detection.

3   Community Definition for a HeMLN

3.1   Multilayer Networks: Notations Used in the Paper

We start with a formal multilayer network definition that covers both homogeneous and heterogeneous networks.

Table 1. Notations used in this paper

$G_i(V_i, E_i)$	Simple Graph for layer i
$X_{i,j}(V_i, V_j, L_{i,j})$	Bipartite graph of layers i and j
$MLN(G, X)$	Multilayer Network of layer graphs (set G) and Bipartite graphs (set X)
Ψ	Analysis function for G_i (community)
$\Theta_{i,j}$	**Proposed** Maximum Weighted Bipartite Coupling (**MWBC**) function
$CBG_{i,j}$	Community Bipartite Graph for G_i and G_j
U_i	Meta nodes of layer i 1-community
$L'_{i,j}$	Meta edges between U_i and U_j
c_i^m	m^{th} community of G_i
$v_i^{c_i^m}, e_i^{c_i^m}$	Vertices and Edges in community c_i^m
H_i^m	Hubs in c_i^m
$H_{i,j}^{m,n}$	Hubs in c_i^m connected to c_j^n
$x_{i,j}$	{Expanded (meta edge $< c_i^m, \ c_j^n >$)}
0 and ϕ	null community id and empty $x_{i,j}$
$\omega_e, \omega_d, \omega_h$	Weight metrics for meta edges (see Sect. 5)

Formally, a **multilayer network**, $MLN(G, X)$, is defined by two sets of graphs: (i). The set $G = \{G_1, G_2, \ldots, G_N\}$ contains graphs of N individual layers $L = \{L_1, L_2, \ldots, L_N\}$ each of which is a simple graph, where $G_i(V_i, E_i)$ is defined by a set of vertices, V_i and a set of edges, E_i. An edge $e(v, u) \in E_i$, connects vertices v and u, where $v, u \in V_i$ and (ii). A set $X = \{X_{1,2}, X_{1,3}, \ldots, X_{M-1,M}\}$ of bipartite graphs. Each bipartite graph $X_{i,j}(V_i, V_j, L_{i,j})$ is defined by two sets of vertices V_i and V_j, and a set of edges (also called links or inter-layer edges) $L_{i,j}$, such that for every link $l(a, b) \in L_{i,j}$, $a \in V_i$ and $b \in V_j$, where V_i (V_j) is the vertex set of graph G_i (G_j.) For a HeMLN (the focus of this paper), X is explicitly specified. Without loss of generality, we assume unique numbers for nodes across layers and disjoint sets of nodes across layers.

Our definition of community for a HeMLN uses communities from each layer and inter-layer edge connection strength between communities across layers expressed as a weight. One of the weights (number of inter-layer edges) is compatible with the simple graph definition of a community. In addition, coupling alternatives can be formulated for the same Θ to provide multiple (or a *family* of) community definitions that can be used for different analysis objectives as needed. Finally, the framework is extensible in that it allows one to propose additional parameters (or weights) to customize for a specific data set or a set of analysis objectives. Furthermore, it also preserves the structure and semantics due to composition using the decoupling approach which is also shown to be computationally efficient (see Sect. 7). Table 1 lists notations used in the paper for quick reference.

3.2   Definition of HeMLN Community

A **Community Bipartite Graph** or $CBG_{i,j}(U_i, U_j, L'_{i,j})$ is defined between two disjoint and independent communities U_i and U_j. Each element of U_i (U_j) is a community from G_i (G_j) that is represented as a single meta node. $L'_{i,j}$ is the set of meta edges between the nodes of U_i and U_j (or bipartite graph edges.) For any two meta nodes, a *single edge* is used for $L'_{i,j}$, if there is *at least one inter-layer edge* between any pair of nodes from the corresponding communities (acting as meta nodes in CBG) in layers G_i and G_j. The strength (or weight) component of the meta edges is elaborated in Sect. 5.

For a layer graph, a **1-community** is the same as the traditional communities identified using any of the community detection algorithms. A **HeMLN community for 2 layers** (termed **2-community**) is formally defined using the community bipartite graph $CBG_{i,j}(U_i, U_j, L'_{i,j})$ corresponding to layers G_i and G_j. A 2-community is a set of tuples each with a pair of elements $< c_i^m, c_j^n >$, where $c_i^m \in U_i$ and $c_j^n \in U_j$, that satisfy the *Maximum Weighted Bipartite Coupling (MWBC)* (composition function Θ) for the bipartite graph of U_i and U_j, along with the set of inter-layer edges between them (denoted by $x_{i,j}$.) The idea is to obtain the group(s) of nodes that are tightly coupled within and across layers. The pairing is done from left-to-right (as it is *not commutative*) and a single community from the left layer can pair with *zero or more communities* from the right layer. That is, *one-to-many or many-to-one pairings* are possible, unlike traditional bipartite matching.

A **HeMLN community of k connected layers**, termed **k- community** is defined as the application of *2-community* definition recursively to compute k-community. The 2-community definition can be applied to t1-community and t2-community to generate a (t1+t2)-community. The base case corresponds to applying the definition of 2-community for 2 layers t1 and t2. This applies to any expression with precedence. For sinplicity, we discuss the left-to-right computation of k-community.

For a left-to-right computation, the base case is applied to the first 2 layers. For each recursive step, there are two cases for the 2-community under consideration:

i) the U_i is from a layer G_i *already in the t-community* and the U_j is from a *new layer G_j*. This bipartite graph match is said to **extend** a t-community (t < k) to a *(t+1)-community*, or ii) **both** U_i and U_j from layers G_i and G_j, respectively are *already in the t-community*. This bipartite graph match is said to **update** a t-community (t < k), **not extend** it.

For both cases i) and ii) above, two outcomes are possible. A meta node from U_i either, a) matches one or more meta nodes in U_j resulting in one (or many) **consistent match**, or b) does not match a meta node in U_j resulting in a **no match**. However, for case ii) a third possibility exists which can be characterized as c) matches a node in U_j that is not consistent with a previous match, termed **inconsistent match**. Since both communities have already been matched, a previous match exists (either consistent or no match). If the current match is not the same, then it is an **inconsistent** match.

Table 2. Cases and outcomes for MWBC (Algo. 1)

(G_{left}, G_{right}) outcome	Effect on tuple t
case (i) – one processed and one new layer	
a) consistent match	Copy & Extend t with paired community id and $x_{i,j}$
b) no match	Copy & Extend t with 0 and ϕ
case (ii) – both are processed layers	
a) consistent match	Copy & Update t only with x
b) no match	Copy & Update t only with ϕ
c) inconsistent match	Copy & Update t only with ϕ

Structure preservation is accomplished by retaining, for each tuple of t-community, either a matching community id (or 0 if no match) and $x_{i,j}$ (or ϕ for the empty set) representing inter-layer edges corresponding to the meta edge between the meta nodes (termed **expanded(meta edge).**) The *extend* and *update* carried out for each of the outcomes on the representation is listed in Table 2. Note that due to multiple pairing of nodes during any composition, the number of tuples (or t-communities) may increase. Copying is used to deal with multiple pairings. In general, each element of a k-community can be total or partial. A **partial k-community element** has *at least one* ϕ or 0 as part of the tuple. Otherwise, it is a **total k-community element**. Any k-community that is **total** reflects a stronger coupling as it includes all inter-layer edges for those communities (as is the case of M-A-D-M in Fig. 6 (b) in Sect. 7.)

3.3 Characteristics of k-Community

To clearly understand, a HeMLN can be viewed as a simple graph (termed HeMLN-graph) with each *HeMLN layer being a node* and the inter-layer edges between layers denoted by a weighted edge between corresponding nodes. Then, a k-community can be specified over any connected subgraph of this HeMLN-graph. Case i) above corresponds to a k-community of *an acyclic* subgraph, and case ii) to a k-community of a *cyclic* subgraph of the HeMLN-graph. For both, the number of compositions will be determined by the number of edges in the connected subgraph and can be more than the number of layers (or nodes). Also, for both cases, MWBC results in one of the 3 outcomes: a *consistent match, no match*, or an *inconsistent match (only for case (ii))* as shown in Table 2.

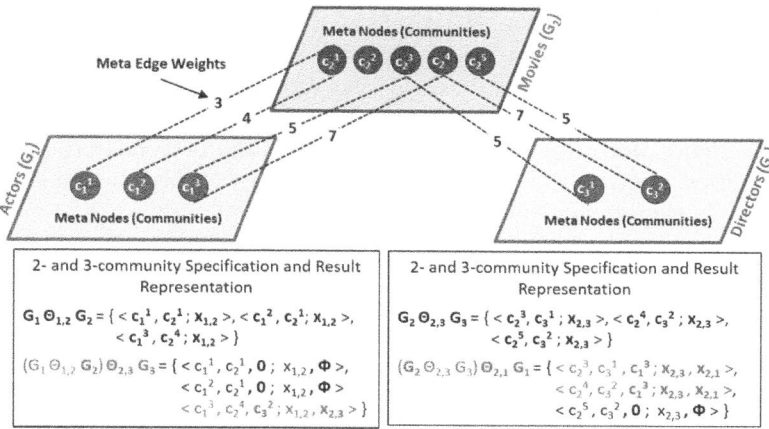

Fig. 3. Illustration of order dependence on a k-community

The above definition when applied to a specification (such as the one shown in Fig. 3 generates *progressively strong coupling of communities between layers* (due to left-to-right precedence of Θ) using MWBC. *Thus, our definition of a k-community is characterized by dense connectivity within the layer (community definition) and strong coupling across layers (comparable to community definition captured by MWBC semantics.)* Hence, we believe, that this definition of k-community matches or comes close to the original modularity intuition [19] of a community[1] for a simple graph (see Table 3). By refining the edge weight using participating community characteristics, a family of community definitions is supported that can be customized.

For the evaluation purpose, we used the IMDb (layer details are shown in Table 4 of Sect. 7) and DBLP HeMLNs. For IMDb, we have used the Actor and Director layers with their inter-layer edges. For DBLP, we have used the Author and Paper layers with their inter-layer edges. For compo-

Table 3. Community modularity comparison: Type-independent vs. Proposed definitions

HeMLN	Type-Independent	Decoupling
IMDb	0.77696	0.643508
DBLP	0.694208	0.694208

sition, we have used the metric ω_e that takes into account the number of inter-layer edges between the layer-wise communities while performing the match-

---

[1] Modularity is a measure of the structure of a network or a graph which measures the strength of division of a network into modules (also called groups, clusters or communities). Networks with high modularity have dense connections between the nodes within modules but sparse connections between nodes in different modules. Similarly, in our definition, we pair the communities between two layers based on the inter-layer edge strength connected to that pair (of all pairs), and hence the pairs produced have dense connections within and across two layers. Here the modularity of the HeMLN is based on the dense coupling between the dense communities of layers as compared to all possible inter-layer couplings.

ing. This metric is closest to the traditional definition as type-independent aggregation does not consider any other layer-wise community characteristics. Table 3 shows the modularity values[2] for the *decoupling approach and the type-independent approach*. As can be seen they are identical for DBLP and very close for IMDb. Hence, our community definition, apart from preserving the structure and semantics, generates communities whose quality, based on modularity *is comparable to* the type-independent communities.

Need for a New Pairing Algorithm. In a traditional bipartite graph (used for dating, hiring, etc.), each node is atomic. The goal is to find the maximum number of matches (bipartite edges) such that no two matches share the same node. Hence, a node from one set is paired with *at most* one node from the other set. To accommodate multiple edges, weights are needed without changing the pairing semantics [9].

In contrast, each node of our bipartite graph is a meta node (non-atomic and corresponds to a community) and the bipartite edge is also a meta edge (set of edges between two communities). Each meta node, in our case, is a community representing a group of entities (layer nodes) with additional characteristics. Each meta edge needs to, at the least, capture the number of edges in that meta edge (i.e., inter-layer edges.) The number of edges between the meta nodes is one of the edge weights (ω_e) proposed, which corresponds to the traditional intuition behind a community.

Since edge weights play a significant role in the matching and are also used as a mechanism for determining the strength of the coupling of communities across layers, edge weights can be leveraged to include participating community characteristics. In addition to ω_e, it is possible to bring in participating community characteristics to capture additional aspects of coupling. This can be done by defining different edge weights to capture different characteristics of the participating communities. We have used hub participation from communities and the density of participating communities as weight alternatives.

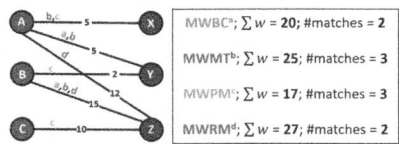

Fig. 4. Illustration of Traditional and Relaxed Pairings on a weighted bipartite graph

For pairing nodes of the bipartite graph, since traditional approaches are not suited for our coupling, we propose an edge weight-based coupling that reflects the semantics of the community. Each node from the first set is paired with *zero or more nodes* from the second set solely based on the outgoing edge weights of that node. This is repeated for each node from the first set. *Most importantly, unlike current alternatives in the MLN community literature, there is no information loss or distortion or hiding the effect of different entity types or relationships in our definition.*

---

[2] A modularity value greater than 0.5 is considered acceptable. Modularity value close to 1 indicates strong community structure, whereas a value close to 0 indicates that the community structure is not better than random.

Figure 4 provides an example of a bipartite graph to illustrate multiple types of pairings appropriate for MLN communities. MWM (Maximum Weight Matching); MWMT (MWM with Ties); MWPM (Maximum Weight Perfect Match); MWRM (Maximum Weight with Relaxed Matching).

4   HeMLN k-Community Computation

Algorithm 1 accepts a linearized specification of a k-community and computes the result as described earlier. The output is a *set* whose *elements are tuples corresponding to distinct, single HeMLN k-community* for that specification. Figure 3 shows 2- and 3-community example results computed using this algorithm.

Algorithm 1. HeMLN k-community Detection Algorithm

Require: -
    INPUT: HeMLN, $(G_{n1} \; \Theta_{n1,n2} \; G_{n2} \; ... \; \Theta_{ni,nk} \; G_{nk})$, and a weight metric ($\omega$).
    OUTPUT: Set of distinct k-community tuples
1: **Initialize:** k=2, $U_i = \phi$, $U_j = \phi$, $result' = \emptyset$
    $result \leftarrow \text{MWBC}(G_{n1}, G_{n2}, \text{HeMLN}, \omega)$
    left, right $\leftarrow$ left and right subscripts of second Θ
2: **while** *left* $\neq$ null && *right* $\neq$ null **do**
3:     $U_i \leftarrow$ subset of 1-community$(G_{left}, result)$
4:     $U_j \leftarrow$ subset of 1-community$(G_{right}, result)$
5:     $MP \leftarrow \text{MWBC}(U_i, U_j, \text{HeMLN}, \omega)$ //a set of comm pairs $< c^p_{left}, c^q_{right} >$
6:     **for** each tuple $t \in result$ **do**
7:         kflag = false
8:         **if** *both* c^x_{left} *and* c^y_{right} are part of t and $\in$ MP [case ii (processed layer): consistent match] **then**
9:           Update *a copy of* t with $(x_{left, right})$ and append to result$'$
10:        **else if** c^x_{left} is part of t and $\in$ MP and G_{right} layer has been processed [case ii (processed layer): no and inconsistent match] **then**
11:           Update *a copy of* t with ϕ and append to result$'$
12:        **else if** c^x_{left} is part of t and for each $c^x_{left} \in MP$ [case i (new layer): consistent match] **then**
13:           copy and Extend t with paired $c^y_{right} \in$ MP and $x_{left, right}$ and append to result$'$; kflag = true
14:        **else if** c^x_{left} is part of t and $\notin$ MP [case i (new layer): no match] **then**
15:           copy and Extend t with 0 (community id) and ϕ and append to result$'$; kflag = true
16:        **end if**
17:     **end for**
    left, right = next left, right subscripts of Θ or null
    if kflag k = k + 1; result = result$'$; result$'$ = $\emptyset$
18: **end while**

The bipartite graph for the base case and each iteration is constructed for the participating layers (either one is new or both are from the t-community for some t) and MWBC algorithm is applied. The result obtained is used to either extend or update the tuples of the t-community and add new tuples as well. All cases are described in Table 2.

The algorithm iterates (**lines 2 to 18**) until there are no more compositions to be applied. Line 5 computes the first 2-community. **Lines 6 to 17** apply the results of the MWBC (**line 5**) to generate tuples of the k-community using the

Table 2. Care is taken in the composition to make sure either the tuple is updated or extended by keeping a flag and checking it after **line 17**. The order of checking inside the for loop (**lines 6 to 17**) is important to generate the correct k-community tuples.

5 Customizing the Bipartite Graph

Without including the characteristics of meta nodes and edges for the match, we cannot argue that the pairing obtained represents analysis based on participating community characteristics. Hence, it is important to identify how qualitative community characteristics can be mapped quantitatively to a weight metric (that is, the weight of the meta edge in the community bipartite graph) to influence the bipartite matching. Out of the three developed weight metrics based on (number of inter-community edges (ω_e), density (ω_d), and hub participation (ω_h)), we detail only one weight metric due to space constraints below.

Hub Participation (ω_h): For many analyses, we are interested in knowing whether highly influential nodes within a community also interact across the community. This can be translated to the *participation of influential nodes within and across each participating community* for analysis. This is modeled by using the notion of **hub[3] participation** within a community and their interaction across layers. In this paper, we have used degree centrality for this metric to connote higher influence. The ratio of participating hubs from each community and the edge fraction is multiplied to compute ω_h. Formally,

For every $(u_i^m, u_k^n) \in L'_{i,k}$, where u_i^m and u_k^n are the meta nodes denoting the communities, c_i^m and c_k^n in the CBG, respectively, the weight,

$$\omega_h(u_i^m, u_k^n) = \frac{|H_{i,k}^{m,n}|}{|H_i^m|} * \frac{|x_{i,k}|}{|v_i^{c^m}| * |v_k^{c^n}|} * \frac{|H_{k,i}^{n,m}|}{|H_k^n|},$$

where, $x_{i,k} = \{(a,b) : a \in v_i^{c^m}, b \in v_k^{c^n}, \text{ and } (a,b) \in L_{i,j}\}$; H_i^m and H_k^n are set of hubs in c_i^m and c_k^n, respectively; $H_{i,k}^{m,n}$ is the set of hubs from c_i^m that are connected to c_k^n; $H_{k,i}^{n,m}$ is the set of hubs from c_k^n that are connected to c_i^m.

6 Expressing Analysis Objectives on HeMLNs

We demonstrate how analysis objectives can be expressed as k-community computations on HeMLNs. Also, depending on the analysis, appropriate weight metrics can be chosen. Due to space constraints, we are not discussing the results for other real-world (like DBLP) and synthetic data sets.

---

[3] High centrality nodes (or hubs) have been defined based on different metrics, such as degree centrality (vertex degree), closeness centrality (mean distance of the vertex from other vertices), betweenness centrality (fraction of shortest paths passing through the vertex), and eigenvector centrality.

IMDb Data Set [1]: The IMDb data set captures movies, TV episodes, actors, directors, and other related information, such as rating. Some IMDb detailed analysis objectives are listed below,

(A1) Based on the similarity of genres, for each director group which are the actor groups whose *majority of the most versatile members interact?*
2-community: D $\Theta_{A,D}$ A; ω_h

(A2) For the *most popular* actor groups, for each movie rating class, find the director groups with which they have *maximum interaction* and who also direct movies with similar ratings.
Cyclic 3-community: M $\Theta_{M,A}$ A $\Theta_{A,D}$ D $\Theta_{D,M}$ M; ω_e

To address the IMDb analysis requirements, three layers for the IMDb data set are generated. Layer A and Layer D connect actors and directors who act in or direct *similar genres frequently* (intra-layer edges), respectively. Layer M connects movies within the same rating range. The inter-layer edges depict *acts-in-a-movie* $(L_{A,M})$, *directs-movie* $(L_{D,M})$ and *directs-actor* $(L_{A,D})$ relationships. There are multiple ways of quantifying the similarity of actors (directors) based on the movie genres they have worked in. A vector was generated with the number of movies for each genre he/she has acted in (directed). To take into account the *frequency of genres* in the similarity calculation, two actors (directors) are connected if the Pearsons' Correlation between their corresponding genre vectors is ≥ 0.9[4]. Moreover, 10 movie rating ranges are considered - [0–1), [1–2), ..., [9–10].

For a specific analysis, the characteristics of the communities connected in the bipartite graph need to be used as meta edge weight to get the desired coupling.

For example, *most popular* in (A2) is interpreted as the higher number of edges between the participating communities. In contrast, *versatility* is mapped to the participation of hub nodes in each group as in (A1).

To compute a k-community, k needs to be identified. (A1) requires 2-community. analysis:IMdbHespsmadm requires a cyclic 3-community using inter-layer relationships between all layers in a particular order. Note that the analysis objectives have been chosen carefully to cover the weights discussed in the paper. The limitation on the number of analysis objectives is purely due to space constraints.

7   HeMLN Community Analysis on Real-World Data Sets

We would like to point out that the choice of data sets and sizes was mainly to demonstrate the versatility of analysis using the k-community detection and its efficiency as well as drill-down capability based on structure- and -semantics

---

[4] The choice of the coefficient is not arbitrary as it reflects relationship quality. The choice of this value can be based on how actors (directors) are weighted against the genres. We have chosen 0.9 for connecting actors (directors) in their top genres.

preservation. We are not trying to demonstrate scalability in this paper. Also, instead of presenting communities, we present a few important drill-down results to showcase the structure and semantics preservation, and the general applicability of our approach.

7.1   Experimental Setup and Data Sets

Due to the lack of real-world HeMLNs, we generated HeMLNs from data collected/crawled from some well-known real-world domains. For IMDb HeMLN, we extracted, for the top 500 actors, the movies they have worked in (7500+ movies with 4500+ directors). The actor set was repopulated with the co-actors from these movies, giving a total of 9000+ actors. For this set of actors, directors, and movies, the HeMLN with 3 layers described in Sect. 6 was built. Widely used Louvain method [3] was used to detect layer-wise 1-communities[5]. The k-community detection algorithm 1 was implemented in Python version 3.6 and was executed on a quad-core 8th generation Intel i5 processor Windows 10 machine with 8 GB RAM.

Table 4. IMDB HeMLN Statistics

	Actor	Director	Movie
#Nodes	9,485	4,510	7,951
#Edges	996,527	250,845	8,777,618
#Communities (Size > 1)	63	61	9
Average Community Size	148.5	73	883.4

	#Inter-layer Edges
Actor-Director	32,033
Actor-Movie	31,422
Director-Movie	8,581

Individual Layer Statistics: Table 4 shows the IMDb HeMLN statistics. 63 Actor (A) and 61 Director (D) communities based on similar genres are generated. Out of the 10 ranges (communities) in the movie (M) layer, most of the movies were rated in the range [6–7], while the least popular rating was [1–2]. No movie had a rating in the range [0–1].

7.2   Analysis Results and Discussion

(A1)
Analysis Results: 34 D-A (Director-Actor) similar genre-based community pairs were obtained, where *the majority of the most versatile members interact.* Intuitively, a group of directors that prominently makes movies in some genre

---

[5] Louvain numbers all communities from 1 and we only consider communities having *at least two members* for this paper. The numbering used in the paper has the layer name followed by the Louvain-generated community ID (e.g. A91).

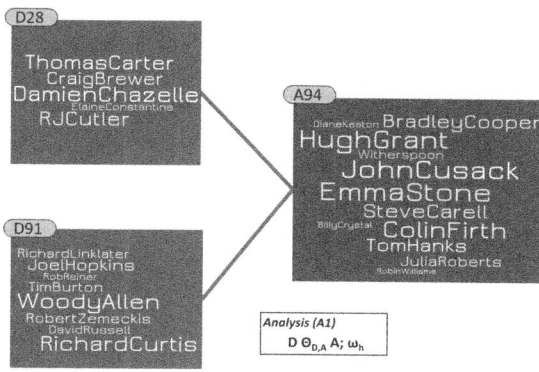

Fig. 5. Sample (A1) Result for *Romance, Comedy, Drama*

(say, Drama, Action, Romance, ...) must pair up with the group(s) of actors who primarily act in similar kinds of movies. Moreover, a *director group may work with multiple actor groups and vice-versa*. For example, in Fig. 5, the sample result shows that the director groups, D28 and D91, with academy award winners like **Damien Chazelle and Woody Allen, respectively, pair up with the actor group with members like Diane Keaton, Emma Stone, and Hugh Grant**. Members from these groups are primarily known for movies from the **Romance, Comedy, and Drama** genre.

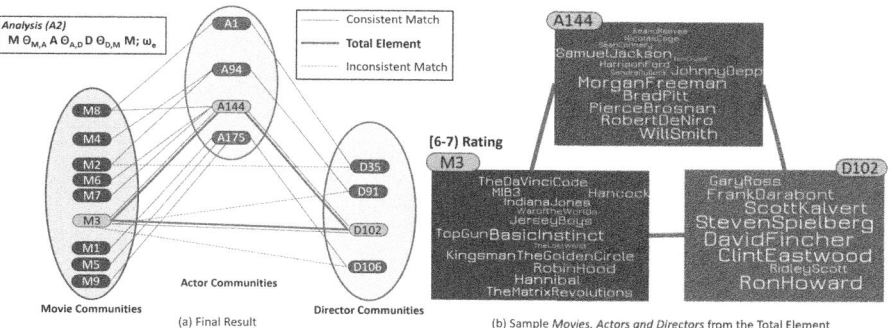

Fig. 6. (A2) Result: **1 Total, 9 Partial Elements**. (Color figure online)

(A2) **Analysis Results:** Here, the *most popular actor groups for each movie rating class are further coupled with directors*. These *director groups are coupled again with movies to check whether the director groups also have similar ratings*. Results of each successive pairing (there are 3) are shown in Fig. 6 (a) using the same color notation. The coupling of movie and actor communities (first composition) results in 10 consistent matches. When the base case is extended to the director layer (second composition) using all director communities and the matched 4 actor communities, we get 4 consistent matches. The final composition to complete the cycle uses 4 director communities and 9 movie communities as left and right sets of community bipartite graph, respectively. **Only one consistent match is obtained to generate the total element (M3-A144-D102-M3) for the cyclic 3-community** (bold blue triangle.) The resulting total element is from the **Action, Drama genre** as can be seen from the sample members shown in Fig. 6 (b). It is interesting to see 3 inconsistent matches (red broken lines) between the communities, which clearly indicate that all couplings are not satisfied by these pairs. These result in 9 partial elements. **The inconsistent matches also highlight the importance of mapping an analysis objective to a k-community specification for computation.** If a different order had been chosen (viz., director and actor layer as the base case), the result could have included the inconsistent matches.

7.3   Efficiency of Decoupling Approach

The goal of the decoupling approach was to preserve the structure as well as improve the efficiency of k-community detection. We illustrate that with the largest k-community we have computed which uses 3 iterations (including the base case.) Fig. 7 shows the execution time for the one-time and iterative costs discussed earlier for (A2). The difference in one-time 1-community cost for the 3 layers follows their density shown in Table 4. We can also see how the iterative cost is insignificant as compared to the one-time cost (by an order of magnitude.) Iteration cost includes creating the bipartite graph, computing ω_e for meta edges, and MWBC cost. **The cost of all iterations together (0.515 sec) is still almost** *an order of magnitude less than the largest one-time cost* **(5.21 sec for Movie layer.)** We have used this case as this subsumes all other cases. The **additional incremental cost for computing a k-community is extremely small validating the efficiency of decoupled approach.**

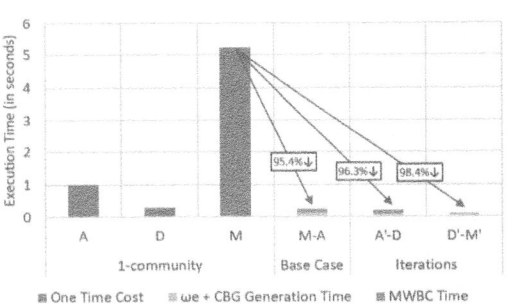

Fig. 7. Performance Results for cyclic 3-community in (A2)

8   Conclusions

In this paper, we have provided a community definition for HeMLNs that is consistent with the traditional definition and is structure preserving. This definition can be applied to an arbitrary number of layers of a HeMLN. *In fact, with ω as a customizable parameter, this supports a family of definitions that are customizable to analysis needs.* We proposed a new bipartite match-based composition function (MWBC algorithm) for the decoupling approach. We have compared our results with the traditional ground truth using modularity to show their compatibility. Finally, we used the proposed approach to demonstrate its analysis versatility using the IMDb data set. In the future, we plan to extend this work to weighted MLNs.

Acknowledgment:. This work was supported by NSF awards CCF-1955798, and CCF-1956373.

References

1. The internet movie database. ftp://ftp.fu-berlin.de/pub/misc/movies/database/
2. Berenstein, A., Magarinos, M.P., Chernomoretz, A., Aguero, F.: A multilayer network approach for guiding drug repositioning in neglected diseases. PLOS (2016)
3. Blondel, V.D., Guillaume, J., Lambiotte, R., Lefebvre, E.: Fast unfolding of community hierarchies in large networks. CoRR arXiv:abs/0803.0476 (2008)
4. Boccaletti, S., et al.: The structure and dynamics of multilayer networks. Phys. Rep. **544**(1), 1–122 (2014)
5. Boden, B., Günnemann, S., Hoffmann, H., Seidl, T.: Mining coherent subgraphs in multi-layer graphs with edge labels. In: Proceedings of the 18th ACM SIGKDD International Conference on Knowledge Discovery and Data Mining, pp. 1258–1266 (2012)
6. Bohlin, L., Edler, D., Lancichinei, A., Rosvall, M.: Community detection and visualization of networks with the map equation framework (2014). http://www.mapequation.org/assets/publications/mapequationtutorial.pdf
7. Brandes, U., Gaertler, M., Wagner, D.: Experiments on graph clustering algorithms. In: Di Battista, G., Zwick, U. (eds.) ESA 2003. LNCS, vol. 2832, pp. 568–579. Springer, Heidelberg (2003). https://doi.org/10.1007/978-3-540-39658-1_52
8. Domenico, M.D., Nicosia, V., Arenas, A., Latora, V.: Layer aggregation and reducibility of multilayer interconnected networks. CoRR arXiv:abs/1405.0425 (2014)
9. Edmonds, J.: Maximum matching and a polyhedron with 0, 1-vertices. J. Res. Natl. Bureau Stand. B **69**(125–130), 55–56 (1965)
10. Fang, Y., Yang, Y., Zhang, W., Lin, X., Cao, X.: Effective and efficient community search over large heterogeneous information networks. Proc. VLDB Endow. **13**(6), 854–867 (2020)
11. Fortunato, S., Castellano, C.: Community structure in graphs. In: Encyclopedia of Complexity and Systems Science, pp. 1141–1163 (2009)
12. Huang, X., Chen, D., Ren, T., Wang, D.: A survey of community detection methods in multilayer networks. Data Min. Knowl. Disc. **35**, 1–45 (2021)
13. Jin, D., et al.: A survey of community detection approaches: from statistical modeling to deep learning. IEEE Trans. Knowl. Data Eng. **35**(2), 1149–1170 (2021)
14. Kim, J., Lee, J.: Community detection in multi-layer graphs: a survey. SIGMOD Rec. **44**(3), 37–48 (2015)
15. Kivelä, M., Arenas, A., Barthelemy, M., Gleeson, J.P., Moreno, Y., Porter, M.A.: Multilayer networks. CoRR arXiv:abs/1309.7233 (2013)
16. Leskovec, J., Lang, K.J., Dasgupta, A., Mahoney, M.W.: Community structure in large n/ws: natural cluster sizes and absence of large well-defined clusters (2008)
17. Magnani, M., Hanteer, O., Interdonato, R., Rossi, L., Tagarelli, A.: Community detection in multiplex networks. ACM Comput. Surv. (CSUR) **54**(3), 1–35 (2021)
18. Mukunda, K., Roy, A., Santra, A., Chakravarthy, S.: Stress centrality in heterogeneous multilayer networks: heuristics-based detection. In: IEEE 9th International Conference on Big Data Computing Service and Applications, pp. 103–110 (2023)

19. Newman, M.E.: Modularity and community structure in networks. Proc. Natl. Acad. Sci. **103**(23), 8577–8582 (2006)
20. Santra, A., Bhowmick, S.: Holistic analysis of multi-source, multi-feature data: modeling and computation challenges. In: Big Data Analytics - Fifth International Conference, BDA 2017 (2017)
21. Santra, A., Bhowmick, S., Chakravarthy, S.: Efficient community re-creation in multilayer networks using Boolean operations. In: International Conference on Computational Science, ICCS 2017 (2017)
22. Santra, A., Bhowmick, S., Chakravarthy, S.: Hubify: efficient estimation of central entities across multiplex layer compositions. In: IEEE International Conference on Data Mining Workshops, ICDM Workshops 2017 (2017)
23. Santra, A., Irany, F.A., Madduri, K., Chakravarthy, S., Bhowmick, S.: Efficient community detection in multilayer networks using boolean compositions. Front. Big Data **6** (2023)
24. Shi, C., Li, Y., Zhang, J., Sun, Y., Philip, S.Y.: A survey of heterogeneous information network analysis. IEEE Trans. Knowl. Data Eng. **29**(1), 17–37 (2017)
25. Su, X., et al.: A comprehensive survey on community detection with deep learning. IEEE Trans. Neural Netw. Learn. Syst. (2022)
26. Sun, Y., Han, J.: Mining heterogeneous information networks: a structural analysis approach. ACM SIGKDD Explor. Newsl. **14**(2), 20–28 (2013)

i-QLS: Quantum-Supported Algorithm for Least Squares Optimization in Non-linear Regression

Supreeth Mysore Venkatesh[1,2]([✉])[iD], Antonio Macaluso[2]([✉])[iD],
Diego Arenas[3][iD], Matthias Klusch[2][iD], and Andreas Dengel[1,3][iD]

[1] University of Kaiserslautern-Landau (RPTU), Kaiserslautern, Germany
`supreeth.mysore@rptu.de`
[2] German Research Center for Artificial Intelligence (DFKI), Saarbruecken, Germany
`{supreeth.mysore,antonio.macaluso,matthias.klusch}@dfki.de`
[3] German Research Center for Artificial Intelligence (DFKI), Kaiserslautern,
Germany
`{diego.arenas,andreas.dengel}@dfki.de`

Abstract. We propose an iterative quantum-assisted least squares (i-QLS) optimization method that leverages quantum annealing to overcome the scalability and precision limitations of prior quantum least squares approaches. Unlike traditional QUBO-based formulations, which suffer from an exponential qubit overhead due to fixed discretization, our approach refines the solution space iteratively, enabling exponential convergence while maintaining a constant qubit requirement per iteration. This iterative refinement transforms the problem into an anytime algorithm, allowing for flexible computational trade-offs. Furthermore, we extend our framework beyond linear regression to non-linear function approximation via spline-based modeling, demonstrating its adaptability to complex regression tasks. We empirically validate i-QLS on the D-Wave quantum annealer, showing that our method efficiently scales to high-dimensional problems, achieving competitive accuracy with classical solvers while outperforming prior quantum approaches. Experiments confirm that i-QLS enables near-term quantum hardware to perform regression tasks with improved precision and scalability, paving the way for practical quantum-assisted machine learning applications.

Keywords: Quantum Annealing · Least Squares Optimization · Non-Linear Regression · Quantum Machine Learning

1 Background

A key distinction in machine learning methodologies lies in their optimization strategies, which are shaped by the assumptions that models make about the relationship between input features and the target variable. Models as Neural networks rely on non-convex optimization, often requiring extensive datasets and

M. H. Lees et al. (Eds.): ICCS 2025, LNCS 15904, pp. 19–34, 2025.
https://doi.org/10.1007/978-3-031-97629-2_2

prolonged training due to the difficulty of escaping local minima [7]. In contrast, parametric models based on least squares (LS) optimization [10], such as splines [19] and support vector machines [2], benefit from convex optimization, ensuring global optimality and theoretical robustness. However, the polynomial complexity of matrix operations in LS optimization introduces significant computational constraints as the number of features grows, limiting scalability.

Recently, several quantum machine learning models have been proposed for supervised learning tasks [14]. However, relatively few studies have explored the potential of leveraging near-term quantum computing exclusively for training and parameter estimation while maintaining a classical model for inference [12,20]. In particular, quantum annealing has been proposed as a method for reformulating LS optimization into a Quadratic Unconstrained Binary Optimization (QUBO) problem for execution on quantum hardware [1,3,4]. Despite these advancements, quantum annealing approaches face two fundamental bottlenecks. First, scalability remains a major challenge, as a tradeoff exists between the number of qubits and the precision of estimates of the optimized weights. Second, these methods are primarily to linear models, which fail to capture the complexities of real-world problems.

In this work, we introduce i-QLS, an iterative for quantum-assisted least squares optimization that overcomes the precision-scalability tradeoff inherent in prior quantum annealing-based methods. Instead of attempting to solve the problem in a single QUBO formulation with a fixed discretization, our approach iteratively refines the search space over multiple annealing cycles, exponentially converging to the optimal solution. This iterative approach is an anytime algorithm, meaning that at any point during execution, we can extract the best solution found, allowing for flexible computational tradeoffs. Furthermore, our method enables scalable adaptation to the constraints of existing quantum hardware by adjusting the granularity of discretization dynamically across iterations. In addition, we test i-QLS to non-linear regression via spline-based approximations, further demonstrating the versatility of our method beyond conventional linear regression. Our objective is to establish a practical and scalable framework for quantum-enhanced least squares optimization that leverages quantum hardware efficiently while remaining competitive with classical solvers. To summarize, this paper makes the following key contributions:

- **Iterative QUBO Refinement for Least Squares Optimization:** We introduce a novel iterative quantum-assisted least squares optimization strategy that mitigates the precision-qubit tradeoff inherent in prior single-shot QUBO formulations.
- **Scalability and Anytime Computation:** Our approach enhances scalability by enabling computation-limited devices to obtain progressively refined solutions throughout execution, supporting flexible computational tradeoffs.
- **Extension to Nonlinear Regression:** We extend our iterative framework beyond linear regression by incorporating spline-based function approximation, demonstrating its adaptability to complex modeling tasks.

- **Empirical Validation on Quantum Hardware:** We implement and evaluate our method on the D-Wave quantum annealer, showing that iterative refinement facilitates exponential convergence in optimization precision.

2   Related Works

Quantum annealing has been recently explored as an alternative paradigm for solving combinatorial optimization problems that naturally map onto a QUBO formulation. Although the least squares (LS) problem in regression does not inherently conform to this framework, few studies have attempted to adapt the LS formulation to harness quantum annealing for regression purposes.

For instance, [4] and [1] discretize the solution space by encoding each continuous weight as a series of binary variables, thereby formulating a corresponding QUBO problem that is amenable to solution via a quantum annealer. A significant limitation of these approaches is that increasing the precision of the weights necessitates an exponential increase in the number of qubits, rendering the method infeasible for large-scale problems on current quantum hardware. Consequently, classical LS solvers—such as direct matrix inversion [22], QR decomposition [6], or iterative gradient-based methods [13]—often outperform quantum approaches due to their capacity to achieve high precision without incurring the substantial overhead associated with quantum hardware.

Moreover, existing quantum annealing methods for LS have predominantly focused on linear regression, thereby neglecting non-linear problems that are more representative of real-world applications. In contrast, gate-based quantum computation has addressed non-linear approximation through the development of quantum splines [11,15]. Splines, which are piecewise polynomial functions with smoothness constraints, provide an effective tool for modeling non-linear relationships within a structured regression framework. In the initial formulation of quantum splines [15], a specific LS formulation based on B-splines [5] was employed to exploit the computational advantages of the HHL algorithm [9] for solving sparse linear systems. Although this approach offers a theoretical computational advantage, its reliance on fault-tolerant quantum computation limits its near-term applicability. Consequently, a variational counterpart has been proposed to leverage near-term quantum devices [11], although it has not yet demonstrated a robust advantage over classical methods.

Our work builds on these methodologies but fundamentally differs in two key aspects. First, rather than formulating a fixed QUBO problem with a predetermined discretization, we introduce an iterative refinement process. In our approach, each QUBO solution serves as the input for the subsequent iteration, progressively narrowing the weight search space while maintaining a fixed qubit requirement. This iterative refinement accelerates convergence exponentially and overcomes the scalability limitations encountered in previous methods. Our method is inspired by classical iterative refinement techniques commonly used in numerical optimization—such as Newton's method [17], gradient descent [8], and expectation-maximization [16]. To the best of our knowledge, apart from combinatorial optimization problem like graph-clustering [18], previous QUBO-based quantum regression methods have not incorporated an iterative refinement mechanism, rendering our approach unique in this context.

Second, our approach explicitly extends quantum-assisted regression to non-linear function approximation by integrating quantum spline formulations. In our framework, the advantages of splines—namely, their ability to model non-linear relationships through piecewise polynomial fitting with smoothness constraints—are harnessed within the iterative QUBO process to determine optimal spline coefficients. This integration enables the capture of non-linear dynamics while mitigating the fixed discretization issues of prior methods. By unifying iterative refinement with spline-based modeling, our method addresses the limitations of both conventional quantum annealing approaches and the earlier quantum spline formulations that rely on fault-tolerant computation.

3   Preliminaries

Least squares optimization is a technique in linear regression, where the goal is to identify the best-fitting linear relationship between input variables and a target by minimizing the sum of squared differences between observed and predicted values. This approach underpins many statistical models due to its simplicity and efficiency [10]. To extend these ideas to more complex, non-linear relationships, spline functions are often employed [2]. Spline functions are smoothing methods suitable for modelling the relationships between variables, typically adopted either as a visual aid in data exploration or for estimation purposes [10]. The underlying idea is to use linear models in which the input features are augmented with the so-called *basis expansions*. These consist of transformations of the original variables and serve to introduce non-linearity. Technically, splines are constructed by dividing the sample data into sub-intervals delimited by breakpoints, also referred to as knots. A fixed degree polynomial is then fitted in each of the segments, thus resulting in a piecewise polynomial regression. Formally, in the case of a 1-dimensional input vector x, we can express its relationship with a target variable y in terms of an order-M spline with knots $\{\xi_k\}_{k=1,\ldots,K}$:

$$y_{n_{\mathrm{obs}} \times 1} = N_{n_{\mathrm{obs}} \times (M+K)} \theta_{(M+K) \times 1} + \epsilon_{n_{\mathrm{obs}} \times 1}, \tag{1}$$

where θ is the vector of coefficients attached to the basis expansions, n_{obs} is the sample size, ϵ is a random error term and the design matrix N contains $M + K$ basis functions defined as follows:

$$h_j(x) = x^{j-1}, \qquad\qquad j = 1, \cdots, M \tag{2}$$

$$h_{M+k} = (x - \xi_k)_+^{M-1}, \qquad\qquad k = 1, \cdots, K. \tag{3}$$

Notice that the formulation above includes M basis functions that determine the order-M polynomial to be fitted in each segment. The additional K basis introduce continuity constraints on the spline and its derivatives up to order $M-2$. The goal is then to find the optimal set of parameters $\boldsymbol{\theta}$ that minimises the residual sum of squares (RSS), with a *ridge regularisation* of the curvature acting as a roughness penalty:

$$\text{Score}\,(\boldsymbol{\theta}, \eta) = (\boldsymbol{y} - \boldsymbol{N}\boldsymbol{\theta})^{T}\,(\boldsymbol{y} - \boldsymbol{N}\boldsymbol{\theta}) + \eta\boldsymbol{\theta}^{T}\boldsymbol{\Omega}_{(M+K)\times(M+K)}\boldsymbol{\theta}, \tag{4}$$

where $\boldsymbol{\Omega}$ is a diagonal matrix containing the partial second derivatives. The solution to (4) can easily seen to be:

$$\hat{\boldsymbol{\theta}} = (\boldsymbol{N}^{T}\boldsymbol{N} + \eta\boldsymbol{\Omega})^{-1}\boldsymbol{N}^{T}\boldsymbol{y}. \tag{5}$$

4   Methods

In this section, we present the mathematical formulation and derivation of our proposed quantum-classical hybrid algorithm for performing iterative least squares optimization, named i-QLS. We begin by posing the standard linear regression model as a discretized optimization problem, then describe how to embed it into a QUBO form suitable for a quantum annealer. Subsequently, we extend the iterative scheme to refine the precision of the solution at each iteration without requiring a large number of qubits from the outset. Finally, we outline how the approach generalizes to non-linear regressions via spline representations.

4.1   Problem Formulation

Let
$$\mathbf{X} \in \mathbb{R}^{N \times d}, \quad \mathbf{y} \in \mathbb{R}^{N},$$

where N is the number of data points and d is the number of features (or variables). The goal of linear regression is to find a weight vector

$$\mathbf{w}^{*} = (w_{1}^{*}, w_{2}^{*}, \dots, w_{d}^{*})^{\top}$$

that minimizes the sum of squared errors (SSE):

$$\mathbf{w}^{*} = \arg\min_{\mathbf{w} \in \mathbb{R}^{d}} S(\mathbf{w}) = \sum_{n=1}^{N} (y_{n} - \mathbf{w}^{\top}\mathbf{x}_{n})^{2}, \tag{6}$$

where $\mathbf{x}_{n}$ is the n-th sample. The least squares solution can be found in closed form via $\mathbf{w} = (\mathbf{X}^{\top}\mathbf{X})^{-1}\mathbf{X}^{\top}\mathbf{y}$ in the non-singular case. However, our method operates on a discretized search space to allow a quantum annealer to sample candidate $\mathbf{w}$ values.

4.2 Discretization and QUBO Construction

Representing each weight parameter $w_i \in \mathbb{R}$ by m binary variables and initializing a sufficiently large interval $\Delta_i^{(0)}$ with lower bound $\ell_i^{(0)}$ and the upper bound $u_i^{(0)}$, we find the discretization step size:

$$\delta_i^{(0)} \;=\; \frac{\Delta_i^{(0)}}{2^m - 1} \quad \text{for} \quad i = 1, \dots, d, \qquad \text{where} \qquad \Delta_i^{(0)} \;=\; (u_i^{(0)} - \ell_i^{(0)}) \quad (7)$$

Initializing $w_i^{(0)} \leftarrow \frac{(u_i^{(0)} + \ell_i^{(0)})}{2}$, our goal is to find $w_i^{(1)}$ restricted to 2^m equally spaced values in the interval $[\ell_i^{(0)}, u_i^{(0)}]$ that is closest to the optimal weight w_i^*. Mathematically,

$$w_i^{(1)*} \;=\; \arg \min_{w_i^{(1)}} |w_i^{(1)} - w_i^*|,$$
$$(8)$$
$$\text{where} \quad w_i^{(1)} \in \{\ell_i^{(0)} + \delta_i^{(0)} p \mid p \in \{0, 1, \dots, 2^m - 1\}\}$$

$$\mathbf{b} \;=\; \big(b_{1,0}, b_{1,1}, \dots, b_{1,m-1}, \dots, b_{d,0}, \dots, b_{d,m-1}\big)^{\top} \quad (9)$$

be the binary encoding vector representing the discretized weights. The weight vector $\mathbf{w}$ can now be expressed as a linear function of $\mathbf{b}$:

$$\mathbf{w}^{(1)}(\mathbf{b}) \;=\; \boldsymbol{\ell}^{(0)} + \mathbf{D}^{(0)} \mathbf{B}(\mathbf{b}), \quad (10)$$

$$\text{where} \quad \boldsymbol{\ell} = (\ell_1, \dots, \ell_d)^{\top},$$

$\mathbf{D}$ is a diagonal matrix containing the discretization step sizes δ_i,

$$\mathbf{D} = \operatorname{diag}(\delta_1, \dots, \delta_d),$$

and $\mathbf{B}(\mathbf{b})$ is the vector whose entries are sums of powers of two weighted by the bits i.e.,

$$\mathbf{B}(\mathbf{b}) = (B_1(\mathbf{b}_1), \dots, B_d(\mathbf{b}_d))^{\top},$$

$$B_i(\mathbf{b}_i) \;=\; \sum_{p=0}^{m-1} 2^{m-1-p} b_{i,p}, \quad (11)$$

$$\mathbf{b}_i \;=\; \big(b_{i,0}, b_{i,1}, \dots, b_{i,m-1}\big)^{\top}.$$

Quadratic Cost Function. The sum of squared errors for the discretized weights is

$$S^{(1)}(\mathbf{b}) \;=\; \sum_{n=1}^{N} \Big(y_n - \mathbf{w}^{(1)}(\mathbf{b})^{\top} \mathbf{x}_n\Big)^2. \quad (12)$$

Expanding this, we get

$$S^{(1)}(\mathbf{b}) \;=\; \sum_{n=1}^{N} \Big(y_n - \boldsymbol{\ell}^{(0)\top} \mathbf{x}_n - \mathbf{B}(\mathbf{b})^{\top} \mathbf{D}^{(0)\top} \mathbf{x}_n\Big)^2. \quad (13)$$

Since each component of $\mathbf{B}(\mathbf{b})$ is linear in the binary variables, $S^{(1)}(\mathbf{b})$ becomes a polynomial (up to quadratic order) in those bits:

$$S^{(1)}(\mathbf{b}) \;=\; \alpha^{(1)} \;+\; \sum_r \gamma_r^{(1)} \, b_r \;+\; \sum_{r<s} \Gamma_{r,s}^{(1)} \, b_r \, b_s, \tag{14}$$

where the summations over r, s run over all binary variables in $\mathbf{b}$, and $\alpha^{(1)}, \gamma_r^{(1)}, \Gamma_{r,s}^{(1)}$ are real coefficients derived from $\mathbf{X}, \mathbf{y}, \boldsymbol{\ell}^{(0)}, \mathbf{D}^{(0)}$.

This polynomial is in fact strictly quadratic, because each $b_r^2 = b_r$ as $b_r \in \{0, 1\}$. Thus we have effectively mapped the least squares cost onto a QUBO:

$$\mathbf{b}^{(1)*} \;=\; \min_{\mathbf{b} \in \{0,1\}^{d \times m}} \; \alpha^{(1)} + \sum_{r=1}^{dm} \gamma_r^{(1)} \, b_r \;+\; \sum_{1 \le r < s \le dm} \Gamma_{r,s}^{(1)} \, b_r \, b_s. \tag{15}$$

Given Eq. 15, a quantum annealer can directly solve QUBO problems by mapping them onto its native hardware architecture, where the cost function is minimized adiabatically through quantum tunneling. The annealer attempts to find a low-energy configuration of the binary variables that corresponds to the optimal solution of the given problem. However, the total number of binary variables that can be encoded is constrained by the available qubits, making it challenging to achieve high precision (i.e., large m) in naive discretization, as each additional bit per weight significantly increases qubit requirements.

4.3   i-QLS Algorithm

To address the qubit-precision trade-off, we propose an iterative zoom-in approach that uses a small number of bits $m \ge 1$ per weight but refines the solution region over multiple iterations. Let $\ell_i^{(0)}$ and $u_i^{(0)}$ be the initial bounds for weight w_i. At iteration $k = 1, \ldots, K$, each weight w_i is represented in $[\ell_i^{(k-1)}, u_i^{(k-1)}]$ with m bits. With $\Delta_i^{(k-1)} = u_i^{(k-1)} - \ell_i^{(k-1)}$, the discretization step size is evaluated as

$$\delta_i^{(k-1)} \;=\; \frac{\Delta_i^{(k-1)}}{2^m - 1}.$$

Thus,

$$\mathbf{w}^{(k)}(\mathbf{b}) \;=\; \boldsymbol{\ell}^{(k-1)} \;+\; \mathbf{D}^{(k-1)} \mathbf{B}(\mathbf{b})$$

Construct the QUBO from

$$S^{(k)}(\mathbf{b}) \;=\; \sum_{i=1}^{N} \left(y_i - \mathbf{w}^{(k)}(\mathbf{b})^\top \mathbf{x}_i \right)^2$$

with the updated bounds. Simplify the expression so that powers of binary variables are reduced via $b_{j,r}^2 = b_{j,r}$. Solve the QUBO using a quantum annealer. Retrieve the optimal binary solution $\mathbf{b}^{(k)*}$. Compute

$$w_i^{(k)*} \;=\; w_i(\mathbf{b}_i^{(k)*}), \quad i = 1, \ldots, d.$$

For each i:

$$\ell_i^{(k)} = w_i^{(k)*} - \left(\frac{\delta_i^{(k-1)}}{2f(m)}\right). \quad u_i^{(k)} = w_i^{(k)*} + \left(\frac{\delta_i^{(k-1)}}{2f(m)}\right),$$

$$\text{where} \quad f(x) = \begin{cases} 2, & \text{if } x = 1, \\ 1, & \text{otherwise.} \end{cases} \tag{16}$$

This ensures the next iteration's search space is an interval of width $\Delta_i^{(k)}$ centered at the best estimate $w_i^{(k)*}$. Observe that, for $m = 1$, $\delta_i^k = \Delta_i^k \implies w_i^{(k)} \in \{\ell_i^{(k-1)}, u_i^{(k-1)}\}$, thus the step function $f(x)$ helps in shrinking the search space exponentially. Repeat until the maximum number of iterations K is reached, or until a convergence criterion is met (e.g., changes in the loss below a threshold).

This procedure scales linearly with d in qubit usage (only $d \times m$ qubits are needed), yet allows an effective exponential zoom-in on the candidate solution region over multiple iterations, thereby improving precision without globally increasing the qubit count. The pseudocode of the algorithm is given in 1.

Algorithm 1: i-QLS: Iterative Quantum-Assisted Least Squares

Input: $\mathbf{X} \in \mathbb{R}^{N \times d}$ (feature matrix), $\mathbf{y} \in \mathbb{R}^N$ (target values), bits per weight m, max iterations K, initial bounds $\ell_i^{(0)}, u_i^{(0)}$ for each $i = 1, \ldots, d$.

for $k = 1$ **to** K **do**

  Compute step size: $\delta_i^{(k-1)} \leftarrow \frac{u_i^{(k-1)} - \ell_i^{(k-1)}}{2^m - 1}$, $i = 1, \ldots, d$.
  Construct weight representation $w_i^{(k)}(\mathbf{b})$ (Eq. 10)
  Formulate QUBO cost function $S^{(k)}(\mathbf{b})$.
  Solve $\min S^{(k)}(\mathbf{b})$ using a quantum annealer.
  Extract optimal solution: $\mathbf{b}^{(k)*} \leftarrow \arg\min_{\mathbf{b}} S^{(k)}(\mathbf{b})$.
  Compute updated weight estimates: $w_i^* \leftarrow w_i(\mathbf{b}^{(k)*})$.
  Update bounds:
  $\ell_i^{(k)} \leftarrow w_i^{(k)*} - \frac{1}{2f(m)}\delta_i^{(k)}, \quad u_i^{(k)} \leftarrow w_i^{(k)*} + \frac{1}{2f(m)}\delta_i^{(k)}$.

Output: $\mathbf{w}^{(K)*} = (w_1^{(K)*}, \ldots, w_d^{(K)*})$ (optimal weights).

Lemma 1. *If the underlying least squares problem is well-posed (i.e., there exists a unique optimal weight vector w^* such that $y = Xw^*$) and $w_i^* \in [l_i^{(0)}, u_i^{(0)}]\forall i$, then the mean squared error evaluated from the weights estimated at each iteration exponentially converges to 0 as $k \to \infty$.*

Proof. At iteration $k = 0$, the search interval for weight w_i is given by $[l_i^{(0)}, u_i^{(0)}]$ with width $\Delta_i^{(0)}$. In the first iteration, the algorithm discretizes this interval into 2^m equally spaced values. Let the discretization step size be

$$\delta_i^{(0)} = \frac{\Delta_i^{(0)}}{2^m - 1}.$$

The QUBO formulation selects the discrete value closest to the true optimum, denoted by $w_i^{(1)}$. The algorithm then updates the search interval to be centered at $w_i^{(1)}$ with new bounds

$$l_i^{(1)} = w_i^{(1)} - \frac{\delta_i^{(0)}}{2f(m)}, \quad u_i^{(1)} = w_i^{(1)} + \frac{\delta_i^{(0)}}{2f(m)},$$

so that the new interval width is

$$\Delta_i^{(1)} = \frac{\delta_i^{(0)}}{f(m)} = \frac{\Delta_i^{(0)}}{f(m)(2^m - 1)}.$$

By repeating this process iteratively, the width of the interval after k iterations is

$$\Delta_i^{(k)} = \frac{\Delta_i^{(k-1)}}{f(m)(2^m - 1)} = \frac{\Delta_i^{(0)}}{(f(m)(2^m - 1))^k}. \tag{17}$$

Since $f(m)(2^m - 1) > 1$ for any $m \geq 1$, it follows that

$$\lim_{k \to \infty} \Delta_i^{(k)} = 0.$$

Because the search interval shrinks to a single point, the weight estimate $w_i^{(k)}$ converges to a unique value, which must coincide with the optimal weight w_i^* provided the model is correctly specified.

Furthermore, since the mean squared error is a continuous function of the weights, it follows that

$$\lim_{k \to \infty} E^{(k)} = \lim_{k \to \infty} \|Xw^{(k)} - y\|^2 = \|Xw^* - y\|^2.$$

Under the assumption that the least squares problem is consistent (or that w^* minimizes the error), we have $\|Xw^* - y\|^2 = 0$, which implies

$$\lim_{k \to \infty} E^{(k)} = 0.$$

5   Experiments

In this section, we present a series of experiments to evaluate various aspects of i-QLS. Specifically, the first set of experiments examines the convergence rate and the accuracy of thee algorithm, analyzing how the number of bits used for weight representation and the number of iterations influence these two factors.

The second set of experiments evaluates the performance of i-QLS in terms of scalability on a multivariate linear regression problem. We demonstrate that i-QLS overcomes the scalability limitations of existing quantum approaches by successfully handling datasets with up to 175 features. Our results indicate that the performance of i-QLS is comparable to classical methods while outperforming all previously proposed quantum annealing-based techniques.

Finally, we extend our analysis beyond linear regression by applying i-QLS to nonlinear function estimation using the least squares formulation of spline functions. In this case, the algorithm achieves significantly higher precision in weight estimation compared to existing gate-based quantum spline methods, even with a relatively low number of iterations.

5.1   Convergence Rate and Accuracy

In this section, we demonstrate our iterative quantum-assisted least squares algorithm on a synthetic linear regression problem with two features. The goal is to illustrate the exponential rate of convergence of the mean squared error (MSE) across iterations, along with the exponential refinement of the weights' search space. We generate synthetic data (Fig. 1a) with two input features, sampling each feature value x_i uniformly from $[-5, 5]$. The corresponding target values are computed using a linear model ($y_i = w_1^* x_{i1} + w_2^* x_{i2}$) where the true weights w_1^* and w_2^* were set to fixed values within $(-1, 1)$.

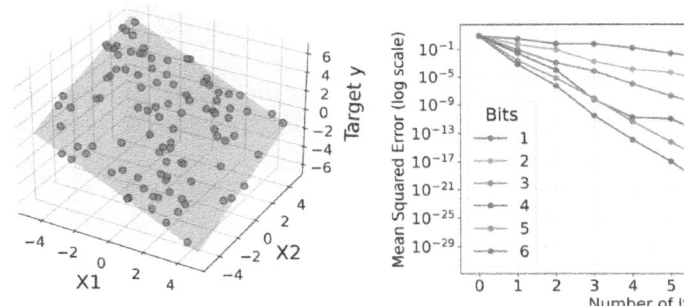

(a) **Two-Feature Linear Data.** 100 data points (x_i, y_i) where $x_i \in [-5, 5]^2$ sampled from a linear model.

(b) **MSE Convergence Over Iterations.** Exponential decay of MSE with iterations, improving with increasing bits per weight.

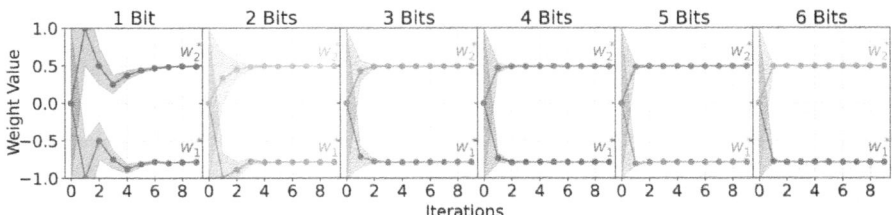

(c) **Exponential Shrinking of the Search Space.** The weight search space for w_1 and w_2 narrows exponentially with iterations. Higher bit precision enables finer updates, accelerating convergence.

Fig. 1. Assessment of convergence rate and accuracy of i-QLS

We use i-QLS to find the optimal set of weights w_1^* and w_2^*. Initially, each weight is discretized into 2^m equally spaced values in a sufficiently broad range (we initialized $[-1, 1]$). At each iteration k, we solve the corresponding QUBO using a D-Wave Advantage annealer, extract the best candidate $(w_1^{(k)*}, w_2^{(k)*})$, and then refine the search interval around it (Eq. 16). This process ensures that, the search space shrinks exponentially by a factor of $f(m)(2^m - 1)$ (Eq. 17).

Figure 1b plots the MSE per iteration for different bit-precisions b. The logarithmic scale on the vertical axis reveals a linear decline, which corresponds to exponential convergence in the original scale. Increasing m (i.e., using more qubits per weight) accelerates convergence. For example, at iteration $k = 9$, the 6-bit case achieves MSE ≈ 0 (within machine precision using `float64`), whereas the 2-bit case remains around 10^{-9}. This follows our theoretical prediction that finer discretization improves solution precision but at the cost of additional quantum resources.

Figure 1c shows how the search space bounds for w_1 and w_2 evolve over iterations. The shaded regions indicate the possible value ranges, which shrink exponentially with each iteration. Higher bit precision results in faster contraction of the search space, leading to earlier convergence to w_1^* and w_2^*. This aligns with our theoretical framework (Lemma 1), which predicts that the bounds contract by a factor of $f(m)(2^m - 1)$ per iteration.

5.2 Scalability Analysis

In this experiment, we investigate how i-QLS scales with the number of features in a linear regression setting. We fix the number of bits per weight to 1 and the maximum number of iterations to 10. We generate synthetic data in the same manner as Sect. 5.1, but now the number of features d ranges from 1 to 175. Specifically, for each d, we draw each feature value x_{ij} uniformly from $[-5, 5]$ and compute the target values via

$$y_i = \sum_{j=1}^{d} w_i^* x_{ij},$$

where the true weights w_i^* are sampled from $(-1, 1)$. At each iteration of i-QLS, a QUBO problem of size proportional to the number of features d is defined, discretizing each weight into $2^1 = 2$ equally spaced values in a broad interval (here we chose $[-1, 1]$). This QUBO is solved on a D-Wave Advantage quantum annealer accessed remotely, employing a default embedding strategy, which is heuristic and may produce different embeddings for repeated runs on identical problems. For a classical QUBO solver, we use Gurobi, a state-of-the-art branch-and-bound mixed-integer optimizer. Additionally, to compare with classical linear regression, we employ scikit-learn's `LinearRegression`, which by default uses an efficient singular value decomposition (SVD)-based method. This makes it highly competitive for moderate to large d (Fig. 2).

The results in Fig. 1 highlight the trade-off between quantum hardware constraints and classical solver limitations. D-Wave Advantage (Quantum Annealer)

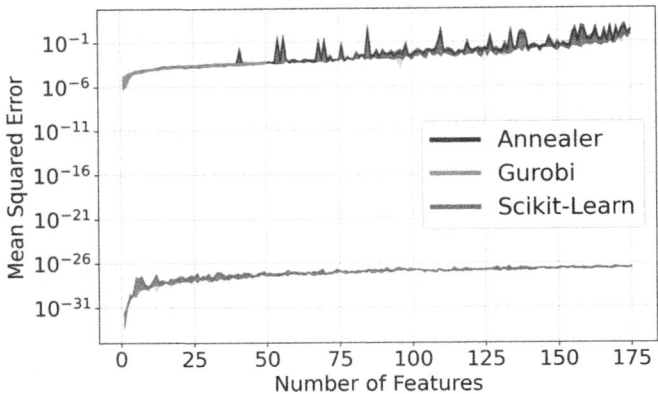

Table 1. R^2 values for the fit generated by our quantum-assisted approach across different numbers of features.

d	R^2
1	0.999943
25	0.999988
50	0.999990
75	0.999987
100	0.999987
125	0.999923
150	0.999627
175	0.998592

Fig. 2. MSE at iteration $k = 10$ for i-QLS executed on D-Wave Advantage, Gurobi, and scikit-learn's `LinearRegression`. Shaded regions indicate variance (darker) and min-max range (lighter) over three random seeds.

efficiently handles problems up to $d = 175$, yielding MSE values near 10^{-1} for larger d. This is, to our knowledge, the first demonstration of linear regression on a real quantum annealer scaling to 175 features while maintaining high accuracy. The inconsistencies of minor embedding and remote access on the performance of the quantum annealer for certain instances can be seen as spikes in Fig. 2. Gurobi's branch-and-bound algorithm struggles beyond $d = 50$ due to the fully connected QUBO (the number of pairwise interactions is equal to $\frac{d(d-1)}{2}$). Consequently, Gurobi's runtime becomes prohibitively large, and the solver occasionally terminates with suboptimal solutions or runs out of memory. Scikit-learn remains highly accurate (MSE $\approx 10^{-28}$) and efficiently solves these linear regressions for d beyond 175.

5.3 Non-linear Function Approximation Using Splines

In this section, we use i-QLS to non-linear function approximation by leveraging the LS optimization of splines provided in Eq. (4). We perform similar experiments to existing quantum splines [11,15] generating synthetic datasets from five widely used non-linear functions in deep learning: (1) the sine function $\sin(\pi x)$, which is periodic and frequently appears in Fourier analysis and wave modeling; (2) the sigmoid function $\sigma(\pi x) = (1 + e^{-\pi x})^{-1}$, commonly used in neural networks for binary classification; (3) the hyperbolic tangent $\tanh(\pi x)$, which serves as a popular activation function for hidden layers; (4) the ReLU (Rectified Linear Unit) $\max(0, \pi x)$, a piecewise linear function that underpins many modern deep networks; and (5) the ELU (Exponential Linear Unit), an extension of ReLU that smooths the transition for negative inputs, thereby improving optimization stability. For each function, we generated 100 data points uniformly in the range $x \in [-1, 1]$. Given these target highly non-linear functions, we define a

LS optimization problem using spline functions (Sect. 3 with 20 knots. Thus, we estimate the optimal parameters of the non-linear regression using i-QLS. The optimization is performed by discretizing the solution space, assigning one qubit per parameter, and refining the weight estimates iteratively for up to 10 iterations.

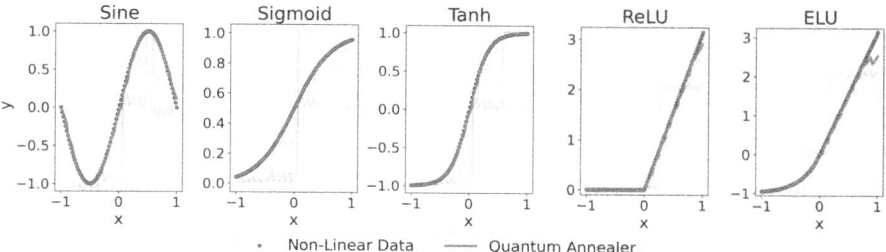

Fig. 3. Spline-Based Approximation of Non-Linear Functions. The green curves illustrate the iterative refinement of the regression fit obtained using our quantum-assisted least squares approach with *linear splines* (20 knots), one qubit per parameter, and up to 10 iterations. The lighter green lines correspond to intermediate fits across iterations, while the darker green curve represents the final approximation after 10 iterations. (Color figure online)

Figure 3 presents the results of our spline-based approximations. The blue scatter points indicate the ground-truth function values, while the green curves represent the fits obtained by estimating the parameters of the splines using i-QLS. Due to the choice of a linear function in each interval of the splines, the approximation is piecewise linear, generating a recurrent angular pattern in the function representation. This effect arises from the limited number of knots (20), which results in a segmented approximation of the underlying function.

While we demonstrated the capability of i-QLS to model non-linear functions, more expressive spline models—such as *quadratic* or *cubic splines*—could further improve smoothness and accuracy. However, these higher-degree splines introduce additional computational overhead, requiring more qubits, which must be accounted for in practical quantum implementations.

By integrating splines with our iterative refinement strategy, we demonstrated how quantum-assisted optimization can be extended beyond traditional linear models, offering a scalable path towards non-linear function learning on quantum hardware.

6   Discussion and Conclusion

In this work, we proposed i-QLS, a novel iterative quantum-assisted least squares optimization algorithm that exhibits favorable scaling compared to prior single-shot QUBO formulations. The key advantage stems from the iterative refinement mechanism, which systematically narrows the search space around the best-found weight values at each step. Given that the search space contracts by a factor of $f(m)(2^m - 1)$ per iteration, the number of iterations K required to achieve a given precision ϵ scales as:

$$K = \mathcal{O}\left(\log_{f(m)(2^m-1)} \frac{1}{\epsilon}\right),$$

where m is the number of bits allocated per weight. This implies that, with higher bit precision, convergence is achieved in fewer iterations. However, as observed in Fig. 1b, increasing m requires additional quantum resources, introducing a trade-off between iteration count and hardware feasibility. Nevertheless, the approach theoretically guarantees to find the optimal solution even with $m = 1$ as $k \to \infty$, making it inherently more scalable.

As a well-rounded analysis of the approach, considering practical hardware constraints, we acknowledge that if an iteration selects a weight far from the true optimal, restricting the subsequent shrunk search space from containing the true optimal weight, successive iterations may struggle to recover. This issue is particularly relevant when practical quantum hardware noise or solver limitations prevent finding the best candidate in an iteration. Also, our method demonstrates strong scalability in terms of problem size, the practical runtime depends heavily on hardware access constraints. As the annealer is remotely accessed and shared among many users, the time to solution for each QUBO problem is primarily dominated by network latency and queue wait times rather than raw annealing time [21]. Consequently, having the annealer on-site could significantly reduce overhead and make i-QLS a highly competitive method for real-time applications.

The weight chosen at step k might be nearer to the optimal weight than the weight chosen at step $k + 1$, but the search space would have shrunk by a factor of $f(m)(2^m - 1)$. This effect is exacerbated when fewer qubits per weight are used, as demonstrated in Fig. 1b for the 1-bit case. From iteration 8 to 9, the MSE temporarily increases instead of decreasing. This occurs because the selected weight at iteration 8 was closer to the optimal than any available discretized value at iteration 9. However, the MSE resumes decreasing at iteration 10, illustrating that while local fluctuations may arise, the overall convergence trend remains intact. Thus, one can eventually mitigate this issue by running a sufficiently large number of iterations.

Empirical validation on the D-Wave quantum annealer demonstrated the effectiveness of i-QLS, showing that it scales efficiently to problems with up to 175 features while maintaining high accuracy. Additionally, we extended our framework to non-linear regression using spline-based modeling, demonstrating its adaptability beyond linear problems.

While our approach significantly improves existing quantum-assisted regression, practical implementations against classical solvers must carefully balance bit precision, iteration count, and hardware limitations to ensure robust performance. Future work will explore adaptive bit precision per iteration and weight to further enhance efficiency and accuracy.

Acknowledgments. This work has been partially funded by the German Ministry for Education and Research (BMB+F) in the project QAIAC-QAI2C under grant 13N17167.

Code Availability. The code associated with this paper is available at: https://github.com/supreethmv/i-QLS.

References

1. Borle, A., Lomonaco, S.J.: Analyzing the quantum annealing approach for solving linear least squares problems. In: Das, G.K., Mandal, P.S., Mukhopadhyaya, K., Nakano, S. (eds.) WALCOM 2019. LNCS, vol. 11355, pp. 289–301. Springer, Cham (2019). https://doi.org/10.1007/978-3-030-10564-8_23
2. Cortes, C., Vapnik, V.: Support-vector networks. Mach. Learn. **20**, 273–297 (1995)
3. Cruz-Santos, W., Venegas-Andraca, S.E., Lanzagorta, M.: A qubo formulation of minimum multicut problem instances in trees for d-wave quantum annealers. Sci. Rep. **9**(1), 17216 (2019). https://doi.org/10.1038/s41598-019-53585-5
4. Date, P., Potok, T.: Adiabatic quantum linear regression. Sci. Rep. **11**(1), 21905 (2021). https://doi.org/10.1038/s41598-021-01445-6
5. De Boor, C., De Boor, C.: A Practical Guide to Splines, vol. 27. Springer, New York (1978)
6. Francis, J.: The qr transformation a unitary analogue to the lr transformation–part 1. Comput. J. **4**(3), 265–271 (1961). https://doi.org/10.1093/comjnl/4.3.265
7. Goodfellow, I., Bengio, Y., Courville, A.: Deep Learning. MIT Press, Cmbridge (2016). http://www.deeplearningbook.org
8. Haji, S.H., Abdulazeez, A.M.: Comparison of optimization techniques based on gradient descent algorithm: a review. PalArch's J. Archaeol. Egypt/Egyptology **18**(4), 2715–2743 (2021)
9. Harrow, A.W., Hassidim, A., Lloyd, S.: Quantum algorithm for linear systems of equations. Phys. Rev. Lett. **103**(15), 150502 (2009)
10. Hastie, T., Tibshirani, R., Friedman, J.: The Elements of Statistical Learning. Springer Series in Statistics, Springer New York Inc., New York (2001)
11. Inajetovic, M.A., Orazi, F., Macaluso, A., Lodi, S., Sartori, C.: Enabling nonlinear quantum operations through variational quantum splines. In: International Conference on Computational Science, pp. 177–192. Springer, Heidelberg (2023). https://doi.org/10.1007/978-3-031-36030-5_14
12. Jerbi, S., Gyurik, C., Marshall, S.C., Molteni, R., Dunjko, V.: Shadows of quantum machine learning. Nat. Commun. **15**(1), 5676 (2024)
13. Lanczos, C.: Solution of systems of linear equations by minimized iterations. J. Res. Nat. Bur. Stand. **49**(1), 33–53 (1952)
14. Macaluso, A.: Quantum supervised learning. KI-Künstliche Intelligenz, pp. 1–15 (2024)
15. Macaluso, A., Clissa, L., Lodi, S., Sartori, C.: Quantum splines for non-linear approximations. In: Proceedings of the 17th ACM International Conference on Computing Frontiers, pp. 249–252 (2020)
16. Moon, T.K.: The expectation-maximization algorithm. IEEE Signal Process. Mag. **13**(6), 47–60 (1996)
17. Moré, J.J., Sorensen, D.C.: Newton's method. Technical report. Argonne National Lab.(ANL), Argonne, IL (United States) (1982)

18. Mysore Venkatesh, S., Macaluso, A., Klusch, M.: GCS-Q: quantum graph coalition structure generation. In: Mikyška, J., de Mulatier, C., Paszynski, M., Krzhizhanovskaya, V.V., Dongarra, J.J., Sloot, P.M. (eds.) Computational Science - ICCS 2023, pp. 138–152. Springer, Cham (2023). https://doi.org/10.1007/978-3-031-36030-5_11

19. Reinsch, C.H.: Smoothing by spline functions. Numerische mathematik **10**(3), 177–183 (1967)

20. Sinha, A., Macaluso, A., Klusch, M.: NAV-Q: quantum deep reinforcement learning for collision-free navigation of self-driving cars. Quant. Mach. Intell. **7**(1), 1–20 (2025)

21. Venkatesh, S.M., Macaluso, A., Nuske, M., Klusch, M., Dengel, A.: Q-SEG: quantum annealing-based unsupervised image segmentation. IEEE Comput. Graph. Appl. **44**(5), 27–39 (2024). https://doi.org/10.1109/MCG.2024.3455012

22. Woodbury, M.A.: Inverting modified matrices. Department of Statistics, Princeton University (1950)

Uncovering and Verifying Optimal Community Structure in Complex Networks: A MaxSAT Approach

Carlos Ansótegui[1], Vaidyanathan Peruvemba Ramaswamy[2], Stefan Szeider[2], and Hai Xia[2(✉)]

[1] Logic and Optimization Group, University of Lleida, Lleida, Spain
carlos.ansotegui@udl.cat
[2] Algorithms and Complexity Group, TU Wien, Vienna, Austria
{vaidyanathan,sz,hxia}@ac.tuwien.ac.at

Abstract. Network modularity is central to understanding phenomena in diverse domains, from biology and social science to engineering and computational physics. However, computing the optimal modularity—an NP-hard measure quantifying community strength—has remained computationally intractable at large scales. Most approaches resort to heuristics without formal optimality guarantees.

This paper contributes to the computational science of complex systems by introducing a novel MaxSAT-based framework that can compute optimal network modularity values for larger networks than previously possible. Leveraging this new capability, we extensively evaluate heuristic solutions and, for the first time, include the state-of-the-art memetic graph clustering heuristic VieClus. Remarkably, VieClus identifies optimal modularity values for all tested networks, ranging from 103 previously studied instances to 52 new, larger ones, and does so in seconds. This result contrasts with earlier conclusions that heuristics frequently fail to find the optimal modularity.

By combining a powerful MaxSAT encoding, which supports proof logging for verification, with a fast and effective heuristic, we demonstrate that even intricate network structures can be tackled efficiently. This synergy brings us closer to making complex network analysis and community detection tractable, robust, and verifiable—a goal firmly aligned with the core mission of computational science.

Keywords: Complex Networks · Modularity Optimization · Maximum Satisfiability · Graph Clustering Heuristics · Computational Verification

1 Introduction

In network science, the concept of modularity has emerged as a cornerstone for understanding the intricate structure of complex networks. This foundational measure was introduced by Newman and Girvan [25], building on earlier

© The Author(s), under exclusive license to Springer Nature Switzerland AG 2025
M. H. Lees et al. (Eds.): ICCS 2025, LNCS 15904, pp. 35–49, 2025.
https://doi.org/10.1007/978-3-031-97629-2_3

work by Freeman [15] (for a more in-depth discussion, we refer to Newman's book [26]). Modularity provides a quantitative lens through which the subtle yet significant patterns of interaction within networks can be discerned, revealing insights pivotal for theoretical exploration and practical application. At its core, modularity measures the strength of division within a network, distinguishing the dense interconnections within communities from the sparser connections between them. This metric has helped to significantly enlarge our understanding of network topology, offering a robust framework to unravel the community structure inherent in various types of networks arising in physics, sociology, biology, telecommunications, and other areas. In propositional satisfiability (SAT), modularity has been used to link the performance of CDCL-based SAT solvers with the structure in industrial instances [1]. MaxSAT is an optimization version of SAT, finding an assignment that maximizes the number of satisfied clauses, and the clauses can be weighted, with a natural extensibility to solving network modularity optimization.

Computing modularity is NP-hard [10], even for trees [12]. As a result, most research focuses on developing heuristics [7–9,11,22,30,31,34], while exact ILP-based methods have been proposed but do not scale to large networks [10,12, 32]. Recent studies by Aref et al. [4,5] found that near-optimal partitions from heuristics are often disproportionately dissimilar to optimal partitions.

Contributions. We challenge these findings through two main contributions. Firstly, we develop an efficient MaxSAT-based approach for computing optimal modularity values, scaling to larger networks than previous exact methods. Our approach provides mathematical guarantees of optimality through its MaxSAT foundation, and additionally enables independent verification of these guarantees through recent advances in proof logging techniques implemented in the Pacose solver. This is in contrast to heuristic algorithms [21,24] that cannot provide optimality guarantees. Secondly, we analyze the memetic network clustering heuristic VieClus [6], overlooked by previous surveys [4,5,21,24]. We surprisingly found that VieClus finds optimal modularity values for all test networks within seconds—both the 103 networks from their study and 52 additional larger networks.

Our findings have significant implications for computational science: applications in physics, biology, social sciences, and other domains can use VieClus for rapid optimal clustering. When additional verification is needed, our MaxSAT approach provides guaranteed optimality at the cost of longer computation times. For the highest level of trust—crucial for validating scientific results—the proof logging implementation generates independently verifiable certificates for optimality. This portfolio of methods, with increasing levels of verification at corresponding computational cost, makes complex network analysis tractable and trustworthy. Our MaxSAT approach also provides the means for analyzing whether optimal clusterings are unique.

2   Preliminaries

2.1   Graphs and Networks

We use the terms *graph* and *network* synonymously. All graphs considered are finite, undirected, and simple (i.e., without parallel edges or self-loops). We denote the set of vertices of a graph G by $V(G)$ and the set of edges by $E(G)$. Further, we denote an edge between vertices u and v by uv or, equivalently, vu. We denote the degree of a vertex u by $d(u)$. For subsets $C, C' \subseteq V(G)$, $e(C, C')$ denotes the number of edges uv of G with $u \in C$ and $v \in C'$; we use $e(C)$ as a shorthand for $e(C, C)$.

2.2   Modularity

A *clustering* $\mathcal{P}$ of a graph G is a partition of $V(G)$ into nonempty, mutually disjoint sets, called *clusters*. We will write $u \equiv_{\mathcal{P}} v$ if the vertices u and v belong to the same cluster of $\mathcal{P}$. Following Newman and Girvan [25], the *modularity* $Q(\mathcal{P})$ of a clustering of a graph G with m edges is given by

$$Q(\mathcal{P}) := \sum_{C \in \mathcal{P}} \left[\frac{e(C)}{m} - \left(\frac{e(C) + \sum_{C' \in \mathcal{P}} e(C, C')}{2\,m} \right)^2 \right]. \tag{1}$$

A clustering $\mathcal{P}$ is *optimal* if $Q(\mathcal{P})$ is maximal. The value $Q(\mathcal{P})$ for an optimal clustering $\mathcal{P}$ of G is called the *modularity* of G. Finding an optimal clustering is an NP-hard optimization problem [10].

2.3   MaxSAT

A (weighted partial) MaxSAT instance is a propositional formula Φ in conjunctive normal form (CNF), whose clauses are partitioned into hard and soft clauses. Each soft clause has a non-negative integer *weight*. A *solution* τ of a MaxSAT instance Φ is a total truth assignment to Φ that satisfies all the hard clauses. The weight $w(\tau)$ of a solution τ is the sum of weights of all the soft clauses τ satisfies. A solution of maximal weight is *optimal*.

2.4   Graphs with Multiple Optimal Clusterings

It has been empirically observed that most real-world graphs have a unique optimal clustering [5]. However, some graphs do have multiple optimal clusterings. We distinguish between optimal clusterings that are *essentially the same* and those that are *essentially different*. If they are essentially the same, one optimal clustering can be transformed into the other by a symmetry of the graph To define these terms, we need some additional graph-theoretic notions. An *isomorphism* from a graph G to a graph G' is a bijective mapping $\varphi : V(G) \to V(H)$ such that for all pairs $u, v \in V(G)$ we have $uv \in E(G)$ if and only if $\varphi(u)\varphi(v) \in E(H)$. G and H are *isomorphic* if there exists an isomorphism from

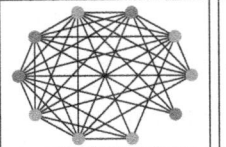 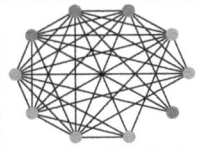 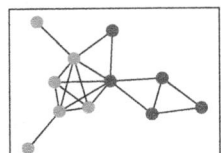 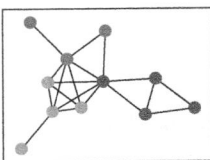

(a) Essentially same optimal clusterings for the **dom** network, with modularity value 0.017. In this case, the vertices which switch clusters happen to be so-called twin vertices.

(b) Essentially different optimal clusterings for the **ambassador** network, with modularity value 0.231. In this special case, even the number of clusters is different.

Fig. 1. Two examples of networks with multiple optimal solutions. Nodes which have the same color belong to the same cluster.

G to H. An *automorphism* of a graph G is an isomorphism from G to itself. An automorphism is *non-trivial* if it is not the identity mapping.

We say that two clusterings $\mathcal{P}$ and $\mathcal{P}'$ of a graph G are *essentially the same* if there exists an automorphism φ of G such that $\mathcal{P} = \{\,\{\,\varphi(v) \mid v \in C\,\} \mid C \in \mathcal{P}'\,\}$; otherwise $\mathcal{P}$ and $\mathcal{P}'$ are *essentially different*. Clearly, if $|\mathcal{P}| \neq |\mathcal{P}'|$ then $\mathcal{P}$ and $\mathcal{P}'$ must be essentially different; however, $|\mathcal{P}| = |\mathcal{P}'|$ does not imply that they are essentially the same. We also note that even if a graph has a non-trivial automorphism, it might have a unique optimal clustering since the automorphism might map vertices to vertices of the same cluster.

We can check whether two clusterings $\mathcal{P} = \{C_1, \ldots, C_k\}$ and $\mathcal{P}' = \{C'_1, \ldots, C'_k\}$ of a graph G are essentially different by using a standard graph isomorphism check. From G, we obtain a graph $G(\mathcal{P})$ by adding k new vertices $v_1, \ldots, v_k$ and all the edges uv_i for $u \in C_i$, $1 \leq i \leq k$. If $\mathcal{P}$ and $\mathcal{P}'$ are essentially the same, then $G(\mathcal{P})$ and $G(\mathcal{P}')$ must be isomorphic. Conversely, if we have found an isomorphism φ from $G(\mathcal{P})$ to $G(\mathcal{P}')$ which induces bijection between $V(G(\mathcal{P})) \setminus V(G)$ and $V(G(\mathcal{P}')) \setminus V(G)$, then $\mathcal{P}$ and $\mathcal{P}'$ are essentially the same.

3   MaxSAT Encoding of Modularity

We now reformulate the definition of modularity as given in (1) so that it can easily be turned into a MaxSAT encoding. Again, let G be the input graph and $m = |E(G)|$. We define the *gain* of a pair $u, v \in V(G)$ as

$$g(u, v) = \begin{cases} 2\,m - d(u)d(v) & \text{if } uv \in E(G); \\ -d(u)d(v) & \text{otherwise.} \end{cases}$$

For a clustering $\mathcal{P}$ of G and a pair $u, v \in V(G)$, we define the *offset gain of u, v relative to $\mathcal{P}$* as

$$g_{\mathcal{P}}^{+}(u, v) = \begin{cases} g(u, v) & \text{if } g(u, v) \geq 0 \text{ and } u \equiv_{\mathcal{P}} v; \\ |g(u, v)| & \text{if } g(u, v) < 0 \text{ and } u \not\equiv_{\mathcal{P}} v; \\ 0 & \text{otherwise,} \end{cases} \tag{2}$$

and

$$Q^+(\mathcal{P}) = \sum_{u<v\in V(G)} g_{\mathcal{P}}^+(u,v). \tag{3}$$

Theorem 1. *For every graph G, there are integers r and s such that for every clustering $\mathcal{P}$ of G, we have $Q(\mathcal{P}) = rQ^+(\mathcal{P}) + s$. Consequently, $\mathcal{P}$ maximizes $Q^+(\mathcal{P})$ if and only if $\mathcal{P}$ is optimal.*

Proof. Let Ω be the sum of $|w(u,v)|$ over all pairs $u < v \in V(G)$ with $w(u,v) < 0$ and $A_{u,v} \in \{0,1\}$ such that $A_{u,v} = 1$ if and only if $uv \in E(G)$. Then, we can rewrite (3) as

$$Q^+(\mathcal{P}) = \Omega + \sum_{C\in\mathcal{P}} \sum_{u<v\in C} g(u,v)$$

$$= \Omega + \sum_{C\in\mathcal{P}} \sum_{u<v\in C} \left(2\,mA_{u,v} - d(u)d(v)\right)$$

$$= \Omega + \sum_{C\in\mathcal{P}} \left(2\,m\cdot\mathrm{e}(C) - \sum_{u<v\in C} d(u)d(v)\right).$$

We have

$$\sum_{u<v\in C} d(u)d(v) = \frac{1}{2}\left(\sum_{u,v\in C} d(u)d(v) - \sum_{u\in C} d(u)^2\right)$$

$$= \frac{1}{2}\left(\sum_{u\in C} d(u)\sum_{v\in C} d(v) - \sum_{u\in C} d(u)^2\right)$$

$$= \frac{1}{2}\left(\sum_{u\in C} d(u)\right)^2 - \frac{1}{2}\sum_{u\in C} d(u)^2.$$

Setting $\Omega' = \Omega + \frac{1}{2}\sum_{u\in V(G)} d(u)^2$ and observing that $\mathrm{e}(C) + \sum_{C'\in\mathcal{P}} \mathrm{e}(C,C') = \sum_{v\in C} d(v)$, we can further rewrite (3) as

$$Q^+(\mathcal{P}) = \Omega' + \sum_{C\in\mathcal{P}} \left[2\,m\cdot\mathrm{e}(C) - \frac{1}{2}\left(\sum_{u\in C} d(u)\right)^2\right]$$

$$= \Omega' + 2\,m^2 \sum_{C\in\mathcal{P}} \left[\frac{\mathrm{e}(C)}{m} - \left(\frac{\sum_{u\in C} d(u)}{2\,m}\right)^2\right]$$

$$= \Omega' + 2\,m^2 Q(\mathcal{P}).$$

Hence, the claim of the theorem holds for $r = 1/2m^2$ and $s = \Omega'$.   □

This theorem gives rise to a MaxSAT encoding. First, we introduce variables and hard clauses that represent a clustering $\mathcal{P}$ of G. Then, we introduce weighted soft unit clauses to maximize modularity.

Let G be the given graph and $V(G)$ linearly ordered by $<$. For any pair $u, v \in V(G)$ with $u < v$, we introduce a variable $c_{u,v}$, indicating whether u and v are in the same cluster (i.e., $u \equiv_{\mathcal{P}} v$); $c_{v,u}$ denotes the same variable as $c_{u,v}$.

To achieve this, we only need to add clauses that enforce the transitivity of $\equiv_{\mathcal{P}}$, i.e., $u \equiv_{\mathcal{P}} v$ and $v \equiv_{\mathcal{P}} w$ implies $u \equiv_{\mathcal{P}} w$. We thus add the following clause for all triples of distinct vertices $u, v, w \in V(G)$

$$\neg c_{u,v} \vee \neg c_{v,w} \vee c_{u,w}. \tag{4}$$

We call u, v and v, w the *premise pairs*, u, w the *forced pair*, and u and w the *end vertices* of the above transitivity clause.

For each pair $u, v \in V(G)$ with $u < v$ and $g(u, v) > 0$ we introduce a soft clause $c_{u,v}$ with weight $g(u, v)$; for each pair $u, v \in V(G)$ with $u < v$ and $g(u, v) < 0$ we introduce a soft clause $\neg c_{u,v}$ with weight $|g(u, v)|$; we do not need to consider pairs with $g(u, v) = 0$.

This concludes the definition of the MaxSAT instance $\Phi(G)$ representing the modularity of G.

Theorem 2. *A solution τ to $\Phi(G)$ is optimal if and only if τ corresponds to an optimal clustering $\mathcal{C}$ of G.*

Proof. By construction of $\Phi(G)$, the weight of an optimal solution τ to $\Phi(G)$ equals $Q^+(G)$, and by Theorem 1, the clustering corresponding to τ is optimal. ☐

A drawback of the MaxSAT encoding defined above is that it introduces a cubic number of clauses for enforcing transitivity. Adapting ideas from Din and Thai [12] to our MaxSAT setting, we can omit a large fraction of the transitivity clauses without affecting optimal solutions, thus scaling the MaxSAT encoding to larger graphs. For a graph G and vertices $u, v \in V(G)$ with $u \neq v$, let $K_G(u, v)$ be the smallest set of vertices such that u and v belong to different connected components of the graph $G_{u,v}$ obtained by (i) removing all vertices that belong to $K_G(u, v)$ from G and (ii) removing the edge uv in case $uv \in E(G)$. We will later explain how to efficiently compute the sets $K_G(u, v)$ for all pairs $u, v \in V(G)$.

The *Sparse MaxSATencoding* $\Phi^*(G)$ for modularity is now obtained from $\Phi(G)$ by limiting the transitivity clauses to a subset; namely, we add the clause (4) only if $v \in K_G(u, w)$.

Theorem 3. $\Phi(G)$ *and* $\Phi^*(G)$ *have the same optimal solutions.*

Proof. We observe that any optimal solution of $\Phi(G)$ is a solution of $\Phi^*(G)$ of the same weight. We will show that also the converse holds, i.e., any optimal solution of $\Phi^*(G)$ is a solution of $\Phi(G)$ of the same weight, which, together with the previous statement, establishes the theorem. Let τ be an optimal solution of $\Phi^*(G)$ and let G_τ be the graph with $V(G_\tau) = V(G) =: V$ and $E(G_\tau) = \{ uv \mid \tau(c_{u,v}) = 1 \}$. We say that a vertex $v \in V$ is τ-*reachable* from u if G_τ

contains a path between u and v, and we say v is $\tau + G$-*reachable* from u if G_τ and G contain the same path between u and v; we call such a path a $\tau + G$-*path*.

Claim 1: For any $u, v \in V$, if v is τ-reachable from u then v is also $\tau +$ G-reachable from u. To show the claim, suppose to the contrary that there are $u, v \in V$ such that v is τ-reachable from u but not $\tau + G$-reachable. Let $X \subseteq V$ be the set of all vertices that are $\tau + G$-reachable from u, and let $Y \subseteq V$ be the set of all vertices that are τ-reachable from u. Clearly $X \subseteq Y$, and by assumption $v \in Y \setminus X$. Since v is τ-reachable from u, there must be at least one edge $xy \in E(G_\tau)$ with $x \in X$ and $y \in Y \setminus X$. We obtain a new assignment τ' from τ by setting $\tau'(c_{x,y}) = 0$ for all $x \in X$ and $y \in Y \setminus X$. Note also that no transitivity clause forces such $c_{x,y}$ to true since for any such transitivity clause where x, y is the forced pair, at least one of its premise pairs x', y' would have $x' \in X$ and $y' \in Y \setminus X$ and $c_{x',y'}$ set to true. However, since $xy \notin E(G)$, $g(x, y) < 0$, and so the encoding contains the soft clause $\neg c_{x,y}$ of weight $|g(x, y)|$; hence switching the truth value of $c_{x,y}$ from false to true increases the overall weight of the assignment, i.e., $w(\tau') > w(\tau)$, contradicting our assumption that τ is optimal. Hence, the claim is shown to be true.

Claim 2: Any $v \in V$ that is τ-reachable from some $u \in V$ is a neighbor of u in G_τ. Consider pairs $u, v \in V$ such that v is τ-reachable from u. Let $d(u, v)$ denote the length of a shortest $\tau + G$-path between u and v (such a path exists by Claim 1). We show Claim 2 by induction on $d(u, v)$. If $d(u, v) = 1$, then u, v are adjacent in G_τ, and we are done. Now assume Claim 2 holds for all pairs of distance $d - 1 \geq 0$ and consider a τ-connected pair u, v with $d(u, v) = d$. Consider a shortest $\tau + G$ path P of length d between u and v. P runs through a vertex $x \in K_G(u, v)$. We have $d(u, x), d(x, v) \leq d(u, v) - 1$, hence $ux, xv \in E(G_\tau)$ by the induction hypothesis, i.e., $\tau(c_{u,x}) = \tau(c_{x,v}) = 1$. By construction, $\Phi^*(G)$ contains the transitivity clause $\neg c_{u,x} \vee \neg c_{x,v} \vee c_{u,v}$, consequently $\tau(c_{u,v}) = 1$, and so $uv \in E(G_\tau)$ as required. Hence Claim 2 holds.

By Claim 2, G_τ is a disjoint collection of cliques. Consider any transitivity clause $\neg c_{u,v} \vee \neg c_{v,w} \vee c_{u,w}$ of $\Phi(G)$, with end vertices u, w, and v not necessarily from $K_G(u, w)$. If u, w belong to the same clique of G_τ, then $\tau(c_{u,w}) = 1$ and the clause is satisfied. If u, w belong to different cliques of G_τ, v cannot be in the same clique with u and in the same clique with w, hence $\tau(c_{u,v}) = 0$ or $\tau(c_{v,w}) = 0$, and the clause is satisfied. We conclude that τ is indeed a solution of $\Phi(G)$. This concludes the proof of the theorem. □

4 Accuracy of Heuristically Computed Modularity

As mentioned above, it is NP-hard to determine the modularity of a network exactly [10], even for tree-structured networks [12]. Since exact methods for computing modularity do not scale to large networks, scientists rely on heuristic methods to compute modularity and associated clusterings.

Aref et al. [4,5] presented a systematic analysis of heuristic methods for modularity computation. In particular, they considered Clauset-Newman-Moore (CNM) [11], Louvain [7], Reichardt-Bornholdt with the configuration model as the null model (LN) [20], Combo [30], Belief [34], Paris [8], Leiden [31], EdMot-Louvain [22], recurrent graph neural network (GNN) [29], and Bayan [2]. They tested all these approaches on 104 networks (54 real-world and 50 synthetic) and computed the exact modularity via an ILP encoding. Their key findings were quite negative: Combo had the highest success rate among the eight heuristics, returning an optimal partition for 90.4% of the networks; the average success rate of all 8 heuristics combined was only 43.9%. On average, GNNs and Bayan achieve optimality for 68.7% and 82.3% of networks, respectively. Aref et al. [4,5] used several partition similarity metrics (AMI, RMI, ECS) to quantify differences between partitions and found that suboptimal clusterings tend to be disproportionately dissimilar to the optimal clustering. In summary, their results suggest significant limitations in commonly used heuristics for modularity.

However, Aref et al. did not consider the *Vienna graph clustering* framework (VieClus) by Biedermann et al. [6], an omission that turns out to be significant. VieClus provides a general *memetic algorithm* [23] for various graph clustering problems, including modularity computation. A key innovation in VieClus is using recombine operators that leverage ensemble clusterings and multi-level techniques. These operators create an overlay clustering from two input clusterings, determining whether pairs of vertices should be in the same cluster based on their groupings in the input clusterings. This recombination is then enhanced with local search algorithms to improve the clustering quality further. The algorithm also employs a multi-level approach to find even better clusterings. VieClus emphasizes randomized tie-breaking throughout the process to diversify the search and improve solutions. Additionally, the algorithm incorporates a scalable communication protocol, enabling it to compute high-quality solutions efficiently. This combination of techniques results in a powerful and versatile clustering tool that can reproduce or improve upon previous benchmark results for various graph clustering instances.

In the sequel, we extend the study by Aref et al. by comparing the modularity computed by VieClus with the optimal modularity. Our powerful MaxSAT approach for computing exact modularity allows us to extend the benchmark set with significantly larger networks.

5   Checking Optimality of VieClus

5.1   Benchmark Networks

Our benchmark set includes 103 networks[1] that were considered in the previous work [5]. 50 of these networks were synthetic LFR and ABCD graphs generated by Aref et al. [3], while the remaining 53 graphs are real-world benchmarks drawn from the well-known Netzschleuder repository [28]. Aref et al. selected these networks to be (optimally) solved within a reasonable time by their sparse

---

[1] Previous work [5] used a benchmark set consisting of 104 networks. However, MaxHS cannot find an optimal solution for the network `physician_trust` within 48 h, so we exclude it from our experimental analysis.

IP formulation and all the 8 heuristics considered. We denote this as the *base* benchmark set. Further, we extend this set by adding another 52 networks from the Netzschleuder repository [28], and solve the networks optimally with our sparse MaxSAT formulation. We denote these 52 new networks as the *extended* benchmark set. The networks in the extended benchmark set contain up to 1461 vertices and 2812 edges. We show the number of nodes and edges of the networks as a scatter plot in 2, where the blue and orange dots represent the networks from the base and extended benchmark sets, respectively.

The newly added networks in the extended set generally have a larger number of nodes and edges compared to the previously used networks. We also provide detailed information on each network in the supplementary material[2].

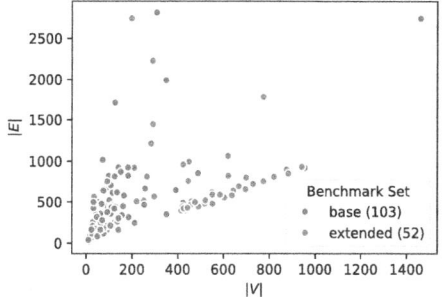
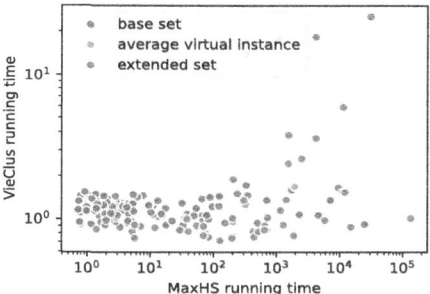

Fig. 2. Scatter plot of the number of nodes ($|V|$) and edges ($|E|$) of the 155 benchmark networks

Fig. 3. The running time of MaxHS on sparse encoding vs VieClus

5.2 Experimental Setup

We perform all our experiments on a compute cluster with nodes equipped with two AMD 7403 processors (24 cores at 2.8 GHz) and 32 GB of RAM per core. We use a timeout of 10 h while running VieClus and a timeout of 48 h while running the MaxSAT solver. We did preliminary testing with different MaxSAT exact solvers participating in the MaxSAT Evaluation 2022[3], including EvalMaxSAT, MaxCDCL, and MaxHS. Then, we identified MaxHS as the most promising solver for our use case.

We generate the encodings using Python 3.10 along with the NetworkX 3.2 graph library [17] and the PySAT 0.1 library [18]. Recall that to construct the sparse encodings, we need a *separator set* for each pair of vertices. We precompute these separator sets for all networks by adapting the basic functions from the NetworkX library (which compute a single $s - t$ cut using flow-based methods [13,14,19]) to run in a parallel setting. The benchmark networks and the

---

[2] https://figshare.com/s/caec9f3c34f5c31488c2.
[3] https://maxsat-evaluations.github.io/2022/descriptions.html.

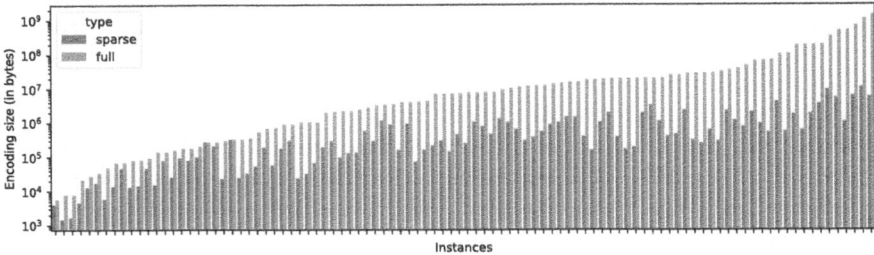

Fig. 4. Comparing the sizes of the sparse vs full encodings

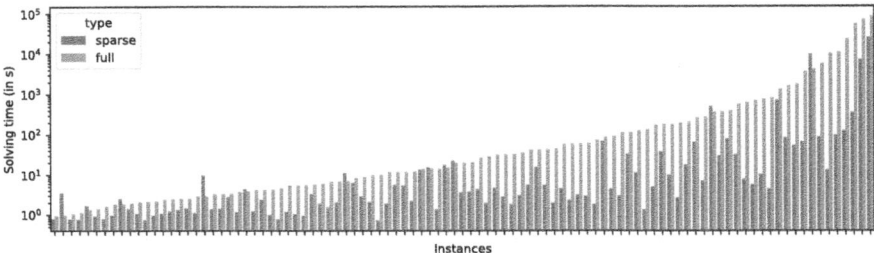

Fig. 5. Comparing the solving times of the sparse vs full encodings

scripts for generating the MaxSAT encodings for those networks are available in the supplementary material.

5.3   Results

Sparse vs. Full Encodings. We use MaxHS to find the optimal modularity of the 103 networks of the base benchmark set with both the full and the sparse encodings. We then compare the file sizes of the two encodings and their respective solving times. In Figs. 4 and 5, it should be noted that the axes are in logarithmic scale. In Fig. 4, we sort the networks according to the file size of their full encodings. We can see that on the larger networks, the sparse encodings result in a more than 10-fold reduction compared to the corresponding full encodings.

Further, we also compare the solving efficiency of MaxSAT with both the full and the sparse encodings on the base benchmark set. From Fig. 5, sparse encodings can speed up the solving time on many large networks by a factor of 5 compared to the full encoding. For several networks, the full encoding is quicker than the sparse encoding; however, most of these networks are too small to take advantage of the sparse encoding. More specifically, all such networks are solved within 20 seconds by MaxHS on both encodings. On average, the MaxSAT sparse encoding can reduce the solving time by 57.59% among the 103 networks. Considering the space and time savings, our first argument is justified: The sparse MaxSAT encoding provides a higher solving efficiency for computing optimal modularity, especially on larger networks.

VieClus vs. MaxHS with Sparse Encoding. We compute the optimal modularity values for all 155 networks in the base benchmark set and the extended benchmark set using MaxHS to solve the sparse MaxSAT encodings. We then compare the modularity values obtained by VieClus with the optimal values obtained by MaxHS. Amazingly, we find that they match on all networks, i.e., VieClus obtains the optimal modularity for all the networks. Furthermore, regarding time consumption, VieClus requires a lot less time than MaxHS (roughly an order of magnitude faster).

Figure 3 compares the time consumption between MaxHS with the sparse encoding (y-axis) and VieClus (x-axis), where each dot represents a network. The blue and green dots represent the networks from the base and extended sets, respectively. We emphasize that this is for reference only, not a direct comparison between the two algorithms: MaxHS finds an optimal clustering and verifies the optimality, whereas VieClus finds a clustering that happens to be optimal. However, the plot indicates which instances are harder and which are easier for the two approaches.

Regarding the average time consumption, as the orange virtual dot (average solving time of the full and sparse encodings) shown in Fig. 3, the MaxHS takes over 2000 seconds on average to get the optimal modularity values, but VieClus achieves the same modularity values within only 2 seconds, although without verifying their optimality.

In general, on all 155 networks from the base set and the extended set (which were selected such that MaxHS can find an optimal solution within 48 h), VieClus always manages to match the optimal modularity value (and it does so in under 25 seconds), even though it is only a heuristic algorithm. Our conclusion significantly differs from recent claims [4,5] that heuristic methods perform poorly and rarely return an optimal partition.

5.4   Analysis of Networks with Multiple Optimal Solutions

A vast majority of the considered networks, 137 out of 155, have a unique optimal clustering; therefore, the optimal clusterings found by VieClus and MaxHS are necessarily the same. As for the remaining 18 networks, we used the MaxSAT encoding to enumerate *all* optimal clusterings. To this end, when MaxHS finds an optimal clustering, the corresponding satisfying assignment is negated and added to the encoding, thereby forbidding the same solution from being discovered again. Now, if MaxHS finds a solution to this new encoding with the same modularity value as before, we will have found another clustering with the optimal modularity. The moment the modularity value drops, we know for certain that all clusterings with the optimal modularity value have been enumerated. Furthermore, by analyzing the isomorphisms between the optimal clusterings of these networks, we found that for only 11 of them, the optimal clusterings are essentially different (details are available in the supplementary material).

Due to its randomized nature, VieClus can find different optimal clusterings when run with different random seeds. We verify this by running VieClus repeatedly on some of the networks that have multiple optimal clusterings. However, in contrast to the MaxSAT approach, we cannot block previously found optimal clusterings for VieClus. Hence, it is up to chance that repeated runs will eventually find all optimal clusterings. In practice, we notice that VieClus has a tendency to gravitate towards the same solution. For example, in several instances, despite trying more than 1000 random seeds, VieClus always returns the same optimal clustering. However, there were also a handful of instances where repeated runs of VieClus were eventually able to discover all optimal clusterings.

6   Verifying Optimality Through Proof Logging

State-of-the-art MaxSAT solvers are complex algorithms with lots of moving parts. If they claim a solution is optimal, we cannot always blindly trust this. This limitation can be significant for computational science applications where the high trustworthiness of results is crucial. We address this gap using *proof logging* in MaxSAT solving, which provides machine-verifiable certificates of optimality. Specifically, we use Pacose [27], a state-of-the-art MaxSAT solver that outputs VeriPB [16] proofs—formal certificates that can be independently verified. These proofs record every step in the solving process, from CNF encoding verification to optimality confirmation, allowing third parties to validate results without trusting the solver implementation. Since the proof logs output by Pacose by default can get large, we modify Pacose to directly output compressed proof logs. For instance, compressing a 65 GB proof log can reduce its size to 12 GB. We include the modified version of Pacose along with some generated proof logs in the supplementary material.

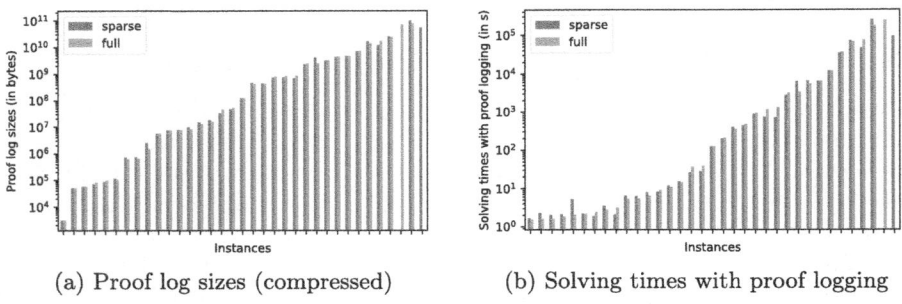

(a) Proof log sizes (compressed)   (b) Solving times with proof logging

Fig. 6. Comparing sparse vs full encoding when solving with proof logging

We run Pacose with proof logging enabled on both sparse and full encodings with a 72-h timeout. Of the 155 benchmark networks, we attempted proof logging on 91 networks. Out of these, 36 networks were successfully solved with verifiable proofs. Of these 36 networks, the network mu_0_01_LFR_0 was solved only by the full encoding, and the network 7th_graders was solved only by the sparse encoding. The largest number of nodes and edges in a network that was solved with proof logging are 141 and 1017, respectively. Figures 6a and 6b compare the proof log sizes and solving times, respectively, of the sparse and full encoding when proof logging is enabled.

Surprisingly, while sparse encodings are consistently faster for regular MaxSAT solving, this advantage disappears with proof logging. For approximately half the instances, full encodings resulted in faster solving times. The proof log sizes were also comparable between sparse and full encodings.

This illustrates the three levels of confidence and effort balance that one can achieve with the different methods we consider. On one end, heuristic methods like VieClus can quickly find optimal solutions but do not provide any guarantees. On the other end, MaxSAT solving with proof logging offers the highest level of rigor but requires significant computational overhead. Finally, MaxSAT solving without proof logging is a good middle ground. It provides optimality guarantees with a reasonable degree of confidence while maintaining scalability to a decent extent.

7   Conclusion and Future Work

In our work, we introduce a novel MaxSAT-based approach for computing optimal modularity and demonstrate that VieClus consistently finds optimal clusterings orders of magnitude faster than exact methods, which challenges previous findings about heuristic unreliability in modularity computations. Our findings provide a variety of options for the modularity analysis:

1. VieClus offers lightning-fast optimal solutions, although without formal guarantees.
2. Our MaxSAT encoding provides optimality guarantees at increased computational cost.
3. For requiring the highest trustworthiness, the proof logging implementation generates independently verifiable certificates of optimality.

The high-efficiency VieClus suggests that network analysis based on optimal modularity is more tractable than previously thought. Simultaneously, our advances in MaxSAT solving with proof logging provide a rigorous foundation when verification is crucial. Future work should explore extending proof logging capabilities to larger networks while maintaining reasonable computational overhead, like integrating the tuning techniques into the (Max)SAT solving [33].

Acknowledgements. This work was funded by the European Union's Horizon 2020 research and innovation program under the Maria Skłodowska-Curie grant agreement No. 101034440, project PID2022-138506NB-C21 (Ministerio de Ciencia e Innovación), and project 10.55776/P36420 (Austrian Research Funds). The full MaxSAT encoding is based on an encoding developed in Rupert Ettrich's Bachelor's Thesis, TU Wien, 2020. The authors also thank Jakob Nordström and Andy Oertel for assisting with VeriPB proof logging.

References

1. Ansótegui, C., Levy, J.: On the modularity of industrial SAT instances. In: CCIA 2011. FAIA, vol. 232, pp. 11–20. IOS Press (2011)
2. Aref, S., Chheda, H., Mostajabdaveh, M.: The Bayan algorithm: detecting communities in networks through exact and approximate optimization of modularity. CoRR arxiv:2209.04562 (2022). https://doi.org/10.48550/ARXIV.2209.04562
3. Aref, S., Mostajabdaveh, M.: Dataset of synthetic modular graphs from LFR and ABCD benchmark models for community detection. Figshare (2023). https://doi.org/10.6084/m9.figshare.24257293.v1
4. Aref, S., Mostajabdaveh, M.: Analyzing modularity maximization in approximation, heuristic, and graph neural network algorithms for community detection. JCS 78, 102283 (2024). https://doi.org/10.1016/J.JOCS.2024.102283
5. Aref, S., Mostajabdaveh, M., Chheda, H.: Heuristic modularity maximization algorithms for community detection rarely return an optimal partition or anything similar. In: ICCS 2023. LNCS, vol. 14076, pp. 612–626. Springer, Heidelberg (2023). https://doi.org/10.1007/978-3-031-36027-5_48
6. Biedermann, S., Henzinger, M., Schulz, C., Schuster, B.: Memetic graph clustering. In: SEA 2018. LIPIcs, vol. 103, pp. 3:1–3:15. Schloss Dagstuhl - Leibniz-Zentrum für Informatik (2018). https://doi.org/10.4230/LIPICS.SEA.2018.3
7. Blondel, V.D., Guillaume, J.L., Lambiotte, R., Lefebvre, E.: Fast unfolding of communities in large networks. JSTAT 2008(10), P10008 (2008). https://doi.org/10.1088/1742-5468/2008/10/P10008
8. Bonald, T., Charpentier, B., Galland, A., Hollocou, A.: Hierarchical graph clustering using node pair sampling. In: MLG (2018)
9. Bouguessa, M., Missaoui, R., Talbi, M.: A novel approach for detecting community structure in networks. In: ICTAI 2014, pp. 469–477 (2014). https://doi.org/10.1109/ICTAI.2014.77
10. Brandes, U., et al.: On modularity clustering. IEEE TKDE 20(2), 172–188 (2008). https://doi.org/10.1109/TKDE.2007.190689
11. Clauset, A., Newman, M., Moore, C.: Finding community structure in very large networks. Phys. Rev. E 70, 066111 (2004). https://doi.org/10.1103/PhysRevE.70.066111
12. Dinh, T.N., Thai, M.T.: Toward optimal community detection: from trees to general weighted networks. Internet Math. 11(3), 181–200 (2015). https://doi.org/10.1080/15427951.2014.950875
13. Esfahanian, A.H.: Connectivity algorithms. In: Topics in Structural Graph Theory, pp. 268–281 (2013)
14. Even, S.: Graph Algorithms. WH Freeman & Co. (1979)
15. Freeman, L.C.: A set of measures of centrality based on betweenness. Sociometry 40(1), 35–41 (1977). https://doi.org/10.2307/3033543
16. Gocht, S., McCreesh, C., Nordström, J.: Veripb: the easy way to make your combinatorial search algorithm trustworthy. In: CPTAI Workshop at CP (2020)
17. Hagberg, A.A., Schult, D.A., Swart, P.J.: Exploring network structure, dynamics, and function using NetworkX. In: SciPy 2008, pp. 11–15 (2008)
18. Ignatiev, A., Morgado, A., Marques-Silva, J.: PySAT: a python toolkit for prototyping with SAT oracles. In: SAT 2018, pp. 428–437 (2018). https://doi.org/10.1007/978-3-319-94144-8_26
19. Kammer, F., Täubig, H.: Connectivity, pp. 143–177. Springer, Heidelberg (2005). https://doi.org/10.1007/978-3-540-31955-9_7

20. Leicht, E.A., Newman, M.: Community structure in directed networks. Phys. Rev. Lett. **100**, 118703 (2008). https://doi.org/10.1103/PhysRevLett.100.118703
21. Li, J., et al.: A comprehensive review of community detection in graphs. Neuro-computing **600**, 128169 (2024). https://doi.org/10.1016/j.neucom.2024.128169
22. Li, P., Huang, L., Wang, C., Lai, J.: Edmot: an edge enhancement approach for motif-aware community detection. In: KDD 2019, pp. 479–487. ACM (2019). https://doi.org/10.1145/3292500.3330882
23. Moscato, P., Cotta, C.: A gentle introduction to memetic algorithms. In: Handbook of Metaheuristics, ISOR, vol. 57, pp. 105–144. Kluwer/Springer (2003). https://doi.org/10.1007/0-306-48056-5_5
24. Nascimento, M.C., de Carvalho, A.C.: Spectral methods for graph clustering - a survey. EJOR **211**(2), 221–231 (2011). https://doi.org/10.1016/j.ejor.2010.08.012
25. Newman, M., Girvan, M.: Finding and evaluating community structure in networks. Phys. Rev. E **69**, 026113 (2004). https://doi.org/10.1103/PhysRevE.69.026113
26. Newman, M.: Networks, 2nd edn. Oxford University Press, Oxford (2018)
27. Paxian, T., Reimer, S., Becker, B.: Pacose: an iterative SAT-based MaxSAT solver. MaxSAT Eval. **2018**, 20 (2018)
28. Peixoto, T.P.: The Netzschleuder network catalogue and repository (2023). https://doi.org/10.5281/zenodo.7839981
29. Sobolevsky, S., Belyi, A.: Graph neural network inspired algorithm for unsupervised network community detection. Appl. Netw. Sci. **7**(1), 63 (2022). https://doi.org/10.1007/S41109-022-00500-Z
30. Sobolevsky, S., Campari, R., Belyi, A., Ratti, C.: General optimization technique for high-quality community detection in complex networks. Phys. Rev. E **90**, 012811 (2014). https://doi.org/10.1103/PhysRevE.90.012811
31. Traag, V., Waltman, L., van Eck, N.J.: From Louvain to Leiden: guaranteeing well-connected communities. Nat. Sci. Rep. **9**(5233) (2019). https://doi.org/10.1038/s41598-019-41695-z
32. Vinh, N.X., Epps, J., Bailey, J.: Information theoretic measures for clusterings comparison: variants, properties, normalization and correction for chance. JMLR **11**(95), 2837–2854 (2010)
33. Xia, H., Szeider, S.: SAT-Based tree decomposition with iterative cascading policy selection. In: AAAI, vol. 38, no. 8, pp. 8191–8199 (2024). https://doi.org/10.1609/aaai.v38i8.28659
34. Zhang, P., Moore, C.: Scalable detection of statistically significant communities and hierarchies, using message passing for modularity. Proc. Natl. Acad. Sci. U.S.A. **111**(51), 18144–18149 (2014)

WebAA: Website Association Analysis via Multi-resource Similarity Computation

Taiyao Zhang[1,2], Dongzheng Jia[3], Xingyu Fu[1,2(✉)], Zhihao Zhang[1,2], and Qingyun Liu[1,2]

[1] Institute of Information Engineering, Chinese Academy of Sciences, Beijing, China
{zhangtaiyao,fuxingyu,zhangzhihao,liuqingyun}@iie.ac.cn
[2] School of Cyber Security, University of Chinese Academy of Sciences, Beijing, China
[3] National Computer Network Emergency Response Technical Team/Coordination Center of China, Beijing, China
jdz@cert.org.cn

Abstract. The rapid proliferation of websites has posed significant challenges to cyberspace management. To effectively manage websites, we introduce the innovative concept of website association in this paper and propose a website association model, WebAA. Website association seeks to determine whether target websites belong to the same organization based on their features, playing a crucial role in cyberspace management. Unlike website identification, website association enables more accurate organizational alignment through fine-grained analysis of website features. Specifically, malicious website association helps investigators identify the organizations behind them, thereby addressing the root causes of the harm they cause. In simple terms, the proposed website association model in this paper consists of four modules. In the first module, we calculate the similarity score based on external resources by analyzing the dependency relationships of the target websites on these resources. The second module employs the BERT model to analyze the similarity of HTML texts between the target websites and generates a similarity score. The third module focuses on the website domain names, calculating their similarity using Levenshtein distance and obtaining the similarity score. In the fourth module, the final similarity score is obtained by weighting the three similarity scores, which is then used to determine whether the two websites belong to the same organization. Extensive experiments on two real-world datasets demonstrate that our model can efficiently associate thousands of website pairs within milliseconds with an accuracy exceeding 90%. Striving for applicability and replicability, we release ready-to-use raw data from our study.

Keywords: Web computing · Cyberspace governance · Website association · BERT

M. H. Lees et al. (Eds.): ICCS 2025, LNCS 15904, pp. 50–64, 2025.
https://doi.org/10.1007/978-3-031-97629-2_4

1   Introduction

With the rapid development of the Internet, more and more organizations and institutions are expanding their influence by building their own websites. According to statistics from Worldwidewebsize [16], as of January 1, 2025, there are 3.76 billion indexable websites worldwide. The vast number of websites and the complex relationships between them pose challenges to cyberspace management, but this issue cannot be effectively addressed solely at the website level. Focusing management efforts on the organizational level behind the websites and conducting website management at that level is a feasible solution. Organizational-level website management enables management from an organizational perspective, revealing potential connections and interactive relationships, which leads to better management of cyberspace ecology.

Organizational-level website management method is highly valuable for fields like information security and data mining. Malicious attackers often create multiple associated websites to conceal their identities or expand their scope of influence [9]. For example, phishing websites, gambling websites, counterfeit platforms, and malicious advertising networks often operate in a distributed manner, using multiple associated websites to evade security detection and blocking [11]. The organization behind the website often uses a new domain name to quickly establish a replacement site when the current website is attacked [10]. Therefore, combating such malicious services solely from the perspective of individual websites is ineffective. Analyzing organizational associations between websites can help quickly uncover hidden malicious network structures and enhance attack prevention capabilities. In addition, with the explosive growth of internet information, different websites within the same organization may contain complementary resources. By identifying and integrating data from these related websites, a more comprehensive and efficient knowledge network can be built, promoting information sharing and collaborative analysis [5].

Although organizational-level website management is becoming increasingly important, there is currently limited research addressing this issue. Most existing research focuses on website-level management needs, such as malicious website identification [2,10]. Those researches mainly analyze various resources of websites, construct an embedded representation of the website using deep learning and other techniques, and use this representation as the input for a classifier to determine whether the target website is malicious.

However, this type of method has two main issues: (1) It focuses on website-level identification tasks. As mentioned above, once a malicious website is discovered, the malicious organizer can easily switch to a new website to continue their activities. Therefore, website-level identification alone cannot fully address such malicious behavior. (2) The various resources required by the method are not always available or accurate. For example, due to CDN technology [12], the IP address associated with the same website may change. Additionally, due to privacy concerns, the WHOIS information of many websites, especially malicious ones, is often incorrect or even nonexistent. The lack of these resources can severely impact the method's accuracy.

To address the above problems, we propose the concept of website association. Website association is an organizational-level alignment task for websites, aiming to determine whether target websites belong to the same organization by analyzing their resource characteristics. At the same time, we propose a website association model, WebAA, which achieves high-accuracy website association while relying on only a small amount of easily accessible website resources. WebAA consists of four modules that analyze the similarities between target websites using multi-resource and produce the final results through a fusion mechanism. Specifically, in the first module, we first extract the external resources that the target websites rely on. These resources are then categorized into core resources, major resources, and general resources based on their importance, with different weights assigned to each category. Next, inspired by Jaccard similarity, we propose weighted Jaccard similarity and calculate the weighted similarity for each type of resource. Finally, the similarities at each category are combined to produce the overall similarity score S_E based on external resources. In the second module, we first extract and preprocess the HTML text of the target websites. Then, we utilize the BERT model [4] to analyze the similarity between the HTML texts of the target websites, resulting in the similarity score S_T. The third module focuses on the domain name similarity of the websites. We apply the Levenshtein distance algorithm [14] to analyze the domain name similarity between the target websites, resulting in the domain name-based similarity score S_D. In the fourth module, the three similarity scores are weighted to calculate the final similarity score S_F, which is then used to determine whether the target websites belong to the same organization. Extensive experiments on two real-world datasets validate the effectiveness of our model.

In general, the contributions of this paper are as follows:

- **New Concept:** To the best of our knowledge, this is the first time to propose the concept of website association. Website association focuses on the organizations behind websites, enabling regulators to effectively manage legitimate websites and fundamentally combat illicit websites.
- **New Technique:** We innovatively propose a website association model, WebAA, which can accurately and efficiently perform the association task by analyzing only three easily accessible resources of the target websites.
- **New Datasets:** We manually constructed two real-world datasets: one consisting of legitimate websites and their organizational information, and the other comprising illegal websites. We release those datasets to support community researchers in conducting studies related to this field[1]
- **New Promotion:** Extensive experiments on two real-world datasets demonstrate that our model can efficiently associate thousands of website pairs within milliseconds with an accuracy exceeding 90%.

2   Related Work

At present, research on websites mainly focuses on malicious website identification. In this section, we briefly introduce several methods for identifying mali-

---

[1] github: https://github.com/SevenZhang123/WebAA-Datasets.

cious websites. We also present text matching and similarity calculation techniques based on the BERT model to help readers understand the text similarity calculation method used in our model.

2.1 Malicious Website Identification

Existing methods for identifying malicious websites analyze the characteristics of various resources on the website and use deep learning and other techniques to perform the identification task. IDTracker [14] first divides the target websites into different clusters by Third-party Service IDs. Then constructs a heterogeneous graph to learn the embedding representation of the website based on the association relationship between various resources (domain names, IP addresses, whois records, etc.). And finally uses a classifier to determine whether the target website is a malicious website. Different from IDTracker, DRSDetector [18] uses CBOW, LightGBM and HAN to process different resources of the target website respectively. And finally determines whether the target website is a malicious website through a scoring mechanism. Furthermore, Lei et al. [6] constructed a bipartite graph to capture the interactions between the terminal host and the domain name, the IP resolution structure of the domain name, and the DNS query time series pattern of the domain name. They then used graph embedding techniques to automatically learn the dynamic and discriminative feature representations of the domain name. Based on these feature representations, they predicted whether a newly observed domain name was malicious or benign.

The above methods perform well in identifying malicious websites, but some of the resources they rely on (such as IP addresses, WHOIS information, etc.) are difficult to obtain. Additionally, these methods are focused solely on the website level, which is insufficient for effective website governance. The method we propose in this paper is aimed at the needs of organizational-level website management, which can help reviewers better grasp the website ecology. Additionally, the method relies on only a small amount of easily accessible resource information and can achieve high-accuracy association tasks.

2.2 Text Matching Based on BERT Model

The BERT-based text matching task aims to evaluate the semantic similarity or relevance between multiple text fragments. The LTM-B model [7] is a long-text matching method based on BERT. It employs a twin network architecture and adopts a hierarchical approach to divide the text into multiple segments. The BERT model is then used to vectorize the text, generating a matrix representation of the document. Finally, the two document matrices are interacted, pooled, concatenated, and the matching results are output through classification in a fully connected layer. Xia et al. [17] proposed a knowledge-enhanced BERT model, which improved its performance on semantic text matching tasks by directly injecting prior knowledge into BERT's multi-head attention mechanism. The DABERT [15] model takes into account the impact of noise in sentences on model performance, and it is confirmed that subtle noise such as adding, deleting, and changing words in sentences may lead to prediction errors. To solve this problem, the authors proposed a novel dual attention method to enhance BERT's ability to capture fine-grained differences in sentence pairs, which to some extent solves the impact of sentence noise on model performance.

The BERT-based text matching method can make full use of the BERT model's ability to extract text information. It adopts a bidirectional attention mechanism and considers the context of the text at the same time, so as to have a deeper understanding of the semantic relationship. This method is suitable for analyzing the semantic structure of long and complex sentences such as HTML text, and has high matching accuracy. In addition, the BERT-based text matching method offers the advantages of flexible transfer learning and task adaptability. It only requires fine-tuning the pre-trained BERT model according to the specific task context to achieve a high matching rate. In this paper, we only use the training dataset to fine-tune the pre-trained BERT model [13] to achieve excellent performance, which greatly reduces the computational cost.

3   WebAA

In this section, we introduce WebAA in detail. The overall architecture of WebAA is shown in Fig. 1. WebAA consists of four modules. The first three modules obtain similarity scores based on external resources, HTML texts, and domain names, respectively. Finally, the fourth module integrates them to obtain the final score, which is used to determine whether the target websites belong to the same organization.

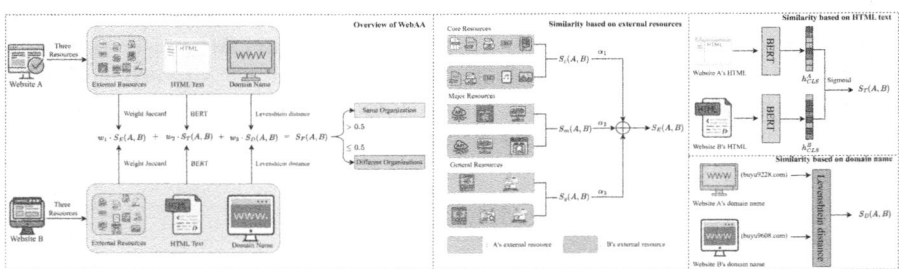

Fig. 1. Overview of WebAA. The left part shows the overall flowchart of the model, while the right part elaborates on the similarity calculation process based on three resources: similarity based on external resources, similarity based on HTML texts, and similarity based on domain names.

3.1   Similarity Based on External Resources

External resources are resources loaded by the website during the parsing process, such as CSS files, JavaScript files, images, audio and video, etc. Due to storage limitations, websites belonging to the same organization often share a large number of external resources, meaning that the external resources loaded by these websites are hosted on the same server.

Taking target websites A and B as an example, to calculate the similarity score between A and B based on external resources, We first obtain the server address where external resources are stored, denoted as $D_A = \{d_1^A, d_2^A, ..., d_n^A\}$, $D_B = \{d_1^B, d_2^B, ..., d_m^B\}$. This process is completed using BeautifulSoup of the bs4 library. Then, the resources are categorized based on their functional types, including core resources (such as images, audio and video, CSS files, etc.), major resources (such as third-party plug-ins), and general resources (such as advertising and analytics tools). Different weights are assigned to resources of different categorized (core resources have a weight of 3, major resources have a weight of 2, and general resources have a weight of 1), expressed as $W_A = \{w_1^A, w_2^A, ..., w_n^A\}$, $W_B = \{w_1^B, w_2^B, ..., w_m^B\}$. Next, inspired by Jaccard similarity, we propose weighted Jaccard similarity that first calculates the weighted similarity of resources at each category. Taking core resources as an example:

$$S_c(A,B) = \frac{\sum_{x \in D_A^{core} \cap D_B^{core}} \min\left(w_x^A, w_x^B\right)}{\sum_{x \in D_A^{core} \cup D_B^{core}} \max\left(w_x^A, w_x^B\right)} \tag{1}$$

Finally, the similarity scores across all categories are integrated to derive the external resources-based similarity score S_E.

$$S_E(A,B) = \alpha_1 S_c(A,B) + \alpha_2 S_m(A,B) + \alpha_3 S_g(A,B) \tag{2}$$

where, α_1, α_2 and α_3 are trainable weight parameters, and we limit $\alpha_1 + \alpha_2 + \alpha_3 = 1$.

3.2   Similarity Based on HTML Texts

We found that although websites under the same organization may serve different functions, their HTML texts show significant similarities. Based on this observation, we calculated the HTML text similarity scores between target websites using the BERT model. BERT has excellent performance in processing long texts and can easily perform transfer learning. For the pre-trained BERT model, only simple fine-tuning on our dataset is required to achieve excellent results.

To calculate the similarity of the HTML texts between target websites A and B, we first use a crawler tool to retrieve the HTML source code of the target websites. Then, we clean the data by removing tags and irrelevant content (such as scripts, style sheets, and other non-essential elements) to obtain clean and meaningful HTML text for input to the BERT model. The input text is first processed by a tokenizer, divided into sub-word units, and mapped into the embedding space. Assuming the input sequence is T, BERT maps it to the initial embedding matrix $E : E = [E_1, E_2, ..., E_n]$, where E_i represents the embedding vector of the i-th word. The input embedding is then processed by a multi-layer

Transformer encoder. Transformer uses a self-attention mechanism to capture contextual information and model the global dependencies of the input sequence:

$$H^l = Transformer(H^{l-1}) \tag{3}$$

where, H^l represents the embedding matrix of the l-th layer, which contains contextual semantic information. The initial state $H^0 = E$.

After multi-layer Transformer encoding, the embedding representation h_{CLS} of the [CLS] tag is extracted from the output matrix of the final layer as the global semantic feature of the input sequence:

$$H = \text{BERT}(T) \quad \Rightarrow \quad h_{\text{CLS}} = H[0] \tag{4}$$

where, h_{CLS} represents the HTML text embedding vector of the target website. For target websites A and B, after obtaining the corresponding embedding vectors h_{CLS}^A and h_{CLS}^B, we calculate the similarity score $S_T(A, B)$ through the sigmoid function:

$$S_T(A, B) = \sigma(W \cdot [h_{CLS}^A \oplus H_{cls}^B] + b) \tag{5}$$

where, $\sigma(x) = \frac{1}{1+e^{-x}}$ is sigmoid function, W and b are trainable parameters, and $\oplus$ represents the embedding vector concatenation operation.

3.3 Similarity Based on Domain Names

We observed that the domain names of websites under the same organization are very similar, so we use the domain name similarity score as one of the features to evaluate whether the websites belong to the same organization. In this section, we compute domain name similarity based on the Levenshtein distance. Specifically, we first calculate the Levenshtein distance $lev_{d_A,d_B}(|d_A|, |d_B|)$ between the domain names d_A and d_B of target websites A and B:

$$lev_{d_A,d_B}(i,j) = \begin{cases} \max(i,j) & \text{if } \min(i,j) = 0 \\ \min \begin{cases} lev_{d_A,d_B}(i-1,j) + 1 \\ lev_{d_A,d_B}(i,j-1) + 1 \\ lev_{d_A,d_B}(i-1,j-1) + 1_{(d_{A_i} \neq d_{B_j})} \end{cases} & \text{otherwise.} \end{cases} \tag{6}$$

where, $1_{(d_{A_i} \neq d_{B_j})}$ is an indicator function, which is equal to 0 when $d_{A_j} = d_{B_j}$ and 1 otherwise.

Since the Levenshtein distance is always an integer greater than or equal to 0, we get the domain name-based similarity score $S_D(A, B)$ by adding 1 to the Levenshtein distance and taking the inverse:

$$S_D(A, B) = \frac{1}{lev_{d_A,d_B}(|d_A|, |d_B|) + 1} \tag{7}$$

3.4 Fusion Module

In the first three modules, we obtained similarities based on external resources, HTML texts, and domain names. In this module, we fuse these similarities using weighted averaging to obtain the final similarity score $S_F(A, B)$ between the target websites:

$$S_F(A, B) = w_1 \cdot S_E(A, B) + w_2 \cdot S_T(A, B) + w_3 \cdot S_D(A, B) \qquad (8)$$

where, w_1, w_2 and w_3 are trainable parameters, represent the weights of the three similarities. $S_F(A, B)$ represents the final similarity score. If $S_F(A, B) > 0.5$, the target websites are judged to belong to the same organization. If $S_F(A, B) \leq 0.5$, the target websites are judged to belong to different organizations.

4 Experiment

In this section, we conduct extensive experiments aimed at answering the following three research questions:

- **(RQ1)** In the real dataset, are there correlations between websites under the same organization in terms of external resources **(RQ1-1)**, HTML text **(RQ1-2)**, and domain names **(RQ1-3)**? Additionally, are there differences in these resources between websites of different organizations?
- **(RQ2)** How does WebAA model perform on real datasets?
- **(RQ3)** How do different modules of the model (such as similarity of external resources, similarity of html text, and similarity of domain names) affect the model performance?

We first describe the dataset composition, experimental environment, and model hyperparameter settings in Sect. 4.1. Then, we address the three questions raised in this section through the experiments presented in Sects. 4.2 and 4.3.

4.1 Experiment Design

Dataset To verify the effectiveness of our model, we manually constructed two datasets based on real-world data. One is the legal website dataset D_{Legal}, and the other is the illegal website dataset $D_{Illegal}$ (including pornographic, gambling, drug, phishing websites, etc.). The detailed information of two dataset is shown in Table 1.

When building D_{Legal}, we used the website registration number as a reference. The registration number is a unique identifier assigned to an organization in mainland China when it registers a website or application. For multiple websites under the same organization, suffixes such as '–1' and '–2' are added to the registration number to distinguish them. Thus, the registration number serves as the identifier of the website owner. If two websites share the same registration number, they can be confirmed to belong to the same organization. When

Table 1. Overview of Datasets

Dataset	Web Num	Org Num	Largest Org	Smallest Org
D_{Legal}	4,746	3,127	281	1
$D_{Illegal}$	8,998	1,606	546	1

constructing the dataset, we first downloaded 51,612 Chinese websites from chinaz [3]. Next, we obtained the registration number for each website. Since the registration number information of some websites could not be found, we retained only those with available registration numbers. Finally, we labeled the websites based on their registration numbers and constructed the groundtruth. After screening, we retained a total of 4,746 websites belonging to 3,127 organizations. Among these, the largest organization included 281 websites, while the smallest organization consisted of just one website.

When constructing $D_{Illegal}$, we could not use the registration number as the basis because illegal websites generally do not contain registration number information. Fortunately, Wang et al. [14] published a dataset of illegal websites in their paper. The dataset contains 18,201 websites, which are divided into organizations. Since the life cycle of illegal websites is generally short, we further analyzed the activity of all websites in the dataset and retained only the currently active websites to construct $D_{Illegal}$. The dataset contains a total of 8,998 websites belonging to 1,606 organizations, among which the largest organization contains 546 websites and the smallest organization contains one website.

D_{Legal} and $D_{Illegal}$ represent legal and illegal website scenarios, respectively. By conducting experiments on these two datasets, we can evaluate the model's performance in different scenarios and further demonstrate its wide applicability and robustness across diverse scenarios.

Experimental Environment and Hyperparameters. When training the model, we set the ratio of the training set, validation set, and test set to 4:3:3. The number of epochs to 100, the learning rate to 0.001. We use the cross-entropy loss function to calculate the loss between the model's output and the true labels, and employ the Adam optimizer to compute gradients and update the parameters. In addition, to balance computational cost and performance, we set the embedding vector of the HTML text to 128 dimensions. We use PyTorch to implement WebAA and conduct experiments on two A100 GPUs with 80 GB of memory, the server operating system is Ubuntu 20.04.

4.2   Website Resource Analysis (answer RQ1)

In this section, we analyze the correlations and differences in external resources, HTML texts, and domain names between websites belonging to the same organization and those of different organizations to answer RQ1.

External Resource Analysis (Answer RQ1-1). In Section 3.1, we mentioned that websites belonging to the same organization share a significant number of external resources, whereas websites from different organizations do not. To verify this hypothesis, we conducted an experiment on the D_{Legal} dataset. We randomly selected two organizations for analysis, and the experimental results are presented in Fig. 2.

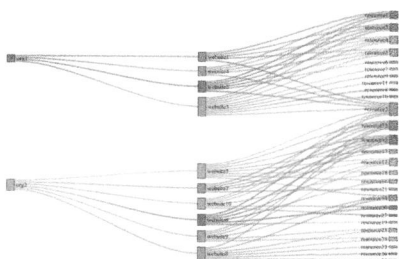

Fig. 2. External resources relied upon by websites under Org1 and Org2.

Considering privacy issues, we have anonymized the organization name, website, and external resource information. The experimental results indicate that websites within the same organization reuse a significant amount of external resources, whereas only a small number of resources are shared between websites from different organizations. For instance, as shown in the figure, only 'resource2' is shared between the websites of org1 and org2. Furthermore, we quantified the external resource reuse rate. Analysis of the two datasets revealed that websites within the same organization had a reuse rate of 66.4%, whereas websites from different organizations had a reuse rate of only 3.8%. These results further validate the accuracy of our idea and the effectiveness of this method for website association analysis.

HTML Text Analysis (Answer RQ1-2). As mentioned in Sect. 3.2, while websites within the same organization may serve different functions, their HTML texts exhibit significant similarities. To validate this, we conducted experiments on two datasets. Specifically, we selected five organizations from each dataset. Using the BERT model, we generated embedding vectors for the websites' HTML texts and used these vectors to represent them. To intuitively analyze the experimental results, we visualized the embedding vectors, as shown in Fig. 3. Since the vector dimension is 128, we used t-SNE [8] to reduce the dimension for easy display.

As illustrated in Fig. 3, the HTML text embeddings of websites belonging to the same organization tend to cluster together, while the HTML text embeddings of websites from different organizations are generally farther apart. This experimental result supports our idea: the HTML texts of websites within the same

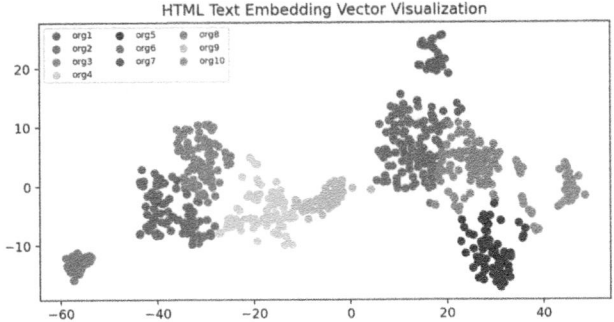

Fig. 3. Visual distribution of HTML text embedding vectors generated based on BERT in 2D space.

organization exhibit significant similarities, while those from different organizations are markedly distinct.

Domain Name Analysis (Answer RQ1-3). As mentioned in Sect. 3.3, the domain names of websites under the same organization are very similar, while those of websites under different organizations are quite different. To verify this, we conducted experiments using two datasets. Specifically, we randomly selected two organizations from each dataset and analyzed the domain name composition of each organization's websites to confirm the validity of our idea.

First, for each organization, we counted the overall distribution of characters in the domain names of all its websites and analyzed the organization-level character distribution. The experimental results are presented in Fig. 4. Among them, Org1 and Org2 are organizations in D_{Legal}, Org3 and Org4 are organizations in $D_{Illegal}$.

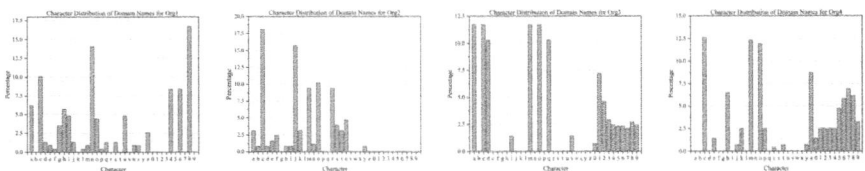

Fig. 4. Overall distribution of characters in the domain names for each organization.

The experimental results show that there are significant differences in character distribution between different organizations. For example, domain names in the Org1 tend to use characters such as 'n' and '8,' while the Org2 does not use numbers in domain names at all. Additionally, compared to legitimate websites, illegal websites are more likely to use numbers when constructing domain names. This phenomenon has also been confirmed by Antonakakis et al. [1].

Next, to further investigate the correlation in character distribution of domain names among websites within the same organization, we analyzed those distribution under four organizations. The experimental results are shown in Fig. 5. Also for privacy reasons, we hide the real domain name of the website and use *domainX* instead.

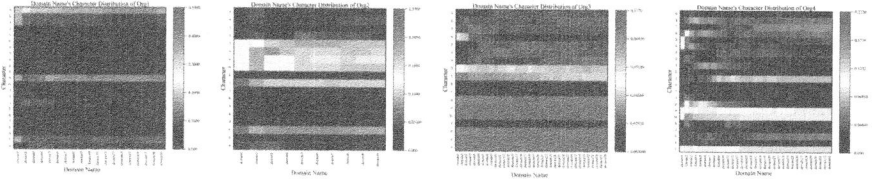

Fig. 5. Correlation heatmap of character distribution in domain names between websites.

The experimental results show that the character distribution of domain names within the same organization is very similar. For example, in Org2, all websites frequently use 'c' and 'j' in their domain names. These results further confirm the validity of our ideas and provide theoretical support for the effectiveness of our model.

4.3   Model Performance

In this section, we answer RQ2 and RQ3 by evaluating the performance of WebAA on two datasets, as well as the performance of each module in WebAA through ablation experiments.

WebAA Performance (Answer RQ2). In this section, we evaluate the performance of the WebAA model using two real-world datasets. Specifically, we randomly selected 2,000 pairs of websites (1,000 pairs of positive samples and 1,000 pairs of negative samples) from the D_{Legal} and $D_{Illegal}$ datasets, respectively, to form the test samples. We employed four metrics—accuracy, recall, F1 score, and the ROC curve—to quantitatively assess the model's performance.

The experimental results are presented in Table 2 and Fig. 6. On both datasets, the accuracy, recall and F1 score of the WebAA model exceed 90%. Specifically, on the D_{Legal} dataset, these three metrics surpass 95%, highlighting the exceptional performance of the WebAA model in the website association task. In addition, the ROC curve in Fig. 6 shows that at a lower false positive rate, the model is able to achieve a higher true positive rate, thus being able to effectively perform the website association task. However, compared to the D_{Legal} dataset, the model performed slightly worse on the $D_{Illegal}$ dataset. We attribute this to deliberate dissimilarity operations performed by the organizations behind illegal websites to avoid detection. These operations include

Table 2. Experimental results on two datasets

Model	D_{Legal}				$D_{Illegal}$			
	Acc	Recall	F1	Time(s)	Acc	Recall	F1	Time(s)
WebAA	**97.15**	95.10	**97.09**	0.45	**92.95**	**92.70**	**92.93**	0.45
WebAA$_r$	91.75	83.70	91.03	0.03	86.90	75.80	85.26	**0.12**
WebAA$_h$	96.80	96.50	96.79	0.38	90.85	88.50	90.63	0.40
WebAA$_d$	82.45	**97.30**	84.72	**0.02**	69.80	89.80	74.83	0.14

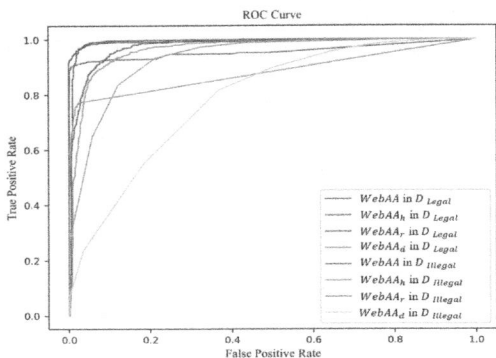

Fig. 6. ROC curves of four models on two datasets.

modifying HTML text, reducing domain name similarity, and minimizing the overlap of external resources. Such modifications introduced slight deviations in our model's performance during website association tasks. Nevertheless, it is worth noting that despite these dissimilarity operations, our model still achieved over 90% on all three metrics, which indirectly demonstrates its robustness and resistance to interference. In terms of time cost, the WebAA model can complete the association of thousands of website pairs within milliseconds on both datasets, fully demonstrating its efficiency and practicality.

Ablation Experiment (Answer RQ3). In this section, we evaluate the impact of different modules through ablation experiments. Specifically, we design three models: WebAA$_r$ (considers only external resource similarity), WebAA$_h$ (focuses solely on HTML text similarity), and WebAA$_d$ (examines only domain name similarity) to analyze the contribution of each module to the website association task. We apply the same samples and evaluation metrics used for WebAA to assess the performance of these three models. The experimental results are presented in Table 2 and Fig. 6.

The experimental results indicate that the performance of different modules varies significantly. Among them, the WebAA$_h$ model demonstrates the best performance, with each metric being only about 1% lower than that of the WebAA.

This suggests that, in the website association task, the similarity of HTML text between target websites is a crucial factor. The performance of the $WebAA_d$ model is relatively poor, with an accuracy of only 69.80% on $D_{Illegal}$. We analyzed that this is because, compared to HTML text and external resources, domain names are simply identifiers for websites and are more prone to dissimilarity. Therefore, using domain similarity alone will not yield the desired results. However, compared to other models, $WebAA_d$ has a higher recall rate and can efficiently recall positive samples.

In general, in the website association task, the three modules each have their own advantages and contribute to the WebAA model to varying degrees. Ultimately, the fusion of these modules enhances the overall performance of the WebAA model.

5   Conclusion

In this paper, we introduce the concept of website association and propose the first model, WebAA, for this task. This model determines whether target websites belong to the same organization by analyzing the similarity of multiple resources. As a further extension of website identification, website association focuses on the organization behind the website, providing a new technical approach for mapping cyberspace ecology and combating illegal websites. Extensive experiments on two manually constructed datasets demonstrate that our model can efficiently associate thousands of website pairs within milliseconds with excellent performance. This fully verifies its significant advantages in terms of accuracy, efficiency, and robustness. Additionally, we will open-source the relevant datasets to assist researchers in the community with their work in this field.

Acknowledgments. This work is supported by the Scaling Program of Institute of Information Engineering, CAS (Grant No. E3Z0041101).

References

1. Antonakakis, M., et al.: From throw-away traffic to bots: detecting the rise of dga-based malware. In: Proceedings of the 21th USENIX Security Symposium, 2012, pp. 491–506. USENIX Association (2012)
2. Chen, J., Zheng, S., Cheng, Y., Zhang, Z.: Data mining based analysis of online gambling sites and illicit financial flows. In: 2024 International Conference on Cloud Computing and Big Data, ICCBD 2024, pp. 205–211. ACM (2024)
3. Chinaz (2024). https://top.chinaz.com/all/
4. Devlin, J., Chang, M., Lee, K., Toutanova, K.: BERT: pre-training of deep bidirectional transformers for language understanding. In: Proceedings of the 2019 Conference of the North American Chapter of the Association for Computational Linguistics: Human Language Technologies, NAACL-HLT 2019, pp. 4171–4186. Association for Computational Linguistics (2019)

5. Ji, S., Pan, S., Cambria, E., Marttinen, P., Yu, P.S.: A survey on knowledge graphs: representation, acquisition, and applications. IEEE Trans. Neural Netw. Learn. Syst. **33**(2), 494–514 (2022)
6. Lei, K., Fu, Q., Ni, J., Wang, F., Yang, M., Xu, K.: Detecting malicious domains with behavioral modeling and graph embedding. In: 39th IEEE International Conference on Distributed Computing Systems, ICDCS 2019, pp. 601–611. IEEE (2019)
7. Long Liu, Xin Liu, L.C.C.T.: Long text matching model based on bert. Comput. Syst. Appl. (2018)
8. Van der Maaten, L., Hinton, G.: Visualizing data using t-sne. J. Mach. Learn. Res. **9**(11) (2008)
9. Sánchez-Rola, I., Dell'Amico, M., Balzarotti, D., Vervier, P., Bilge, L.: Journey to the center of the cookie ecosystem: unraveling actors' roles and relationships. In: 43rd IEEE Symposium on Security and Privacy, SP 2022, pp. 1990–2004. IEEE (2022)
10. Starov, O., Zhou, Y., Zhang, X., Miramirkhani, N., Nikiforakis, N.: Betrayed by your dashboard: Discovering malicious campaigns via web analytics. In: Proceedings of the 2018 World Wide Web Conference on World Wide Web, WWW 2018, pp. 227–236. ACM (2018)
11. Subramani, K., Melicher, W., Starov, O., Vadrevu, P., Perdisci, R.: Phishinpatterns: measuring elicited user interactions at scale on phishing websites. In: Proceedings of the 22nd ACM Internet Measurement Conference, IMC 2022, pp. 589–604. ACM (2022)
12. Subramani, K., Perdisci, R., Skafidas, P., Antonakakis, M.: Discovering and measuring cdns prone to domain fronting. In: Proceedings of the ACM on Web Conference 2024, WWW 2024, pp. 1859–1867. ACM (2024)
13. THUCNews: Bert-chinese-text-classification-pytorch. https://github.com/ 649453932/Bert-Chinese-Text-Classification-Pytorch
14. Wang, C., Li, Z., Yin, J., Liu, Z., Zhang, Z., Liu, Q.: Idtracker: discovering illicit website communities via third-party service ids. In: 53rd Annual IEEE/IFIP International Conference on Dependable Systems and Network, DSN 2023, pp. 459–469. IEEE (2023)
15. Wang, S., Liang, D., Song, J., Li, Y., Wu, W.: DABERT: dual attention enhanced BERT for semantic matching. In: Proceedings of the 29th International Conference on Computational Linguistics, COLING 2022, pp. 1645–1654. International Committee on Computational Linguistics (2022)
16. Worldwidewebsize: The size of the world wide web (the internet) (2025). https:// www.worldwidewebsize.com/
17. Xia, T., Wang, Y., Tian, Y., Chang, Y.: Using prior knowledge to guide bert's attention in semantic textual matching tasks. In: WWW '21: The Web Conference 2021, pp. 2466–2475. ACM/IW3C2 (2021)
18. Zhang, Y., Fu, X., Yang, R., Li, Y.: Drsdetector: detecting gambling websites by multi-level feature fusion. In: IEEE Symposium on Computers and Communications, ISCC 2023, pp. 1441–1447. IEEE (2023)

Exploring the Effect of Spatial Scales in Studying Urban Mobility Pattern

Hoai Nguyen Huynh$^{(\boxtimes)}$ 🆔

Institute of High Performance Computing (IHPC), Agency for Science, Technology and Research (A*STAR), 1 Fusionopolis Way, Connexis (North Tower) #16-16, Singapore 138632, Republic of Singapore
huynhhn@ihpc.a-star.edu.sg

Abstract. Urban mobility plays a crucial role in the functioning of cities, influencing economic activity, accessibility, and quality of life. However, the effectiveness of analytical models in understanding urban mobility patterns can be significantly affected by the spatial scales employed in the analysis. This paper explores the impact of spatial scales on the performance of the gravity model in explaining urban mobility patterns using public transport flow data in Singapore. The model is evaluated across multiple spatial scales of origin and destination locations, ranging from individual bus stops and train stations to broader regional aggregations. Results indicate the existence of an optimal intermediate spatial scale at which the gravity model performs best. At the finest scale, where individual transport nodes are considered, the model exhibits poor performance due to noisy and highly variable travel patterns. Conversely, at larger scales, model performance also suffers as over-aggregation of transport nodes results in excessive generalisation which obscures the underlying mobility dynamics. Furthermore, distance-based spatial aggregation of transport nodes proves to outperform administrative boundary-based aggregation, suggesting that actual urban organisation and movement patterns may not necessarily align with imposed administrative divisions. These insights highlight the importance of selecting appropriate spatial scales in mobility analysis and urban modelling in general, offering valuable guidance for urban and transport planning efforts aimed at enhancing mobility in complex urban environments.

Keywords: Gravity model · Public transport flow · Spatial scales · Urban mobility pattern · Urban modelling

1 Introduction

Urban mobility plays a crucial role in shaping the functionality and efficiency of cities, influencing economic activity, social interactions, and quality of life. An effective transportation system supports accessibility, reduces congestion, and enhances urban sustainability [4]. Public transport, in particular, serves as a

M. H. Lees et al. (Eds.): ICCS 2025, LNCS 15904, pp. 65–77, 2025.
https://doi.org/10.1007/978-3-031-97629-2_5

backbone of mobility in dense urban environments like Singapore, where land constraints necessitate efficient transport planning. Understanding travel patterns within the public transport network is essential for optimising infrastructure, improving service provision, and informing urban development policies [27]. Analysing these patterns requires robust models that can capture the complex dynamics of urban mobility and provide insights into how people move across different spatial scales [23, 25].

A variety of models have been developed to study urban mobility, ranging from agent-based simulations to network-based approaches [2, 3]. Among them, the gravity model has been widely used due to its simplicity and effectiveness in capturing aggregate travel flows [15]. Inspired by Newton's law of gravity, it assumes that the interaction between locations is proportional to their population or activity levels and inversely related to the distance between them [5]. The gravity model has been successfully applied in various urban contexts to estimate mobility patterns and forecast transport demand [18]. However, while the model provides a useful approximation of mobility flows, its accuracy is arguably influenced by the spatial scale at which it is applied.

Spatial aggregation has been shown to play a critical role in urban mobility modelling, affecting both data representation and model performance [6, 9, 20]. At fine spatial scales, such as individual bus stops or train stations, the models may struggle to capture meaningful patterns due to high variability and noise in travel behaviour, leading to overfitting or poor generalisability. Conversely, at very coarse spatial scales, over-aggregation may lead to a loss of critical mobility details, obscuring important underlying dynamics and interaction patterns, and resulting in model underperformance. These challenges reflect the Modifiable Areal Unit Problem (MAUP), a well-known issue in spatial analysis where different zoning schemes or levels of aggregation can lead to significantly different analytical results (sometimes referred to as "Openshaw effect" [10]). In mobility research, this means that both the resolution and method of spatial aggregation must be carefully chosen to ensure accurate model interpretation and policy relevance.

Recent studies have sought to quantify and mitigate the effects of spatial scale in mobility modelling. For instance, it has been showed that while some mobility metrics (e.g. radius of gyration and entropy) remain relatively stable across scales, others vary significantly depending on the spatial resolution, influencing how individual activity spaces are characterised [7, 26]. Similarly, spatial boundaries can be argued to critically affect the predictive power of mobility models [22, 23], which means that spatial scale may not be merely a technical detail but a foundational element of mobility theory. However, while it has been acknowledged that model performance can vary significantly depending on the chosen spatial resolution, there is limited consensus on the optimal level of aggregation [8, 12]. Furthermore, spatial aggregation based on administrative boundaries may not necessarily align with actual urban movement patterns, potentially introducing biases in mobility analysis.

Despite extensive research on urban mobility modelling, several gaps remain in understanding the interaction between spatial scale and model performance using public transport flow data. First, the impact of spatial scale on gravity model performance has not been systematically explored in the context of public transport networks, particularly in highly urbanised environments like Singapore. Second, while administrative boundaries are often used for spatial aggregation, their effectiveness compared to alternative aggregation methods like distance-based clustering remains unclear. Moreover, most of these studies have focused on either individual-level GPS [1], mobile phone records [16] or social media data [7], rather than formal public transport usage data, which reflects structured and policy-relevant travel behaviour. This study addresses these gaps by examining how spatial aggregation affects gravity model performance using public transport flow data in Singapore. Specifically, the influence of different spatial scales on model accuracy is investigated, and comparison between administrative boundary-based aggregation and distance-based methods is made to evaluate which better captures urban mobility patterns. The findings from this study can provide insights into optimal spatial resolutions for mobility analysis and inform urban and transport planning strategies.

The remainder of this paper is organised as follows. Section 2 describes the data and methods used in the study, including details on public transport flow data, spatial aggregation approaches, and gravity model fitting procedures. Section 3 presents the results and discussion, focusing on model performance across different spatial scales and aggregation methods, as well as comparison of mobility pattern between time windows. Finally, Sect. 4 concludes with key findings, implications for urban planning, and potential directions for future research.

2   Data and Methods

2.1   Data

The datasets used in this study were obtained from relevant authorities in Singapore, and can be categorised by public transport and administrative boundaries.

The public transport related data was obtained from the Land Transport Authority (LTA) of Singapore, which provides comprehensive data on public transportation infrastructure and usage across the city-state [14]. The data contains the location of bus stops and train stations and the amount of traffic flow between them. As this study focuses on mobility pattern within Singapore, the bus stops in Johor (Malaysia) that are parts of the cross-border services between Singapore and Malaysia are excluded. The traffic flow data contains information on the number of trips made between a pair of origin and destination transport nodes during hourly time windows (from HH:00 to HH:59) that will be merged to obtain the daily flow on a typical day of a month, for both weekdays and weekends. The public transport flow data used in this study is for October 2024, which was chosen to reflects recent typical mobility patterns without the anomalies and seasonal variations in travel behaviours during major holiday periods in Singapore.

In addition to transport flow data, administrative boundaries delineated in the Master Plan 2019 [24] by the Urban Redevelopment Authority (URA) are also used to compute different levels of spatial aggregation of transport nodes. These boundaries include three hierarchical levels: subzone, planning area, and region. Subzones represent the most granular administrative divisions in Singapore, while planning areas and regions provide broader spatial groupings.

2.2   Spatial Clustering of Nodes

Apart from the spatial aggregation by administrative boundaries, the transport nodes can also be clustered by spatial proximity. In this study, a procedure is devised to identify the clusters of transport nodes given a distance parameter. First, a network is contructed for all public transport nodes in Singapore with links added between pairs of nodes whose Euclidean distance is smaller than a given threshold d_{thr}. The weight of such links is calculated as the ratio between the overlap area of the buffer circles of radius $\rho = d_{thr}/2$ centred at the nodes and their union area (see Fig. 1). This area ratio reflects the strength of relationship between two nodes in terms of how close they are to one another. The network will then be divided into clusters using a procedure of community detection based on modularity (similar to previously employed in [11]). The clustering procedure is described in details in [13], involving multiple runs of the Louvain method for community detection and effective average of clustering patterns to identify the converged clusters of nodes.

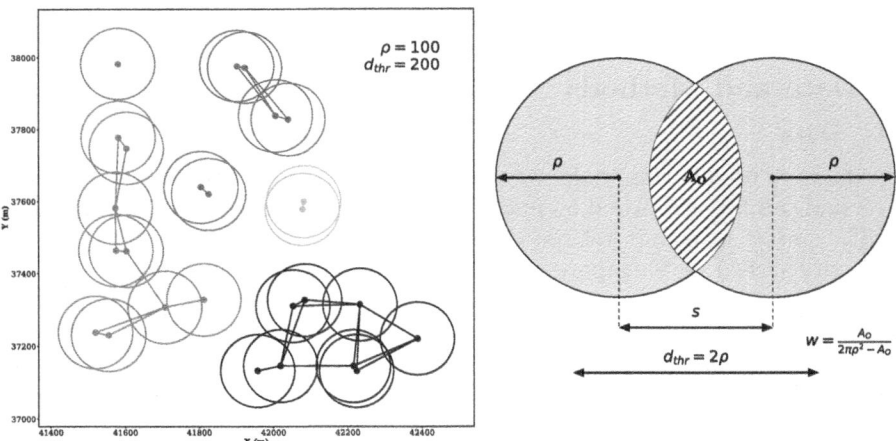

Fig. 1. Construction of network of transport nodes in which links and corresponding weights w are determined by the overlap area (hatched) of buffer circles of radius ρ centred at the nodes (right panel). The threshold distance for a pair of nodes to be considered in the same cluster is $d_{thr} = 2\rho$, beyond which the buffer circles do not overlap (left panel).

Different levels of spatial clustering are obtained by varying the distance threshold parameter d_{thr} from 0 to 6,000 m in steps of 100 m. For every value of d_{thr}, the Louvain community detection algorithm is applied 100 times to yield the clusters of nodes. These distance-based clusters together with administrative boundaries (subzone, planning area and region) will serve as different kinds of spatial aggregation to assess the performance of the urban mobility flow model described in the next section.

2.3 Modelling the Urban Mobility Flow

The gravity model has been widely used in transportation and urban studies to predict mobility flows between locations [18]. It is based on the analogy of Newton's law of gravity, where the interaction between two places is proportional to their population (or activity level) and inversely related to the distance between them. In the context of urban mobility, the model estimates the volume of trips between origin and destination locations and can be expressed as

$$F_{ij} = G \frac{M_i^\alpha M_j^\beta}{D_{ij}^\gamma} \tag{1}$$

in which F_{ij} denotes the traffic volume from location i to j, G is some scaling constant, M_i and M_j denote the corresponding activity level at these locations, and D_{ij} the distance between them, whereas α, β, and γ are associated parameters to be fitted using the mobility data. In this study, the total outflow traffic at location i and inflow traffic at location j are used as proxy for their activity level. The distance between the locations is taken as the Euclidean distance between the centroid of the cluster of transport nodes. It should be noted that in the case of administrative boundaries, the centroid is not the centroid of the polygon but the centroid of the cluster of transport nodes contained within the polygon.

The gravity model is then fitted using linear regression, where the logarithm of observed mobility flows is modelled as a function of explanatory variables including the activity level at origin and destination locations and the distance between them. This is achieved by employing the logarithmic form of Eq. 1

$$\log F_{ij} = \omega + \alpha \log M_i + \beta \log M_j - \gamma \log D_{ij} \tag{2}$$

in which the model parameters α, β, and γ are estimated using ordinary least squares (OLS) regression. Goodness-of-fit is evaluated using the coefficient of determination R^2 to assess how well the model explains variations in urban mobility flows between locations.

As the number of data points varies with different levels of spatial aggregation, the adjusted R^2 [21] is used to characterise the quality of model fitting instead of the usual R^2 to account for the data size and the complexity of the model (i.e. the number of independent variables). The formula for adjusted R^2 is given by $R_{adj}^2 = 1 - (1 - R^2)(n-1)/(n-p-1)$ in which n is the number of data points and p the number of parameters. Given the gravity model has been shown to work very well with urban mobility pattern, aggressive test of the model in this study will be performed by using only 50% of the data for training and the model is tested on the remaining 50%.

3   Results and Discussion

3.1   Patterns of Different Levels of Spatial Aggregation

As the value of the distance threshold d_{thr} varies, different clustering patterns of transport nodes are observed. At $d_{thr} = 0$, every node forms its own cluster, whereas the nodes are grouped into 6 clusters at $d_{thr} = 6,000$ m. The clustering patterns of nodes at different values of d_{thr} are shown in Fig. 2. These patterns are also compared with clustering of nodes by subzones, planning areas and regions to assess their alignment with administrative boundaries. It could be observed that the distance-based aggregation does not necessarily align with imposed administrative boundaries, suggesting nuanced differences in patterns of spatial organisation across scales. The selected distance threshold values of 300 m, 600 m, and 4,400 m in Fig. 2 show the clustering patterns that most closely match the clustering by administrative boundaries, as quantified by the mutual information score, which is commonly used to compare sets of different subset structures [19]. The identified clusters at different spatial aggregation levels will be used to assess the impact of spatial scale on the performance of the gravity model, providing insights into the relationship between spatial aggregation and urban mobility dynamics.

3.2   Performance of Gravity Model Across Spatial Scales

For every clustering structure of the transport nodes, the gravity model is fitted to the corresponding aggregated traffic flow pattern to assess its performance across spatial scales. In order to obtain a reliable measure of the performance, the model fitting is run 100 times with randomisation of 50:50 train-test split. Equation 2 is fitted using 50% of the data to estimate the parameters α, β, and γ, and the model performance is assessed based on its prediction of the remaining 50% of the data. The results for weekday mobility pattern (see Fig. 3, top panel) show that the fitting at the least aggregate level, the transport node, is the worst with R^2_{adj} being only around 0.35, meaning that less than 40% of variance in the traffic flow can be explained by the combination of total outflow at origin, total inflow at destination and the distance between them. The model performance quickly improves as the nodes become spatially aggregated. The same argument as in [12] could be made that the aggregation of nodes better reflects the underlying dynamics of traffic flows where commuters from a particular location may use multiple transport nodes within the vicinity.

As the spatial aggregation increases, the average quality of fitting reaches the peak value at $d_{thr} = 1,500$ m and starts to decline afterwards. This decline

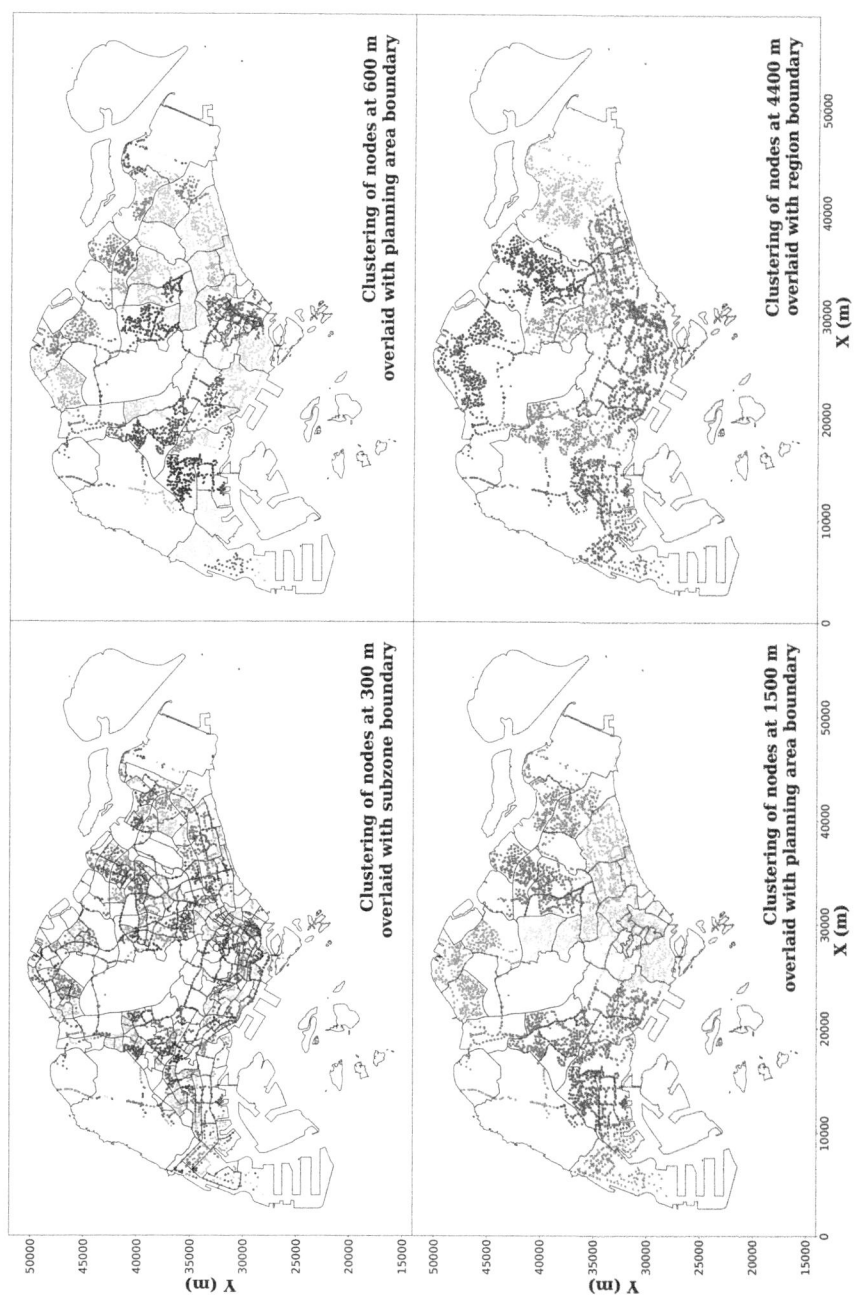

Fig. 2. Clusters of transport nodes at different values of threshold distance d_{thr}.

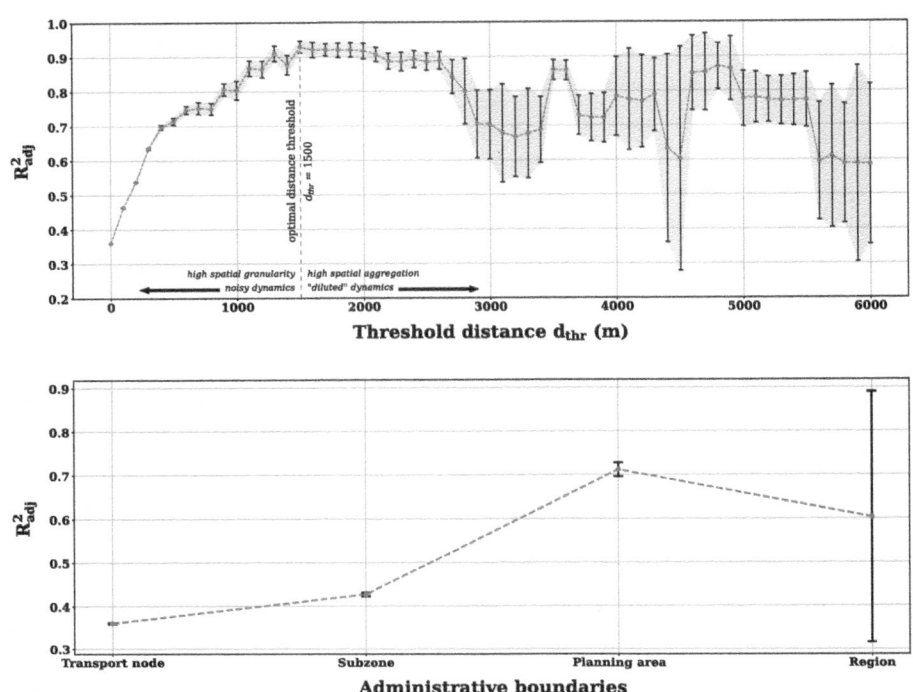

Fig. 3. Quality of fitting the gravity model to weekday mobility flows at different levels of spatial aggregation by distance threshold (top) and administrative boundaries (bottom). At each spatial aggregation, the average R^2_{adj} value and its error bar are computed over 100 runs with randomisation of 50:50 train-test split of the data.

signals that the transport nodes may be over-aggregated and that further congregating nodes may indeed "dilute" the dynamics of traffic flows whereby the true pattern is not captured as well as by a smaller mass of nodes, i.e. the flows are aggregated more than necessary. It is worth pointing out that the quality of fitting at large spatial scales fluctuates significantly compared to smaller ones, indicating low reliability of the fitting. Apart from diluting dynamics, the fact that the number of data points decreases with higher level of spatial aggregation may also contribute to a poorer fitting of the model when its complexity is not justified by the amount of data available.

3.3   Comparison of Mobility Pattern Across Temporal Windows

To further examine the temporal consistency of the gravity model performance, a stratified analysis is conducted based on time of day and weekday versus weekend travel patterns. The model is separately fitted to public transport mobility data from three distinct weekday periods, namely AM peak (6:00 AM to before 10:00 AM), PM peak (4:00 PM to before 8:00 PM), and off-peak hours (10:00

AM to before 4:00 PM), as well as to aggregated flows over the entire day on weekends. Across all time windows, the gravity model consistently shows the best performance when spatial aggregation is applied at 1,500 m (see Fig. 4, top panel). This suggests a robust spatial scale at which urban mobility dynamics in Singapore are optimally captured, regardless of temporal variation in travel behaviour. While minor fluctuations in model fitting are observed between time periods (likely due to differing trip purposes and passenger profiles), the spatial scale of 1,500 m provides a stable balance between granularity and aggregation. These findings reinforce the notion that intermediate spatial scales can effectively reduce noise in fine-grained data without oversimplifying travel patterns, making them suitable for both weekday commuting and weekend leisure mobility analysis.

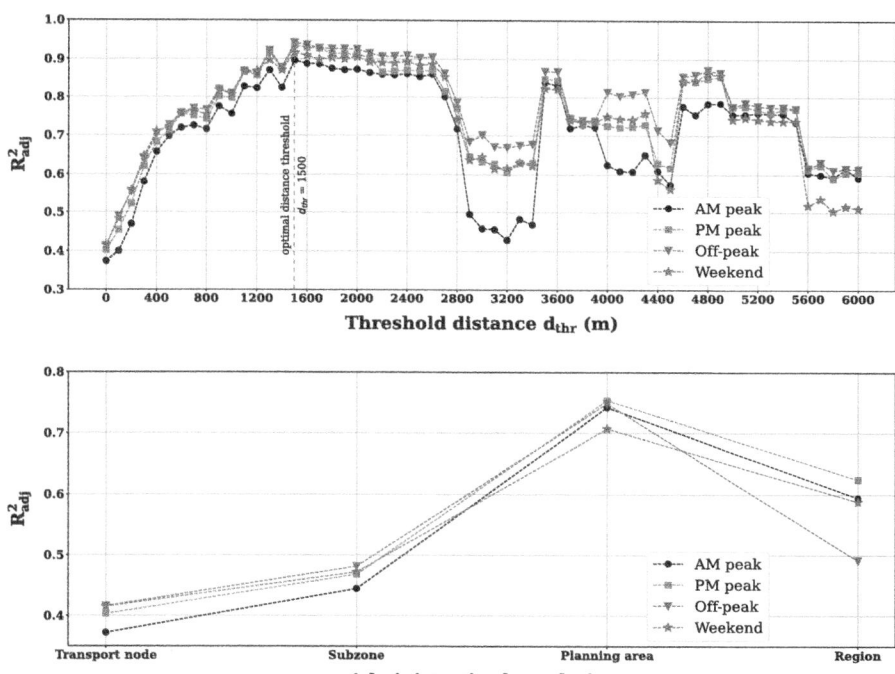

Fig. 4. Comparison of gravity model fitting by diffent periods: weekday AM peak, weekday PM peak, weekday off-peak, and weekend. Quality of fitting the gravity model at different levels of spatial aggregation by distance threshold (top) and administrative boundaries (bottom).

3.4   Effect of Different Spatial Aggregation Methods

A similar trend in the quality of fitting the model is also observed using the administrative boundaries as the method of spatial aggregation (see Fig. 3, bottom panel). At the first level of subzone, the fitting shows some improvement with R^2_{adj} rising above 0.4. The improvement continues at planning area level when the average quality of fitting reaches 0.7. However the trend does not hold beyond planning area when the model performs poorer at region level, indicating over-aggregation of transport nodes. Recalling the corresponding patterns of clusters based on distance threshold and administrative boundaries in Sect. 3.1, grouping of transport nodes at subzone level is closest to the pattern of clusters at 300 m, at planning area level the distance is 600 m, and region level maps best to 4, 400 m (see Fig. 2). The corresponding quality of fitting at these distance also shows comparable values with similar trend as the spatial aggregation level increases. It should be noted that all of these administrative boundary-based clusterings perform worse than the spatial aggregation at 1, 500 m. The poorer performance of spatial aggregation based on administrative boundaries compared to distance-based aggregation indicates that these boundaries may not accurately reflect the true patterns of movement and interaction among transport nodes.

These results are also consistent across temporal windows when fitting mobility data from different periods of weekdays (AM peak, PM peak, and off-peak) as well as on weekends (see Fig. 4, bottom panel). Similar to the results for weekday mobility data, the gravity model exhibits the best performance at planning area level among the administrative boundaries when fitting stratified data from these windows, with the average R^2_{adj} value peaking around 0.7. Notably, while the whole-day data on weekdays (Fig. 3, bottom panel) and weekends (green star marker in Fig. 4, bottom panel) show similar peak R^2_{adj} value of 0.7, the sub-weekday periods all show slightly higher peak value. This hints at the variability in mobility behaviour throughout the day affecting the model performance. In the same vein, the noticeably worse performance of the gravity model at region scale for off-peak period (blue triangle marker in Fig. 4, bottom panel) compared to whole days or peak periods could be due to irregular large-scale movement pattern when long-distance travel appears to be rare outside rush hours. Nevertheless, these observations require further substantiation which is beyond the scope of the current study.

3.5   Contribution to Mobility Research and Future Directions

This study employs the gravity model as a mean to illustrate the effect of spatial scales and units in urban modelling. While the gravity model offers a simple and intuitive framework for modelling urban mobility, it is not without limitations. It assumes that flows between locations depend solely on size and distance, overlooking other influential factors such as land use mix, transport connectivity, and individual travel preferences, which may influence local variations and dynamic behaviours inherent in real-world mobility. Despite these limitations, this study contributes to the field by systematically evaluating how spatial scale affects the model performance, offering practical guidance on appropriate aggregation levels for transport analysis. Furthermore, it highlights the mismatch between administrative boundaries and actual mobility patterns, suggesting that data-driven, distance-based approaches may yield more accurate representations of urban movement. Future research could build on these findings by incorporating additional variables into the model, such as socio-demographic factors or transport service attributes. The role of spatial scale can also be explored in other modelling approaches like radiation [17] or machine learning-based models [25] to provide more comprehensive understanding of complex travel behaviours in urban systems.

Moving toward practical applications of these findings, actionable insights can be derived by combining mobility patterns with contextual knowledge of the actual urban organisation. In the case of Singapore, the clustering pattern observed at the 1,500-m aggregation level (see Fig. 2, bottom left panel) suggests strong functional integration between areas such as Jurong East and Bukit Batok (violet cluster around X=16,000 and Y=35,000). Similarly, northern neighbourhoods like Woodlands, Sembawang, and Yishun, or northeastern areas such as Serangoon, Hougang, Sengkang, and Punggol, may benefit from being considered as cohesive planning units. These spatial patterns highlight the potential for more integrated planning strategies that align with how residents actually move through the city-state. Further analysis, such as incorporating the spatial distribution of amenities, residential density, and public transport infrastructure like bus routes and train stations, could provide deeper insights into mobility demand and service accessibility. While such integration lies beyond the scope of the current study, it represents a promising direction for future research. In other urban contexts, similar approaches could offer powerful tools for urban planning by combining mobility data with additional layers of urban organisation, such as land use and the spatial distribution of services and infrastructure.

4   Conclusion

In this study, a computational method is developed to analyse the urban mobility pattern in Singapore. The findings reveal that while the gravity model can generally capture the flow dynamics, its performance quality varies significantly when different spatial units are used to calculate the amount of traffic between origin and destination locations in the model. It is found that the model fits poorly at the transport node level and performs best at some intermediate level of spatial aggregation corresponding to a threshold distance of $1,500$ m between nodes. Beyond that scale, the model performance decreases, signaling over-aggregation. Similar pattern is observed if the administrative boundaries are used, where the model fits poorly at the lowest level of subzone and improves at the intermediate level of planning area before decreasing at the highest level of region. However, the spatial aggregation at these administrative boundaries perform poorer than the distance-based aggregation, indicating that the administrative boundaries are artificial and not reflective of the actual organisation of mobility patterns on the ground.

The findings here offer valuable insights into the spatial organisation of urban areas in Singapore. The method developed in this study could be used to identify functional urban areas at different scales when combined with other relevant datasets. Additionally, this approach can help reveal latent mobility patterns and interactions between different parts of a city, offering a data-driven lens through which to interpret urban dynamics. Both the methodology and results can be useful for relevant urban and transport planning authorities in understanding the impact of physical infrastructure on urban mobility behaviours so that future land use and transport network can be effectively developed. Future research could build on this work by incorporating additional layers of spatial information, such as the distribution of amenities, residential densities, and the structure of public transport network. This would allow for a more comprehensive analysis of urban function and accessibility, ultimately contributing to the development of more inclusive and efficient urban environments.

Acknowledgments. This research is supported by A*STAR project number CoT-H1-2025-3 under the Cities of Tomorrow Grant 2024.

Disclosure of Interests. The author has no competing interests to declare that are relevant to the content of this article.

References

1. Alessandretti, L., Aslak, U., Lehmann, S.: The scales of human mobility. Nature **587**(7834), 402–407 (2020)
2. Alessandretti, L., Natera Orozco, L.G., Saberi, M., Szell, M., Battiston, F.: Multimodal urban mobility and multilayer transport networks. Environ. Plann. B: Urban Analyt. City Sci. **50**, 2038 (2022)
3. Anda, C., Erath, A., Fourie, P.J.: Transport modelling in the age of big data. Int. J. Urban Sci. **21**, 19 (2017)
4. Banister, D.: The sustainable mobility paradigm. Transp. Policy **15**, 73 (2008)
5. Barbosa, H., et al.: Human mobility: models and applications. Phys. Rep. **734**, 1–74 (2018)
6. Bin Asad, K.M., Yuan, Y.: The impact of scale on extracting urban mobility patterns using texture analysis. Comput. Urban Sci. **3**, 33 (2023)
7. Bin Asad, K.M., Yuan, Y.: The impact of scale on extracting individual mobility patterns from location-based social media. Sensors **24**(12), 3796 (2024)
8. Chen, L., Gao, Y., Zhu, D., Yuan, Y., Liu, Y.: Quantifying the scale effect in geospatial big data using semi-variograms. PLoS ONE **14**, e0225139 (2019)
9. Dabiri, Z., Blaschke, T.: Scale matters: a survey of the concepts of scale used in spatial disciplines. Eur. J. Remote Sens. **52**, 419 (2019)
10. Goodchild, M.F.: The Openshaw effect. Int. J. Geogr. Inf. Sci. **36**(9), 1697–1698 (2022)
11. Huynh, H.N.: Spatial point pattern and urban morphology: Perspectives from entropy, complexity, and networks. Phys. Rev. E **100**(2), 022320 (2019)
12. Huynh, H.N.: Spatial structure of places in Singapore from a complexity perspective of the public transport network. Int. J. Smart Sustain. Cities **01**, 2340003 (2023)

13. Jiang, Z., Huynh, H.N.: Unveiling music genre structure through common-interest communities. Soc. Netw. Anal. Min. **12**, 35 (2022)
14. Land Transport DataMall. https://datamall.lta.gov.sg/. Accessed 17 Jan 2025
15. Li, B., et al.: Estimation of regional economic development indicator from transportation network analytics. Sci. Rep. **10** (2020)
16. Louail, T., et al.: Uncovering the spatial structure of mobility networks. Nat. Commun. **6**(1) (2015)
17. Masucci, A.P., Serras, J., Johansson, A., Batty, M.: Gravity versus radiation models: on the importance of scale and heterogeneity in commuting flows. Phys. Rev. E **88**(2), 022812 (2013)
18. Mepparambath, R.M., Huynh, H.N., Oon, J., Song, J., Zhu, R., Feng, L.: The impact of COVID-19 pandemic on the fundamental urban mobility theories using transit data from Singapore. Transp. Res. Interdisc. Perspect. **21**, 100883 (2023)
19. Nguyen, X.V., Epps, J., Bailey, J.: Information theoretic measures for clusterings comparison: variants, properties, normalization and correction for chance. J. Mach. Learn. Res. **11**, 2837 (2010)
20. Oshan, T.M., Wolf, L.J., Sachdeva, M., Bardin, S., Fotheringham, A.S.: A scoping review on the multiplicity of scale in spatial analysis. J. Geogr. Syst. **24**(3), 293–324 (2022)
21. Raju, N.S., Bilgic, R., Edwards, J.E., Fleer, P.F.: Methodology review: estimation of population validity and cross-validity, and the use of equal weights in prediction. Appl. Psychol. Measur. (4), 291 (1997)
22. Rozenfeld, H.D., Rybski, D., Andrade, J.S., Batty, M., Stanley, H.E., Makse, H.A.: Laws of population growth. Proc. Natl. Acad. Sci. **105**(48), 18702–18707 (2008)
23. Simini, F., González, M.C., Maritan, A., Barabási, A.L.: A universal model for mobility and migration patterns. Nature **484**(7392), 96–100 (2012)
24. Singapore's open data portal. https://data.gov.sg/. Accessed 17 Jan 2025
25. Toch, E., Lerner, B., Ben-Zion, E., Ben-Gal, I.: Analyzing large-scale human mobility data: a survey of machine learning methods and applications. Knowl. Inf. Syst. **58**(3), 501–523 (2018)
26. Wang, X., Yuan, Y.: Modeling user activity space from location-based social media: a case study of Weibo. Prof. Geogr. **73**(1), 96–114 (2020)
27. Zhao, P., Wang, H., Liu, Q., Yan, X.Y., Li, J.: Unravelling the spatial directionality of urban mobility. Nat. Commun. **15** (2024)

NotiCorr: Exposing Social Relationships via Notification Traffic of Instant Messaging Applications

Jiangchao Chen[1,2], Zhuojun Jiang[1,2(✉)], Jiangyi Yin[1,2], Dongfang Hao[1,2], Zhao Li[1,2], Meijie Du[1,2], and Qingyun Liu[1,2]

[1] Institute of Information Engineering, Chinese Academy of Sciences, Beijing, China
{chenjiangchao,jiangzhuojun,yinjiangyi,haodongfang,lizhao,
dumeijie,liuqingyun}@iie.ac.cn
[2] School of Cyber Security, University of Chinese Academy of Sciences,
Beijing, China

Abstract. Instant Messaging (IM) applications, such as Telegram and WeChat, have become indispensable tools for individuals. To protect users' privacy, popular IM applications employ advanced encryption mechanisms. However, we demonstrate that the encrypted traffic of popular IM applications can still leak information about users' social relationships. In this paper, we reveal that the message notification traffic in IM application is exploitable and propose a novel privacy attack called NotiCorr, which allows an adversary to infer the users in the same group based on flow correlation. Specifically, even if the IM application is not running, the client will still instantly receive group message notifications. To this end, we extract robust fingerprints from both message notification and message transmission traffic to enable attacks in more realistic usage scenarios. To the best of our knowledge, this is the first study to highlight the privacy risks posed by message notification traffic in IM applications. Through extensive experiments, we demonstrate that NotiCorr significantly outperforms related methods. Finally, we discuss the mitigation strategies.

Keywords: Social relationships · Privacy · Instant messaging applications · Flow correlation

1 Introduction

Instant Messaging (IM) applications have become essential tools for daily communication, with over 3 billion people use mobile IM applications worldwide [15]. Due to various factors, IM applications are subject to monitoring by governments and corporations. To protect users' privacy, popular IM applications have implemented encryption technologies to secure user communications.

Despite the use of encryption mechanisms, there remain risks of privacy leakage. Some studies have employed flow correlation techniques to reveal users' privacy [2,3,5,12,13]. For example, [5,12,13] utilized flow correlation to perform

M. H. Lees et al. (Eds.): ICCS 2025, LNCS 15904, pp. 78–93, 2025.
https://doi.org/10.1007/978-3-031-97629-2_6

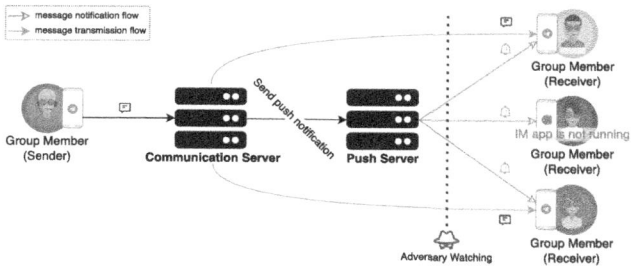

Fig. 1. The adversary correlates message notification flows and message transmission flows to identify users in the same group.

deanonymization on the Tor network. However, these methods fail to extract effective fingerprints for associating group members in IM applications. Bahramali et al. [2] and Bozorgi et al. [3] analyse the MTU-sized packets of message transmission traffic to extract fingerprints from IM application. Then, they identify users within the same group through flow correlation. However, these attacks have some limitations. When users receive messages from multiple groups simultaneously or within a short period, the correlation of message transmission flows become ineffective as the overlapping traffic disrupts the fingerprints. Additionally, these attacks are only effective when the IM application is active, as no message transmission traffic occurs when the application is not running.

To address these limitations, we find that the Push Notification Service in IM applications enables apps to deliver messages to users without requiring the app to be active. Therefore, users can receive message notifications from groups instantly even when the IM application is not running, as shown in Fig. 1. Besides, the interaction patterns of the message notification flows from users within the same group exhibit notable similarities, and the fingerprints extracted from these flows are highly robust. Leveraging these insights, we propose a flow correlation based privacy attack called *NotiCorr*, which combines message notification flows and message transmission flows to identify users in the same group.

To the best of our knowledge, we are the first to leverage message notification flows and message transmission flows to carry out a privacy attack. **Initially**, NotiCorr pre-classifies each user's flows into potential message notification and transmission flows based on the packet size distribution characteristics. **Subsequently**, we propose a PING-PONG based fingerprint extraction algorithm to obtain the message notification sequence and a Packet Inter-Arrival-Time (IAT) based fingerprint extraction algorithm for the message transmission sequence. **Then**, we apply a Spatial-Temporal based Longest Common Subsequence (LCS) algorithm to extract the common message notification sequence (denoted as $cN\text{-}FP$) between clients. Additionally, we utilize merge-Dynamic Time Warping (merge-DTW) algorithm to extract the common message transmission sequence (denoted as $cD\text{-}FP$) between clients. **Ultimately**, NotiCorr extracts features from the $cN\text{-}FP$ and $cD\text{-}FP$ between clients and inputs these features into a well-trained classifier model to obtain the final correlation result.

Extensive experiments are conducted to evaluate our method's performance. Comparisons with related flow correlation attacks demonstrate that NotiCorr achieves a higher True Positive Rate (TPR) and a lower False Positive Rate (FPR), maintaining robust performance across varying positive and negative sample ratios. Additionally, NotiCorr is deployed on an enterprise gateway and the experimental results demonstrate its real-world practicality. Ultimately, we propose mitigation strategies to address that privacy risk.

In summary, the contributions of this paper are as follows:

- This paper is the first study to reveal that the message notification traffic flow of IM application can be used to infer users' social relationships.
- We propose a flow correlation based privacy attack called NotiCorr to correlate users in the same group. By leveraging the immediacy of message notifications, our attack remains effective even when the IM application is not running.
- We perform extensive experiments to demonstrate that NotiCorr[1] outperforms related methods. We also deploy our method at an enterprise gateway, and real-world results show its practicality, achieving a TPR greater than 0.9 and a FPR lower than 3.5×10^{-5}.

Ethics. Our analysis focuses solely on traffic from our own IM application clients, without capturing or analyzing data from others. All attacks are conducted exclusively on our own clients. In real-world experiments, we neither save raw network traffic nor analyze packet payloads. Instead, our method processes gateway traffic in real time to extract features, which are not stored afterward. Thus, our experiments do not compromise the privacy of real-world IM application clients.

2 Background and Related Work

This section first introduces the Push Notification Service, followed by a summary of current research on privacy attacks leveraging encrypted traffic analysis.

2.1 Push Notification Service

The Push Notification Service (PNS) is a widely adopted mechanism in modern IM applications, enabling the timely delivery of messages and updates to users. It comprises a cloud of push-based messaging servers that are responsible for relaying messages from application servers to clients [24]. This allows servers to send notifications directly to users' devices, ensuring that messages and updates are received instantly, even when the application is not launched.

---

[1] We release its source code and the datasets at https://anonymous.4open.science/r/NotiCorr-784A.

2.2 Privacy Attacks Through Network Traffic Analysis

Numerous technologies have been proposed to protect user privacy, such as SSL/TLS protocols, SSH, and anonymous communication systems like Tor. Despite the increasing use of encryption, attackers can still exploit encrypted network traffic to conduct privacy attacks. Some researchers [11,13,17] have focused on website fingerprinting attacks, which can infer the websites a user visits. Other studies [7,9,21] demonstrate that attackers can identify the applications a user is using based on encrypted traffic. Moreover, some researchers [6,8,14] have advanced further, being able to recognize specific user activities within applications. In addition, progress has been made in content identification [1,10,18,19], enabling the detection of videos users are watching or the specific webpages they are accessing.

Recently, researchers have attempted to conduct privacy attacks on users' social relationships based on analysis of the encrypted traffic of IM application. Bahramali et al. [2] and Bozorgi et al. [3] utilize burst patterns in message transmission flow of IM application to extract the fingerprints, identifying users in the same group through flow correlation. These attacks are ineffective in real-world usage scenarios, such as when users receive messages from multiple group or when the application is not running. In this paper, we extract the fingerprints from the message notification flow of the IM application, which can leak more information. By combining these fingerprints with message transmission fingerprints, we implement a practical and effective privacy attack.

3 Analysing IM Application Message Notification and Message Transmission Traffic

In this section, we analyse the message notification traffic and the corresponding message transmission traffic in IM applications. Among various IM applications, we select Telegram for our traffic analysis due to two primary reasons. Firstly, Telegram has over 700 million monthly active users [16], making the analysis of privacy risks in its traffic broadly impactful. Secondly, the Telegram API provides detailed information on group messages, including sending times, message types, and message sizes. Furthermore, programmable bots in Telegram allow us to customize message content, facilitating the analysis.

Initially, we create a group and add two Android clients, running Telegram in the background with message notification enabled. And a bot sends messages of varying sizes, types, and sending intervals within the group. We utilize tcpdump to capture the traffic and export notification logs from phones via Android Debug Bridge (ADB). Ultimately, we collect message notification and transmission traffic for 500 messages. After that, we use the Telegram API to retrieve message timestamps and analyze logs to extract notification timestamps, which help identify message transmission and notification traffic.

After analysing the traffic of IM application, we identify three key insights that guide the development of NotiCorr.

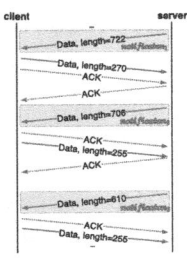   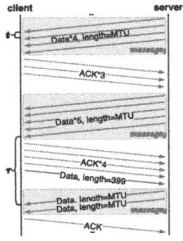

(a) message notification flow   (b) message transmission flow

Fig. 2. Interaction patterns of message notification flow and message transmission flow.

(i): *The message notification flow exhibits a distinct interaction pattern denoted as PING-PONG. Additionally, the downstream MTU-sized packets of the message transmission flow exhibit a distinct pattern denoted as burst.*

Our analysis reveals that message notifications are transmitted in a flow[2] with a distinct interaction pattern, as illustrated in Fig. 2a. Specifically, when a message notification is generated, the server encapsulates its content in a packet payload and transmits it to the client (PING). Upon receiving the packet, the client responds with an acknowledgment packet (PONG). We refer to this packet as a message notification packet and this interaction pattern as PING-PONG. For the message transmission flows, we find that they exhibit burst patterns, as shown in Fig. 2b. Specifically, the server transmits a message to the client by sending multiple MTU-sized packets within a short period. This burst pattern is consistent with findings from studies of Bahramali et al. [2] and Bozorgi et al. [3]. Within each burst, packets have short inter-arrival times (e.g., $t < 0.5$ *seconds*), whereas the intervals between bursts are longer (e.g., $\tau \geq 0.5$ *seconds*).

(ii): *Message notification packets from different users in the same group display highly consistent temporal (timing) and spatial (size) characteristics.*

We calculate the time and size differences of the message notification packets containing the same content between the two clients, plotting the cumulative distribution functions (CDF) in Fig. 3a and Fig. 3b. Figure 3a illustrates that nearly 100% of message notifications are received with a time difference of less than 1000 milliseconds, while over 80% arriving within 200 milliseconds. Figure 3b shows that almost 100% of message notification packets have a size difference of less than 100 bytes, with approximately 85% having identical sizes.

(iii): *There is a strong temporal correlation between a client's message notification packets and their corresponding message transmission packets. Exploiting this relationship enables the extraction of more robust fingerprints.*

We analyse the time distribution between a message notification packet and the corresponding message transmission on a single client. We calculate the time difference between the message notification packet and the corresponding mes-

---

[2] We define a flow as a sequence of packets composed of identical five-tuple(source/destination IP, source/destination port, transport layer protocol).

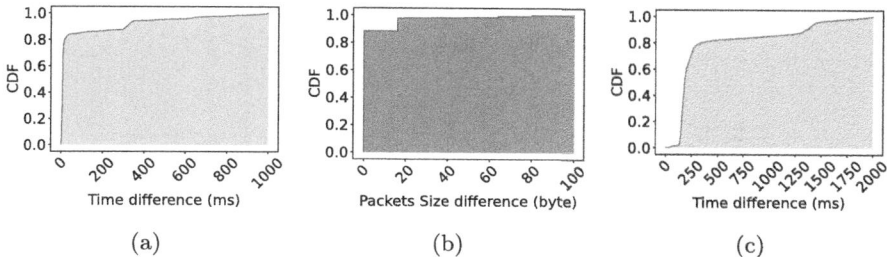

(a)   (b)   (c)

Fig. 3. Characterization of message notification and message transmission from two clients in the same group. 3a shows the CDF of arrival intervals of message notification packets between two clients; 3b presents the CDF of size differences of message notification packets between two clients; 3c illustrates the CDF of time intervals between message notifications and their corresponding message transmissions.

sage transmission, and plot the CDF, as shown in Fig. 3c. Figure 3c demonstrates that approximately 80% of message notifications and their corresponding message transmissions occur within 500 milliseconds, and nearly 100% occur within 2000 milliseconds.

We also analyse other popular IM applications, like Wechat, Whatsapp and QQ, whose message notification and transmission flows exhibit similar characteristics as described above.

4   Threat Model

We assume that all traffic between IM application clients and servers is encrypted and that the message notification function within IM application is enabled. Adversaries need to monitor users' encrypted network traffic. Leveraging Noti-Corr, attackers can achieve their goals without collaborating with IM providers or exploiting vulnerabilities within the application.

The attacker aims to identify the IP addresses of IM application users within the same group. By using NotiCorr, the attacker can detect IP pairs belonging to the same group and construct a social graph, revealing the social relationships of users associated with these IPs. Associating IP addresses with specific users is beyond the scope of this study and has been addressed in prior work [4].

5   Approach

In this section, we describe the methodology of this paper, referred to as Noti-Corr. As shown in Fig. 4, NotiCorr consists of four primary steps: (i) Preprocessing network traffic and pre-categorizing flows. (ii) Fingerprint extraction. (iii) Fingerprint comparison. Then, we extract the comparison features from these sequences. (iv) Identification.

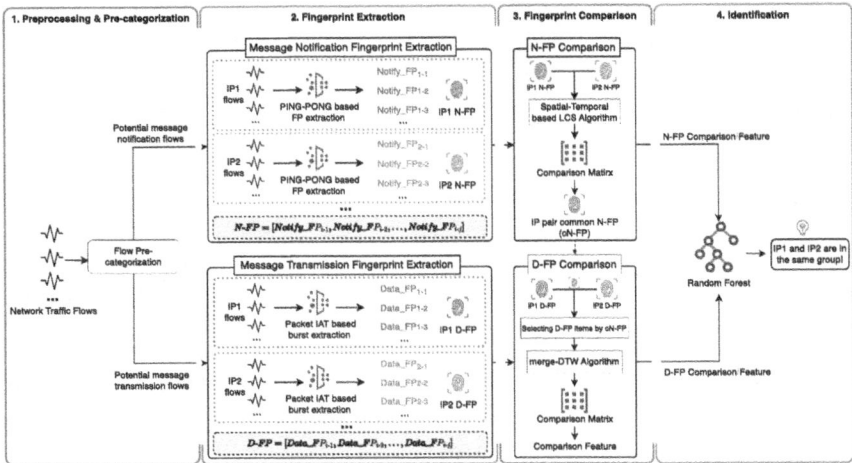

Fig. 4. The framework of NotiCorr.

5.1 Network Preprocessing and Flow Pre-Categorization

In this step, network traffic is initially divided into flows. We then classify these flows into either potential message notification flows or potential message transmission flows based on the proportion of MTU-sized packets in the downstream of each flow. A detailed analysis of the packet size distribution for message notification flows and other flows is provided in Sect. 6.

5.2 Fingerprint Extraction

Based on the analysis in Sect. 3, we develop distinct fingerprint extraction algorithms for message notification and transmission flows.

PING-PONG Based Fingerprint Extraction. For a given flow j of a client IP i, the objective of this algorithm is to extract the message notification sequence as a fingerprint $Notify\_FP_{i\text{-}j} = [(NT_{i\text{-}j_1}, NS_{i\text{-}j_1}), (NT_{i\text{-}j_2}, NS_{i\text{-}j_2}), ..., (NT_{i\text{-}j_n}, NS_{i\text{-}j_n})]$, where $NT_{i\text{-}j_n}$ represents the timestamp of the n-th message notification and $NS_{i\text{-}j_n}$ denotes its size. As detailed in Sect. 3, each message notification packet sent from the server to the client is typically followed by an ACK packet from the client to the server. Consequently, we identify message notification packets by locating downstream packets that arrive between two consecutive upstream packets in the potential message notification flow. We denote the downstream packet time as the message notification time and packet size as the message notification size. If there are multiple downstream packets between two consecutive upstream packets, the message notification size is determined by summing the sizes of these downstream packets, and the message notification time is marked by the arrival time of the first downstream packet. This method allows an adversary to extract potential message notification sequences from the observed message notification flows associated with a given client IP.

Packets IAT Based Fingerprint Extraction. When an IM client receives a message (e.g., text, picture, video, audio, etc.), the downstream packets of message transmission flow exhibit burst pattern. This feature has also been analyzed in previous studies [2,3]. In this paper, the burst sequence extracted from the message transmission flow as message transmission fingerprint is an auxiliary feature to reduce the false positive rate. We extract burst sequence from the MTU-sized packets of message transmission downstream based on packets IAT. After extracting fingerprint based on packets IAT, we get the burst sequence $Data\_FP_{i\text{-}j} = [(DT_{i\text{-}j_1}, DS_{i\text{-}j_1}), (DT_{i\text{-}j_2}, DS_{i\text{-}j_2}), ..., (DT_{i\text{-}j_n}, DS_{i\text{-}j_n})]$ from message transmission flow j of client IP i, where $DT_{i\text{-}j_n}$ represents the timestamp of the first packet in the n-th burst and $DS_{i\text{-}j_n}$ denotes the cumulative packet size in that burst.

After fingerprint extraction, each flow of one client is extracted either a notification fingerprint or a message transmission fingerprint. Ultimately, we construct a list of notification fingerprints (denoted as $N\text{-}FP$) and a list of message transmission fingerprints (denoted as $D\text{-}FP$) for all flows of each client.

5.3   Fingerprint Comparison

This step aims to derive message notification comparison features denoted as $\mathcal{F}_{N\text{-}FP}$, and message transmission comparison features denoted as $\mathcal{F}_{D\text{-}FP}$, by comparing the message notification fingerprint lists and message transmission fingerprint lists between each client IP pair.

Message Notification Fingerprint Comparison. According to Sect. 3, message notification packets of clients from the same group exhibit temporal and spatial similarities. Therefore, given two message notification fingerprint items $(NT_{p\text{-}j_m}, NS_{p\text{-}j_m})$ and $(NT_{q\text{-}k_n}, NS_{q\text{-}k_n})$, we consider these items to be similar if $|NT_{p\text{-}j_m} - NT_{q\text{-}k_n}| < \alpha$ and $|NS_{p\text{-}j_m} - NS_{q\text{-}k_n}| < \beta$, where α and β are the thresholds of time and size. Based on this logic, we design a Spatial-Temporal based LCS algorithm to identify the common message notification sequence between IP pair, as shown in our source code. By comparing the message notification fingerprints between a client IP pair using this algorithm, we finally obtain a comparison matrix and we denote the element in the k-th row and j-th column of this matrix as $cN\text{-}FP_{k\text{-}j} = (cN\text{-}FP_{k\text{-}j_p}, cN\text{-}FP_{k\text{-}j_q})$, where $cN\text{-}FP_{k\text{-}j_p}$ is the common message notification sequence of flow j from client IP p and $cN\text{-}FP_{k\text{-}j_q}$ is the common message notification sequence of flow k from client IP q. We sort all message notification sequence items of each IP in the comparison matrix by time to derive the common message notification sequence denoted as $cN\text{-}FP$. Subsequently, we calculate statistical features from $cN\text{-}FP$ to form the $\mathcal{F}_{N\text{-}FP}$.

These features include the length of $cN\text{-}FP$ and the mean, minimum, and maximum values of $mean(cN\text{-}FP)$. The $mean(cN\text{-}FP)$ is defined as follows:

$$mean(cN\text{-}FP) = [\frac{S_{p1} + S_{q1}}{2}, \frac{S_{p2} + S_{q2}}{2}, ..., \frac{S_{pn} + S_{qn}}{2}]$$

where S_{pn} and S_{qn} are the sizes of the n-th item in $cN\text{-}FP$ for client IPs p and q, respectively.

Message Transmission Fingerprint Comparison. Fig. 3c demonstrates that message notification packets and corresponding message transmissions exhibit a strong temporal correlation. Consequently, we select the burst from message transmission fingerprint based on the time of the item in $cN\text{-}FP$. Specifically, given a message transmission fingerprint item $(DT_{p\text{-}j_m}, DS_{p\text{-}j_m})$, and an item (T_{pn}, S_{pn}) in $cN\text{-}FP$ of client IP p, if $T_{pn} - \tau_- < DT_{p\text{-}j_m} < T_{pn} - \tau_+$, we consider that there is a temporal correlation between $(DT_{p\text{-}j_m}, DS_{p\text{-}j_m})$ and (T_{pn}, S_{pn}), where T_{pn} is time of the n-th item in $cN\text{-}FP$, τ_- and τ_+ are predetermined thresholds. After that, we obtain the selected message transmission fingerprint list $\widetilde{D\text{-}FP}_p$ and $\widetilde{D\text{-}FP}_q$ for a client IP pair p and q, respectively. We then apply our merge-DTW algorithm to identify the most similar burst sequence and calculate the corresponding distance between $\widetilde{D\text{-}FP}_p$ and $\widetilde{D\text{-}FP}_q$. By comparing the message transmission fingerprints between a client IP pair, we construct a comparison matrix and denote the element in the k-th row and j-th column of this matrix as $cD\text{-}FP_{k\text{-}j} = (dis_{k\text{-}j}, cN\text{-}FP_{k\text{-}j_p}, cN\text{-}FP_{k\text{-}j_q})$, where $cN\text{-}FP_{k\text{-}j_p}$ and $cN\text{-}FP_{k\text{-}j_q}$ are the common burst sequences of the client IP pair p and q, $dis_{k\text{-}j}$ is the distance between $cN\text{-}FP_{k\text{-}j_p}$ and $cN\text{-}FP_{k\text{-}j_q}$. Subsequently, we identify a matrix element with the minimum distance dis denoted as $min(cD\text{-}FP)$. We then calculate the total length, maximum value, and mean value of the two common burst sequences in $min(cD\text{-}FP)$. These features serve as the $\mathcal{F}_{D\text{-}FP}$.

Merge-DTW is an optimized version of the standard DTW algorithm that we developed to account for the effects of network conditions such as network latency, which may cause a single message to be split into two bursts in flow. Specifically, when calculating the distance between two burst items $(DT_{p\text{-}j_m}, DS_{p\text{-}j_m})$ and $(DT_{q\text{-}k_n}, DS_{q\text{-}k_n})$, merge-DTW computes the absolute difference between $DS_{p\text{-}j_m}$ and $DS_{q\text{-}k_n}$. Additionally, it calculates the absolute difference between $DS_{p\text{-}j_m}$ and the sum of $DS_{q\text{-}k_n}$ and $DS_{q\text{-}k_{n-1}}$. The minimum value of these two absolute difference values is then taken as the distance between two burst items.

5.4   Identification

In this step, we have derived the comparison features $\mathcal{F}_{D\text{-}FP}$ and $\mathcal{F}_{N\text{-}FP}$ for a given client IP pair. These features are used as inputs to a classifier, which outputs the probability that the IP pair belongs to the same group. We finally select the Random Forest model as the classifier.

Table 1. Group relationships of Telegram clients.

Clients	$group_A$	$group_B$	$group_C$	$group_D$	$group_E$
C_1	yes	yes	no	no	no
C_2	no	yes	yes	yes	no
C_3	no	no	yes	no	yes

6 Evaluation

In this section, we evaluate the performance of NotiCorr under various configurations and compare its effectiveness against related methods.

6.1 Experimental Setup

Dataset. We simulate group relationships using three Telegram clients, where two client pairs share the same groups, and each client also participates in other groups independently, as shown in Table 1. Each client remains active with message notifications enabled. And Telegram bots send messages in these groups, generating the traffic of message notification and message transmission. To simulate realistic group chat scenarios, we add 500 public groups and record message features, including message time, type and size, using the Telegram API. Following the method of Bahramali et al. [2], we generate message sequences for each group, including sending intervals, message types and sizes. In each round, bots send these messages and we capture traffic from each client using tcpdump until all messages are sent, resulting in 8,000 rounds of traffic with 126,798 flows.

Since our evaluation requires ground truth labels, we export Telegram's message notification logs via ADB. We label the message notification flows in each round of traffic using the log information. We divide the traffic corresponding to 8000 rounds into training, testing, and validation datasets using a 7:1:2 ratio.

Baseline. The most relevant methods to NotiCorr are traffic flow correlation. Therefore, we compare NotiCorr with existing privacy attack methods that utilize flow correlation, which either have open source code or provide the source code through contact with the authors. These related methods include Deep-Corr [12], DeepCoffea [13], FlowTracker [5] and Bahramali [2].

Metrics. Similar to previous studies [2,12,13], True Positive Rate (TPR) and False Positive Rate (FPR) are used to evaluate flow correlation performance Due to the TPR and FPR of NotiCorr, DeepCorr, DeepCoffea and Bahramli vary with changes in the threshold. Therefore, we also calculate the area under the curve (AUC) based on their TPR and FPR at different thresholds to measure their overall performance. The closer a method's AUC value is to 1, the higher its TPR and the lower its FPR at a given threshold.

Table 2. Hyperparameters of NotiCorr. (Note: The bold numbers in the search range represent the default values of the parameters)

threshold	search range	optimal value
$ratio_{D-MTU}$	✗	0.4
IAT_{burst}	✗	0.5 s
α	[1000, **2000**, ..., 5000]	1000
β	[0, 100, **200**, ..., 400]	0
τ_-	[1000, **2000**, ..., 5000]	2000
τ_+	[1000, **2000**, ..., 5000]	2000

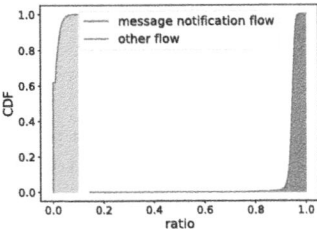

Fig. 5. CDF graph of the proportion of MTU size packets in the downstream of message notification flows and other flows

Fig. 6. Comparison of the effectiveness of using different algorithms for D-FP comparison in NotiCorr.

6.2   NotiCorr Effectiveness

We evaluate the impact of the different settings and ideas behind NotiCorr on the overall effectiveness of the attack.

Hyperparameter Selection. There are multiple thresholds in NotiCorr, including the ratio of the number of MTU-size packets in downstream flow for *Flow Pre-categorization* (denoted as $ratio_{D-MTU}$), the IAT of packets in Sect. 5.2 (denoted as IAT_{burst}), the time threshold α and size threshold β in Sect. 5.3, the time range threshold τ_- and τ_+ in Sect. 5.3. For the threshold $ratio_{D-MTU}$, we conduct a statistical analysis on the proportion of MTU-sized packets in the downstream of message notification flows and the other flows in the training and testing datasets, as shown in Fig. 5. The proportion of MTU-sized packets shows distinct difference between message notification flows and other flows. Therefore, we empirically select 0.4 as the value of threshold $ratio_{D-MTU}$. For threshold IAT_{burst}, we use the value of 0.5 s from the work of Bahramali et al. [2] And for the remaining thresholds α, β, τ_- and τ_+, we search within a certain range for each threshold in the training and testing datasets to find the value that maximize the effectiveness of NotiCorr. Specifically, we iterate through each

Table 3. The performance of NotiCorr and other four methods with different positive and negative sample ratios. The values in brackets are the performance of methods using our features

method	metrics	positive and negative sample ratio				
		1:9	1:49	1:99	1.149	1.199
NotiCorr	AUC	**0.9822**	**0.9820**	**0.9820**	**0.9822**	**0.9822**
DeepCorr		0.7321(0.9700)	0.6758(0.9668)	0.6363(0.9695)	0.5702(0.9690)	0.4894(0.9667)
DeepCoffea		0.6227(0.9507)	0.5979(0.9506)	0.5789(0.9504)	0.5327(0.9492)	0.5259(0.9438)
Bahramali		0.7685(0.9659)	0.6945(0.9599)	0.6689(0.9546)	0.5698(0.9539)	0.5064(0.9467)
FlowTracker	TPR	0.5083(0.9698)	0.4368(0.9584)	0.4232(0.9509)	0.4209(0.9498)	0.4168(0.9550)
	FPR	0.2991(0.0764)	0.2197(0.0662)	0.2099(0.0640)	0.2066(0.0635)	0.2050(0.0635)

parameter, change its value in the search space each time, and train a random forest model in NotiCorr to obtain its best performance in testing dataset. Table 2 shows the search range and optimal value of each parameter in NotiCorr.

The Effectiveness of Message Transmission Fingerprint and Message Notification Fingerprint. To evaluate the effectiveness of the two fingerprint types in NotiCorr, we compare its performance with a version using only the message notification fingerprint, denoted as NotiCorr_only_notify. As shown in Fig. 6, NotiCorr_only_notify achieves an AUC of 0.9707, demonstrating the effectiveness of the message notification fingerprint. When the TPR is 0.9, the FPR of NotiCorr_only_notify is higher than that of NotiCorr, indicating that message transmission fingerprint can reduce the false positive rate.

The Effectiveness of Merge-DTW. To evaluate the effectiveness of merge-DTW, we compare the performance of NotiCorr, NotiCorr using standard DTW algorithm (denoted as NotiCorr_DTW), and NotiCorr using cosine similarity (denoted as NotiCorr_Cosine). The results, shown in Fig. 6, indicate that Noti-Corr using merge-DTW outperforms NotiCorr_DTW. This improvement is due to merge-DTW accounting for scenarios where a message transmission burst may split into two bursts due to network environment, such as network latency. Additionally, NotiCorr using merge-DTW outperforms NotiCorr_Cosine, as merge-DTW allows for non-linear alignment of two sequences, enabling the optimal matching for sequences with unequal lengths or time axis offsets.

6.3   Comparison with Related Methods

In this section, we compare NotiCorr with four related methods on our dataset. Table 3 shows the performance of NotiCorr and four other methods with different positive and negative sample ratios. From the table, we can draw the following conclusions:

NotiCorr achieves a greater TPR with a lower FPR. Across all positive and negative sample ratios, NotiCorr achieves a greater TPR with a lower FPR compared to the four other methods. The success of NotiCorr in achieving such strong experimental results can be attributed to its effective utilization of the PING-PONG interaction pattern in message notification traffic and the burst characteristics of message transmission traffic during the fingerprint extraction phase, enabling the accurate identification of both message notification and message transmission packets.

To further verify the effectiveness of the feature extracted in this paper, we use the common message notification sequence $cN\text{-}FP$ of each IP pair as the input to baseline methods. We then train and evaluate these four methods, and the results are shown in Table 3. According to the result, our features significantly improve the performance of all four methods across different sample ratios. This indicates that the original features used in these methods are not well-suited for the attack scenario described in this paper.

NotiCorr exhibits robustness under different positive and negative sample ratios. As shown in Table 3, even as the proportion of negative samples increases, the AUC for NotiCorr remains consistently high (AUC ≥ 0.98). In contrast, the AUC of the three other methods decreases as the proportion of negative samples rises, and the effectiveness of FlowTracker decreases with the increase in the proportion of negative samples.

6.4   Real-World Experiments

In this section, we evaluate the effectiveness of our method in a real-world network environment. Specifically, we collaborate with an enterprise to deploy NotiCorr at the enterprise gateway and evaluate its effectiveness in a practical setting.

We monitor the traffic continuously, segmenting it into 30-second time windows. Within each window, flows are preprocessed and pre-categorized, and fingerprints are extracted. Next, NotiCorr's Fingerprint Comparison and Identification modules were employed to correlate potential notification flow pairs. Upon correlation, NotiCorr provided the client IPs of the flow pair. Twenty enterprise volunteers are recruited and permitted to freely form groups and join public Telegram groups, with their usage information recorded.

Our method is deployed continuously at the enterprise gateway for 7 days, during which we record the total time taken for correlation within a window. **NotiCorr takes less than 15 s to process each window, ensuring that the current window could be processed before the next one is generated.** NotiCorr's output for each window is then compared with volunteer usage data to evaluate its real-world performance. The results show that NotiCorr achieves a TPR above 0.9 with a FPR below 3.5×10^{-5}.

7   Mitigation Strategies

We propose two types of mitigation strategies for IM application users and providers. For IM application providers, we recommend altering the interaction pattern of the message notification traffic. One approach is to introduce redundant packets into the notification flows, thereby making notification packets indistinguishable. For IM application users, we suggest using proxy or VPN tools (such as V2Ray/V2Fly and shadowTLS) with obfuscation configurations that include multiplexing and random padding. Multiplexing interleaves packets from multiple flows, altering traffic patterns in terms of packet size, timing, and direction [22]. By mixing multiple client flows, multiplexing makes it challenging for attackers to distinguish individual flows. Additionally, random padding involves appending dummy data to payloads to obscure patterns of packet sizes [20, 23]. This technique alters the fingerprints that attackers can extract, thereby complicating successful fingerprint comparison.

8   Conclusion

In this paper, we reveal the message notification traffic in IM applications is vulnerable, which can leak users' privacy. Specifically, we design a privacy attack, called NotiCorr, based on message notification fingerprints and message transmission fingerprints, which allows attackers to identify users in the same group. Through extensive experiments, we demonstrate that the fingerprints extracted by NotiCorr are robust and our method outperforms related methods. We also deploy NotiCorr at an enterprise gateway to evaluate its performance and the results indicate that our method is practical and effective in the wild. Moreover, we propose corresponding mitigation strategies to IM application providers and users for this privacy attack.

Acknowledgments. This work is supported by the Scaling Program of Institute of Information Engineering, CAS (Grant No. E3Z0041101)

References

1. Bae, S., Son, et al.: Watching the watchers: practical video identification attack in LTE networks. In: 31st USENIX Security Symposium, USENIX Security 2022. pp. 1307–1324 (2022)
2. Bahramali, A., Houmansadr, A., Soltani, R., Goeckel, D., Towsley, D.: Practical traffic analysis attacks on secure messaging applications. In: 27th Annual Network and Distributed System Security Symposium, NDSS 2020 (2020)
3. Bozorgi, A., et al.: I still know what you did last summer: inferring sensitive user activities on messaging applications through traffic analysis. IEEE Trans. Dependable Secur. Comput. **20**(5), 4135–4153 (2023)
4. Cui, T., Gou, G., Xiong, G., Li, Z., Cui, M., Liu, C.: SiamHAN: IPv6 address correlation attacks on TLS encrypted traffic via siamese heterogeneous graph attention network. In: 30th USENIX Security Symposium (USENIX Security 21). pp. 4329–4346 (2021)

5. Guan, Z., Liu, C., Xiong, G., Li, Z., Gou, G.: Flowtracker: improved flow correlation attacks with denoising and contrastive learning. Comput. Secur. **125**, 103018 (2023)
6. Hu, X., Shu, Z., Tong, Z., Cheng, G., Li, R., Wu, H.: Fine-grained ethereum behavior identification via encrypted traffic analysis with serialized backward inference. Comput. Netw. **237**, 110110 (2023)
7. Li, J., et al.: Packet-level open-world app fingerprinting on wireless traffic. In: 29th Annual Network and Distributed System Security Symposium, NDSS 2022 (2022)
8. Li, J., et al.: FOAP: fine-grained open-world android app fingerprinting. In: Butler, K.R.B., Thomas, K. (eds.) 31st USENIX Security Symposium, USENIX Security 2022. pp. 1579–1596 (2022)
9. Li, Z., Xu, X.: L2-bitcn-cnn: spatio-temporal features fusion-based multi-classification model for various internet applications identification. Comput. Netw. **243**, 110298 (2024)
10. Mitra, G., Vairam, P.K., Saha, S., Chandrachoodan, N., Kamakoti, V.: Snoopy: a webpage fingerprinting framework with finite query model for mass-surveillance. IEEE Trans. Dependable Secur. Comput. **20**(5), 3734–3752 (2023)
11. Mitseva, A., Panchenko, A.: Stop, don't click here anymore: Boosting website fingerprinting by considering sets of subpages. In: 33rd USENIX Security Symposium (USENIX Security 24). pp. 4139–4156. Philadelphia, PA (2024)
12. Nasr, M., Bahramali, A., Houmansadr, A.: Deepcorr: strong flow correlation attacks on tor using deep learning. In: Proceedings of the 2018 ACM SIGSAC Conference on Computer and Communications Security, CCS 2018. pp. 1962–1976. ACM (2018)
13. Oh, S.E., et al.: Deepcoffea: improved flow correlation attacks on tor via metric learning and amplification. In: 43rd IEEE Symposium on Security and Privacy, SP 2022. pp. 1915–1932. IEEE (2022)
14. Shan, Y., Cheng, G., Chen, Z.: Identifying fine-grained douyin user behaviors via analyzing encrypted network traffic. In: 19th International Conference on Mobility, Sensing and Networking, MSN 2023, December 14-16. pp. 868–875. IEEE (2023)
15. Statista: number of mobile messaging users worldwide. https://www.statista.com/statistics/483255/number-of-mobile-messaging-users-worldwide/ (2023) Accessed 31 Aug 2024
16. Telegram: 700 million users and telegram premium. https://telegram.org/blog/700-million-and-premium (2022) Accessed 31 Aug 2024
17. Wang, T.: High precision open-world website fingerprinting. In: 2020 IEEE Symposium on Security and Privacy, SP 2020. pp. 152–167. IEEE (2020)
18. Wu, H., Li, X., Cheng, G., Hu, X.: Monitoring video resolution of adaptive encrypted video traffic based on HTTP/2 features. In: 2021 IEEE Conference on Computer Communications Workshops, INFOCOM Workshops 2021. pp. 1–6 (2021)
19. Wu, H., Yu, Z., Cheng, G., Guo, S.: Identification of encrypted video streaming based on differential fingerprints. In: 39th IEEE Conference on Computer Communications, INFOCOM Workshops 2020. pp. 74–79 (2020)
20. XTLS: Xtls vision, fixes tls in tls, to the star and beyond · xtls/xray-core · discussion 1295. https://github.com/XTLS/Xray-core/discussions/1295 (2023) Accessed 31 Aug 2024
21. Xu, H., Li, S., Cheng, Z., Qin, R., Xie, J., Sun, P.: Trafficgcn: mobile application encrypted traffic classification based on GCN. In: IEEE Global Communications Conference, GLOBECOM 2022. pp. 891–896. IEEE (2022)

22. Xue, D., Kallitsis, M., Houmansadr, A., Ensafi, R.: Fingerprinting obfuscated proxy traffic with encapsulated TLS handshakes. In: 33rd USENIX Security Symposium, USENIX Security 2024 (2024)
23. Yawning: obfs4. https://gitlab.com/yawning/obfs4 (2023) Accessed 31 Aug 2024
24. Zhao, S., Lee, P.P.C., Lui, J.C.S., Guan, X., Ma, X., Tao, J.: Cloud-based push-styled mobile botnets: a case study of exploiting the cloud to device messaging service. In: 28th Annual Computer Security Applications Conference, ACSAC 2012. pp. 119–128 (2012)

PAWUK: Extensive Annotated Web Corpus of Ukrainian

Witold Kieraś[1]([✉]) [iD], Łukasz Kobyliński[1] [iD], Dorota Komosińska[1] [iD],
Michał Rudolf[1] [iD], Maria Shvedova[2,3] [iD], and Aleksandra Zwierzchowska[1] [iD]

[1] Institute of Computer Science, Polish Academy of Sciences, Warsaw, Poland
{w.kieras,l.kobylinski,d.komosinska,m.rudolf,
a.zwierzchowska}@ipipan.waw.pl
[2] National Technical University "Kharkiv Polytechnic Institute", Kharkiv, Ukraine
Mariia.Shvedova@khpi.edu.ua
[3] University of Jena, Jena, Germany

Abstract. In this paper, we present PAWUK, the Polish Automatic Web corpus of UKrainian language. It is a linguistic corpus containing Ukrainian texts acquired from the Internet (selected web pages and social media accounts) and has been updated daily since 2022. It is automatically annotated with morphosyntactic tags, syntactic dependencies and named entities using Stanza framework with a model custom-built for Ukrainian to produce both Universal Dependencies tags and VESUM morphological tags. Users can interact with the corpus through a publicly available web interface.

Keywords: language corpus · Ukrainian language · natural language processing · PAWUK · regional variation

1 Introduction

Comprehensive balanced and representative linguistic corpora are a primary resource for studying language in its various registers. Building such a multipurpose resource is both time- and cost-consuming and requires regular updating. On the other hand, the Internet has been a large source of text data for many years now and the idea of using these data as linguistic corpora goes back at least to the WaCky corpus [1]. Such corpora proved to be useful resources in various linguistic research in which a large dataset representing the most recent vocabulary is more important than its representativeness, as it is in the case of studying neologisms and neosemantisms. Recently such web-based corpora are also being used as the training data for building Large Language Models.

In this paper we present PAWUK[1], a corpus of contemporary Ukrainian built from various resources collected from the Internet and automatically updated on a daily basis. Our goal was to provide a reliable source of language data

---

[1] https://pawuk.ipipan.waw.pl.

enriched by various layers of linguistic annotation: lemmatization, morphosyntactic annotation, dependency parsing and named entities recognition. Also we intended to provide the researchers with a relatively familiar environment at least with respect to the morphosyntactic annotation. Apart from the Universal Dependencies annotation scheme [7] we provide a morphosyntactic annotation consistent with the VESUM morphological dictionary of Ukrainian [12], allowing the user to query the corpus with the same tagset that was used in GRAC, the large reference corpus of contemporary Ukrainian[2]. To combine the Universal Dependencies annotation scheme and VESUM morphosyntactic scheme in a single natural language processing pipeline, a custom model of Stanza framework [9] needed to be trained. PAWUK's data sources were also manually assigned region labels according to the same scheme that was used in GRAC.

2   Related Work

PAWUK stands out as a significant online corpus of the Ukrainian language, joining resources such as GRAC, Zvidusil, UberText, and the Ukrainian components within multilingual collections like Aranea, ParlaMint, and the Leipzig Corpora Collection, as well as UkTenTen and Ukrainian Trends on Sketch Engine. However, PAWUK possesses key distinctions that make it a unique resource.

Notably, PAWUK adopts the regional annotation scheme pioneered by the large Ukrainian language corpus GRAC. Originally designed for studying regional variations in Ukrainian, GRAC evolved into a universal resource and a cornerstone of Ukrainian linguistics. GRAC v.18 comprises texts primarily from printed sources, spanning from the year 1816 to the present, totaling 1.9 billion tokens. It selectively includes contemporary online media that exemplify standard literary Ukrainian. PAWUK serves as a valuable complement to GRAC by offering a substantially larger volume of contemporary material, encompassing both regional and non-standard language. The shared annotation principles enable researchers to extend linguistic investigations initially based on GRAC using PAWUK. A primary limitation is the partial implementation of regional annotation in PAWUK, currently applied to online media and some Telegram channels, while regional information is unavailable for Twitter and YouTube comments.

The majority of large web-based Ukrainian language corpora are static, containing texts downloaded from the Internet at a specific point in time. While some undergo periodic updates, such as UkTenTen (updated in 2014, 2020, and 2022), Araneum Ucrainicum (2014, 2015, 2021, and 2022), ParlaMint (twice in 2023), and the Ukrainian corpus from the Leipzig Corpora Collection (2011, 2014, 2019 onwards), they lack the capacity to observe daily language evolution in real time, a feature offered by monitoring corpora.

PAWUK and Ukrainian Trends function as monitoring corpora, automatically collecting new Ukrainian texts daily and providing searchable online interfaces. Both have been actively uploading texts since spring 2022. In comparison

---

[2] https://uacorpus.org/.

to Ukrainian Trends, PAWUK is updated more frequently (twice a week/daily), downloads a greater volume of text, and includes social media content alongside standard news. This social media component offers valuable material for studying social media discourse and allows for the quicker identification of new words prevalent in online communication and memes. Furthermore, the social media data within PAWUK provides valuable insights into non-standard linguistic features and their regional distribution. Conversely, Ukrainian Trends, accessible through Sketch Engine, boasts a more sophisticated search interface, including the functionality to track trends for individual words.

For a more comprehensive list of current Ukrainian corpora, see Table 1.

3   Architecture of PAWUK

While designing the architecture of PAWUK, we had to address the challenge of being able to regularly collect large amounts of text from various sources on the Internet and to process them efficiently. More specifically, we had to tackle the issue of creating processing pipelines which could be configured independently for various text sources, solve such problems as proper language recognition, separation of main content from downloaded documents, text tagging, efficient storage and indexing.

Consequently, the main processing module is designed to manage the processes of acquiring, processing, and sending data for indexing from various online sources. The system architecture is built using Apache Airflow[3], which defines processing pipelines for different data sources (channels). The data acquisition process varies depending on the source.

In the case of websites, data is collected based on a list of URLs and specially prepared templates for each site. These templates include rules for parsing HTML, such as which elements to skip. In case of social networks data is gathered using lists of user profiles, with content retrieved through APIs provided by each platform.

Once acquired, the data goes through several processing stages:

- text conversion (extracting text from the source format, removing unnecessary characters, eliminating documents that are too short),
- language verification (for Twitter data, relying on source identification, for other sources, automatic verification using tools like Stanza),
- duplicate elimination (removing texts already present in the database),
- annotation (performing morphological and syntactic analysis, as detailed in Sect. 5),
- indexing (using MTAS [2] for text indexing, making the indexed data available through a web service).

The workflow is constantly monitored and statistics are gathered and stored to signal any issues during various stages of the process, such as data collection, tagging, or indexing. For example, we trigger notifications when the amount of data collected for particular sources drops below a set average value.

---

[3] https://airflow.apache.org/.

Table 1. Selected existing Ukrainian corpora.

Corpus	Size (in tokens)	Composition and annotation	Access
General Regionally Anno tated Corpus of Ukrainian (GRAC)	1.9 B	Various texts (1816–2024), VESUM-based annotation	Searchable
Ukrainian Brown corpus	1M	Balanced, various texts, partly manually annotated, VESUM-based annotation	Downloadable
UD Ukrainian IU Treebank	122 K	Various texts, manually annotated corpus	Searchable, downloadable
UD Ukrainian ParlaMint Treebank	51 K	Parliamentary transcripts, manually annotated corpus	Searchable, downloadable
Ukrainian ParlaMint	51 M	Parliamentary transcripts (2002–2023), contains Ukrainian-Russian code switching; language annotation, UD pos annotation	Searchable, downloadable
UberText 2.0	3.3 B	News, Wikipedia, social, fiction, court	Downloadable
Ukrainian Corpus (Leipzig University)	18 B	News, web, Wikipedia (2011, 2014, 2019, 2020, 2021, 2022, 2023, 2024), without lemmatization	Searchable, samples of up to 1 million words for download
Ukrainian Web 2022 (ukTenTen22)	7.5 B	Web texts (2014, 2020, 2022), MULTEXT-East Ukrainian pos annotation, UD pos annotation	Searchable
Ukrainian Trends	>1 B	News, Wikipedia (since spring 2022), MULTEXT-East Ukrainian pos ann.	Searchable
Zvidusil (Laboratory of Ukrainian)	3 B	Web texts (2018), UD annotation	Searchable
Araneum Ucrainicum	125 M—1.25 B	Web texts (2014, 2015, 2021, 2022), pos annotation	Searchable, registration is required
ParaRook Parallel corpora with German, English, French, Spanish, Chinese, Japanese, Persian	28M	Fiction	Searchable
Laboratory of Ukrainian Parallel corpora with English, Polish, French, German, Spanish, Portuguese	5M	Fiction	Searchable
Parallel Russian-Ukrainian	9 M	Fiction, journalism	Searchable

The overall architecture is presented in Fig. 1.

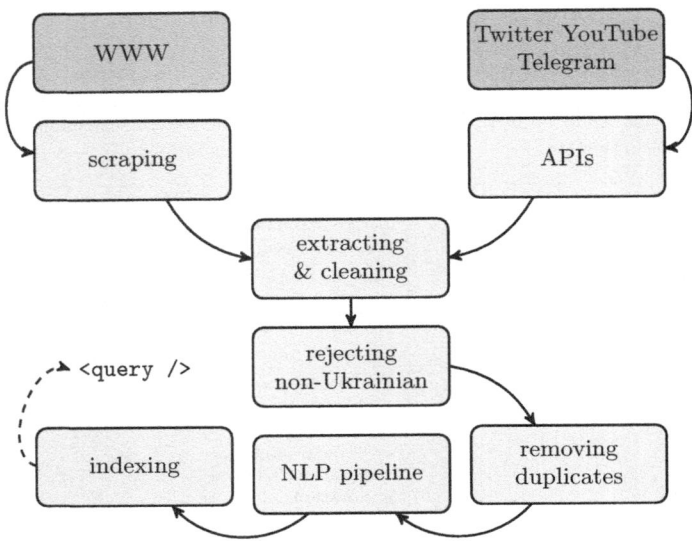

Fig. 1. Architecture of PAWUK.

4   Data Sources and Acquisition

PAWUK monitors 506 selected Ukrainian webpages as well as Ukrainian Twitter[4] (469 accounts), YouTube (294 channels) and Telegram (633 channels). Lists of news sites and Telegram channels were manually selected for the corpus.

The collection includes samples from news websites across all 24 regions of Ukraine as well as Kyiv, and features media outlets published in Ukrainian from other countries. When selecting news sites we aimed to create the most complete, representative, thematically diverse collection, which would include texts from all regions of Ukraine, both large cities and smaller towns. 366 out of 506 news websites are marked by regions.

The principles of selection of Telegram channels are the same as those for news sites. 200 out of 633 channels are marked by regions. There are official channels of institutions and public figures, as well as popular channels covering different topics: music, literature, local city channels.

In the case of Twitter, data was collected from two sources—selected user accounts (targeting 469 accounts) and trends (most popular tweets from Ukraine). The account database initially consisted of several dozen manually selected users and was then automatically expanded based on their connections with other profiles. Accounts were periodically verified and updated.

The list of monitored YouTube accounts is created entirely automatically based on the list of most popular videos from Ukraine. Currently, it includes

---

[4] Data retrieval was discontinued on June 29, 2023, due to Twitter/X's decision to end the free API access.

294 accounts which are periodically verified. Only the text data from comments under the videos are collected.

Each document (a web article or a single post) is assigned metadata that can be used in queries. These consist of: title, author, region, URL, channel, subchannel and date of publication.

Data collection for the corpus occurs daily. Each channel is managed by a separate processing path defined in Apache Airflow. In case of any problems, the system administrator can respond and retry the process at any stage.

The starting point of the corpus collection, March 2022, was at the beginning of the active stage of Russia's war against Ukraine. Each text in the corpus is time-stamped with the day it was created. Thus, the corpus is an accurate testimony of the changes that occurred in the Ukrainian language during the war. This not only concerns the emergence of new words and meanings, but also, for example, a characteristic orthographic feature consisting in beginning proper names associated with Russia with a lowercase letter, also when starting a sentence or headline. On social media, users sometimes employ Unicode subscript characters to render the initial letter of a word visually smaller, as in "$_p$осія".

5   Annotation

The corpus is automatically enriched with linguistic annotation accessible by the user through a corpus search engine. For the annotation process we use Stanza pipeline with a custom model for Ukrainian tweaked to produce VESUM morphosyntactic tags as XPOS values. The tagging and parsing models are trained using the sum of two independent resources:

– standard Ukrainian UD treebank[5] [7] converted to contain VESUM tags instead of existing MULTEXT-East tags as XPOS,
– BrUK[6] corpus manually annotated with VESUM tags (so called Ukrainian Brown corpus) parsed using the standard Stanza model.

The idea behind creating such a heterogeneous training set was to maximize the accuracy of XPOS tagging without impairing the accuracy of parsing as well as to diversify the training set as the Ukrainian UD treebank is rather small. The researchers of Ukrainian are more familiar with the VESUM tagset which is more fine-grained (contains over 1,700 unique tags) and complies with the annotations of Ukrainian large reference corpus GRAC.

Table 2 presents evaluation results for the custom model. We provide the values of standard metrics that are commonly used in evaluation of NLP tagging and parsing frameworks. F1 measure for lemmatization and morphological features (Lem, UPOS, XPOS, UFeats) is provided for testing subsets of both datasets that were used in the training process. For parsing accuracy metrics

---

[5] The treebank was developed by the Institute of Ukrainian and is available at https:// github.com/UniversalDependencies/UD_Ukrainian-IU/.

[6] https://github.com/brown-uk.

(UAS, LAS, CLAS, MLAS, BLEX) only the UD treebank evaluation is provided as the other dataset does not contain a manual dependency annotation layer. The model is publicly available at https://github.com/ipipan/stanza-uk-pawuk. For named entities annotation the standard Stanza model[7] was used.

Table 2. Evaluation results for testing subsets of the two resources used for training the custom model.

Eval. dataset	Lem	UPOS	XPOS	UFeats	AllTags	UAS	LAS	CLAS	MLAS	BLEX
BRuK	98.49	97.29	91.87	92.87	88.92	–	–	–	–	–
UD treebank	98.57	98.96	93.92	95.96	91.30	89.77	87.99	84.71	84.23	84.71
Total:	98.56	98.66	93.55	95.40	90.87	–	–	–	–	–

6   Interacting with the Corpus

The interface of the corpus is very simple and basic. The front page contains the query field through which the user may immediately start working with the corpus. A simple query language cheat sheet may be found in the About subpage.

The simplest corpus query consists of a single word form typed in by the user which should result with the list of all concordances (words within context) containing this word. The multilayer linguistic annotation accessible through Corpus Query Language however allows for much more precise queries regarding lemmatization, morphology, syntax and named entities. The primary annotation scheme used in the corpus is the one provided by the UD scheme which recently became a *de facto* standard for morphosyntactic and dependency annotation. It consists of parts of speech (e.g. noun, adjective, numeral), morphological features, that is morphological categories (e.g. number, case, aspect) and their values (e.g. singular, accusative, imperfective respectively), and the dependency annotation, that is assigning dependency relations between words which are in a direct grammatical relation of certain type such as subject, object, adjunct, modifier etc. Apart from that, the linguistic annotation also includes the morphological tag of every word expressed in the tagset native to a given language (so called XPOS) which usually contains even more precise morphological information and is more familiar to the researchers working with the language on a daily basis. In the case of PAWUK and Ukrainian, the native tagset is the one of VESUM morphological dictionary which was also used in GRAC.

The following examples give an overview of possible corpus queries from basic to more advanced ones, illustrating the expressive power of the linguistic annotation as well as the query language.

– [orth="павуки"] finds all occurrences of a given orthographic word.

---

[7] https://stanfordnlp.github.io/stanza/ner_models.html.

- [lemma="павук"] finds all occurrences of a given lemma regardless of its inflectional form.
- [upos="NOUN"] finds all occurrences of words belonging to a given part of speech (as defined in the UD scheme), here: all nouns.
- [ufeat="fem"] finds all occurrences of words which were assigned a given value of a specific grammatical category, here: all words assigned the feminine gender regardless of their part of speech.
- [deprel="nsubj"] finds all occurrences of words that are dependents in a given grammatical relation, here: words that were assigned as nominal subject in the grammatical structure of a sentence.
- [head.lemma="павук"], [head.upos="ADJ"], [head.ufeat="fem"] work analogously to lemma, upos, ufeat mentioned above but they apply to syntactic head of a given word instead of the word itself.
- [xpos="noun:anim:p:v_naz"] finds all words assigned a given XPOS tag, in this case: animated nouns in plural number and nominative case.
- <ne="PERS" /> finds all occurrences of named entities assigned one of the four categories: person (PERS, as in this query), organization (ORG), place (LOC), other (MISC).

The crucial part of the query language is that all the constraints described above can be combined in one query. For example, the following query will find all occurrences of the word павук ('spider') used as a subject in a sentence and occurring within a proper name referring to a person (see Fig. 2):

```
[lemma="павук" & deprel="nsubj"] within <ne="PERS" />
```

Among the results there could be an instance of Людина-Павук 'Spiderman'.

Additionally the annotation includes an oov (*out of vocabulary*) attribute (used as [oov="true"] in corpus queries) which allows to filter only those interpretations which do not comply with the VESUM dictionary. This feature proved to be useful in researching non-standard grammatical variants. For example, the query [lemma="вухо" & oov="true"] finds word вухи (non-standard variant of the nominative plural form, cf. вуха), which is not a literary norm but is still common in social networks.

The corpus queries may be restricted based on basic metadata, which allows to obtain results only from sources meeting certain specific criteria such as text title, text author, publication date and, most notably, the region of the country from which the source data originates. This is especially useful in researching regional variations of the language.

A graphical query builder allowing the user to construct queries without remembering the CQL syntax and names of all attributes will be available soon.

The multilayer linguistic annotation is a powerful tool for researchers interested in finding rare instances of words or grammatical constructions as well as (ir)regularities in the use of certain types of linguistic phenomena. It is worth noting that users may as well create their own corpus in the Korpusomat application [10] which uses the exact same annotation scheme and provides similar user interface to the one used in PAWUK, which makes the two tools complementary and useful for various linguistics research of modern Ukrainian.

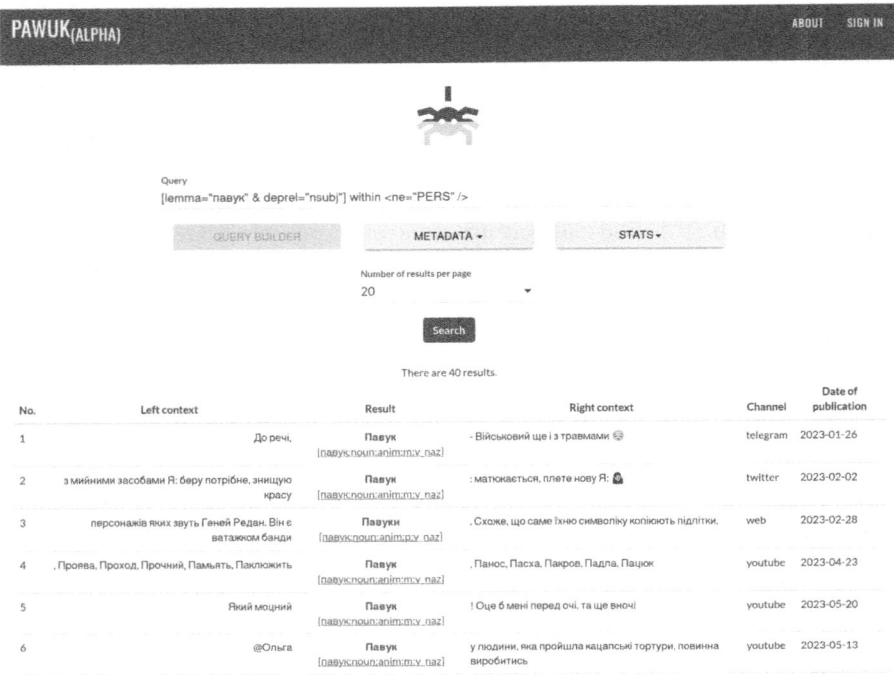

Fig. 2. The screenshot presents search results of the query example.

7　Use-Case Scenario

As a case study of corpus-based research, we analyse herein the distribution of the variant prepositions *vid* and *od*, which share the semantic value 'from' and are historically variant forms of the same preposition. These two variants correlate with distinct Ukrainian dialectal groups. In the South-Western dialectal zone, the variant *vid* prevails, whereas *od* is the predominant form in the Northern dialects. The South-Eastern dialect area represents a more recent and heterogeneous formation, shaped by successive waves of migration and lacking clearly defined dialectal boundaries. Within this mosaic-like region, both variants—*vid* and *od*—are attested with varying frequency, as illustrated in map 269 of the Atlas of the Ukrainian Language, vol. 1 [8]. In the 20th century, the prepositional variant *vid* became established as the main standard form in the literary language, while *od* was largely relegated to spoken language and fictional texts [4].

We used PAWUK to investigate the geographical and quantitative persistence of the older variant preposition *od* in online texts.

The study revealed that the majority of occurrences of the variant *od* within the corpus originate from social media texts. The relative frequency of *od* across different text types is as follows: 3.2 occurrences per 1,000 combined uses of *od* and *vid* in YouTube comments, 1.6 in Twitter, 1.0 in Telegram, and 0.3 in web texts.

To ascertain whether a correlation exists between the distribution of the variant *od* in online texts from various regions and the dialectal areas where this variant is present, we examined web news articles and Telegram channels. While these sources yielded fewer instances of *od* compared to social networks, only this subset of the corpus was annotated for regional origin.

Given that the administrative oblasts represented in the corpus do not precisely align with dialectal distribution areas, we categorized the regions into two groups: those encompassing areas where *od* is used (regions that at least partially include areas of Northern or South-Eastern dialectal distribution) and those where this variant is absent. We excluded all Kyiv-based channels from the dataset due to the capital's diverse population, which renders the linguistic data non-indicative of the region's specific language use. Additionally, we excluded the prominent Lviv portal Zbruch, as it attracts contributions from individuals across Ukraine. Consequently, this analysis yielded a limited number of examples (see Table 3).

Table 3. Distribution of the variants *od* and *vid* in regions with and without *od* in dialects.

Dialectal *od*	*od*	*vid*
Yes	197	558,157
No	21	250,904

The corpus data nonetheless reveal a statistically significant correlation between the regional distribution of the preposition *od* and dialectal areas where this form is traditionally attested. Despite the infrequent overall occurrence of *od* within the corpus, its relative frequency is notably higher in regions with documented dialectal usage. A Fisher's exact test ($p < 0.000001$) and a chi-squared test ($\chi^2 = 45.57, p < 0.00000001$) corroborate that this association is not attributable to random variation.

While the corpus of Internet texts does not constitute dialectal material per se and may not invariably reflect the direct diffusion of a word or feature within a region (as it could appear, for example, in a quotation or a metalinguistic context describing another speaker's language), the influence of local vernacular on written Internet texts is evident.

8   Summary and Future Work

Since the release of its initial version in 2023, PAWUK has been used in researching Ukrainian neologisms [5,6], usage of non-standard language varieties (such as surzhyk) in online communication [3] and teaching corpus linguistics [11].

Table 4. Corpus statistics.

Channel	Documents	Sentences	Tokens	Avg. per day
web	3.9 M	61.3 M	1.1 B	1.3 M
telegram	14.3 M	32.6 M	372 M	310 k
twitter	3.7 M	7.5 M	89 M	200 k
youtube	16.8 M	38.4 M	399 M	360 k
Total:	38.7 M	139.8 M	1.96 B	2.17 M

The corpus is updated daily (approximately 1.5–2.3 million words every day; 2 million on average). Currently it consists of nearly 2 billion tokens (see Table 4).

Although the process of collecting, annotating and indexing daily batches of data is purely automatic, the corpus needs some technical maintenance as some sources of data become unavailable (discontinued websites, social media data policy changes etc.) and new sources need to be added. Therefore, the primary goal is to maintain the corpus updating process and ensure its stability and reliability for users. In the future, we plan to add new data sources as they become available (e.g. new social networks with available APIs and new web sources). We also plan to improve the web interface to make interacting with the corpus more intuitive and user-friendly.

Acknowledgments. The work was partially supported by the Dariah.lab project, as well as the European Regional Development Fund as a part of the 2014-2020 Smart Growth Operational Programme, CLARIN - Common Language Resources and Technology Infrastructure, project no. POIR.04.02.00-00C002/19. The work of Maria Shvedova was partially supported by the visiting program of the Polish Academy of Sciences. We would like to thank everyone who contributed to the collection of channels, including the students of the National Technical University "Kharkiv Polytechnic Institute" - Kateryna Astafieva, Kseniia Luzhna, Daria Lebedenko, Maryna Vozikova, Polina Poltavchenko, Arsenii Lukashevskyi, and Ivan Ahaiev, and participants of the Corpus Linguistics online course (as part of the digital philology program for Ukrainian students, supported by the University of Jena and the DAAD Foundation) - Nazar Kotsur, Nataliia Sheremet, Olha Tochylina, Yurii Petrov, Mariana Martyniv.

Disclosure of Interests. The authors have no competing interests to declare that are relevant to the content of this article.

References

1. Baroni, M., Bernardini, S., Ferraresi, A., Zanchetta, E.: The wacky wide web: a collection of very large linguistically processed web-crawled corpora. Lang. Resour. Eval. **43**(3), 209–226 (2009). https://doi.org/10.1007/s10579-009-9081-4
2. Brouwer, M., Brugman, H., Kemps-Snijders, M.: MTAS: a Solr/Lucene based multi tier annotation search solution. In: Selected papers from the CLARIN Annual Conference 2016. Linköping Electronic Conference Proceedings (2017)

3. Dyka, L., Shvedova, M.: Synonymy in the Ukrainian language and the problems of surzhyk (case study of synonyms with the meaning 'probably') (in Ukrainian). Mova: klasyčne - moderne - postmoderne (9), 50–71 (2023). https://ekmair.ukma.edu.ua/server/api/core/bitstreams/5bd65ed1-25c2-4abf-9403-0776e4c6cea1/content

4. Dyka, L., Shvedova, M.: History and normative status of the Preposition / Prefix од- in Modern Ukrainian (in Ukrainian). Slavia Orientalis **vol. LXXI**(No 4), 797–818 (2022). https://doi.org/10.24425/slo.2022.143220, http://journals.pan.pl/Content/125858/PDF/2022-04-SOR-06.pdf

5. Horoxova, T., Bojko, M.: Dynamics of lexical composition of the ukrainian language in the war period in 2022–2023 (in Ukrainian). Aktual'ni pytannja humanitarnyx nauk: mižvuzivs'kyj zbirnyk naukovyx prac' molodyx včenyx Drohobyc'koho deržavnoho pedahohičnoho universytetu imeni Ivana Franka **67**(1), 213–218 (2023). https://elibrary.kubg.edu.ua/id/eprint/46953/

6. Klym, J.: New invective vocabulary with the component "zet-" in modern internet communication: Based on the web corpus (in Ukrainian). In: Movnyj prostir sučasnoho svitu: tezy dopovidej VII Vseukraïns'koï naukovoï konferenciï studentiv, aspirantiv i molodyx učenyx, pp. 90–94. NaUKMA, Kyiv (2023). https://ekmair.ukma.edu.ua/items/747f8e6f-36c8-4f20-86dd-94ef0788476a

7. de Marneffe, M.C., Manning, C.D., Nivre, J., Zeman, D.: Universal dependencies. Comput. Linguist. **47**(2), 255–308 (2021). https://doi.org/10.1162/coli_a_00402

8. Matvijas, I. (ed.): Atlas of the Ukrainian Language, vol. 1. Naukova dumka, Kyiv (1984)

9. Qi, P., Zhang, Y., Zhang, Y., Bolton, J., Manning, C.D.: Stanza: A Python natural language processing toolkit for many human languages. In: Proceedings of the 58th Annual Meeting of the Association for Computational Linguistics: System Demonstrations (2020). https://nlp.stanford.edu/pubs/qi2020stanza.pdf

10. Saputa, K., Tomaszewska, A., Zawadzka-Paluektau, N., Kieraś, W., Kobyliński, L.: Korpusomat.eu: a multilingual platform for building and analysing linguistic corpora. In: Mikyška, J., de Mulatier, C., Paszynski, M., Krzhizhanovskaya, V.V., Dongarra, J.J., Sloot, P.M. (eds.) Computational Science – ICCS 2023. 23rd International Conference, Prague, Czech Republic, July 3–5, 2023, Proceedings, Part II, pp. 230–237. No. 14074 in Lecture Notes in Computer Science, Springer Nature Switzerland, Cham (2023). https://doi.org/10.1007/978-3-031-36021-3_22

11. Shvedova, M., Pospekhova, A.: Corpus linguistics course for philology students (in Ukrainian). In: Zbirnyk naukovyx prac' I Mižnarodnoï naukovoï konferenciï «Innovacijni texnolohiï v linhvistyci ta perekladi», pp. 60–64. L'viv (2024). https://books.ldubgd.edu.ua/index.php/m/catalog/book/203

12. Starko, V., Rysin, A.: VESUM: a large morphological dictionary of Ukrainian as a dynamic tool. In: Computational Linguistics and Intelligent Systems. vol. 6th Int. Conf, pp. 71–80. COLINS, Gliwice (2022). https://ceur-ws.org/Vol-3171/paper8.pdf

Assessing Physics Parameterizations Using Evolutionary Computation

Iciar Guerrero-Calzas[1,2]($\boxtimes$)(ID), Lorenzo Rossetto[2](ID), Ana Cortés[1](ID), Mauricio Hanzich[2](ID), and Josep Ramón Miró[3](ID)

[1] Computer Architecture and Operating Systems Department, Universitat Autònoma de Barcelona. Carrer de les Sitges, s/n 08193, Cerdanyola del Vallès, Spain
`iciar.guerrero@autonoma.cat`, `ana.cortes@uab.cat`
[2] Mitiga Solutions S.L., Carrer de Julià Portet, 08002 Barcelona, Spain
`{iciar.guerrero,lorenzo.rossetto,mauricio.hanzich}@mitigasolutions.com`
[3] Servei Meteorològic de Catalunya, Carrer del Dr. Roux, 80, 1a planta, 08017 Barcelona, Spain
`jr.miro@gencat.cat`

Abstract. Hailstorms are intense, localized weather phenomena that can severely impact agriculture, infrastructure, and property, making precise forecasting essential for risk management. The Weather Research and Forecasting (WRF) model is widely used for numerical weather prediction, offering numerous physical parameterization options to represent atmospheric processes. However, due to the large number of possible configurations, identifying the most suitable configuration is a challenge. This research uses a genetic algorithm (GA) to systematically refine WRF physics schemes for hail prediction in Central Europe, specifically for the hail events of June 2022. Within this framework, WRF configurations are treated as individuals in a population that evolves through selection, crossover, and mutation over multiple iterations. Fitness is evaluated using the F2 score. This methodology allows to evaluate more than 2.4 million possible setups improving the WRF model's capacity to accurately represent hailstorms. This strategy provides a robust framework for testing a wide range of setups, proving its value in refining parameterizations to better forecast impactful weather phenomena.

Keywords: Hail Modeling · Genetic Algorithm · Numerical Weather Prediction (NWP)

1 Introduction

Hail is a significant weather phenomenon that poses a severe risk to agriculture, infrastructure, and property worldwide. It is a form of precipitation consisting of balls or irregular ice particles that develop within convective clouds under strong upward air currents [2]. Although relatively rare, hailstorms can cause extensive damage due to their localized and intense nature, resulting in billions

© The Author(s), under exclusive license to Springer Nature Switzerland AG 2025
M. H. Lees et al. (Eds.): ICCS 2025, LNCS 15904, pp. 106–120, 2025.
https://doi.org/10.1007/978-3-031-97629-2_8

of US dollars (USD) in damages annually [24], making accurate prediction essential for risk mitigation and disaster preparedness. Hail formation is driven by extreme atmospheric conditions, such as strong updrafts, supercooled water, and freezing-level dynamics. These processes involve complex microphysical interactions, which make accurately predicting hail challenging using traditional numerical weather prediction (NWP) models. Accurate hail forecasting requires high-resolution simulations using advanced models like the Weather Research and Forecasting (WRF) model, which includes multiple physics parameterizations for simulating atmospheric processes such as microphysics, convection, radiation, and planetary boundary layer (PBL) interactions. Choosing the optimal combination of these schemes is critical but challenging due to the large number of possible configurations.

Research on the WRF model has highlighted the importance of optimizing physical parameterization to enhance its ability to accurately simulate severe weather events, including hailstorms. Previous studies have demonstrated the sensitivity of WRF outputs to different parameterization choices. For example, [29] showed that the local rainfall intensity during Typhoon Fitow was highly sensitive to the microphysics schemes chosen, particularly regarding graupel and hail parameterizations. Similarly, [26] highlighted the impact of different combinations of microphysics, cumulus, and PBL schemes on the forecasting of tropical cyclones, affecting their track, intensity, and rainfall. In another study [5], the authors examined the impact of land surface physics on the simulation of horizontal convective rolls in a tropical coastal environment, affecting turbulent structures and the distribution of heat. Furthermore, in [25], it was found that the choice of microphysics greatly affected updraft characteristics during severe thunderstorms in Southeast India, while [20] showed that the model performance was more sensitive to selection of microphysics schemes than to PBL options. Despite these insights, traditional sensitivity tests are limited to a subset of physics parameterization options, providing valuable but incomplete insights, as they do not explore all possible configurations and their interactions, which are crucial for accurately simulating complex phenomena like hail.

To address this challenge, we have applied evolutionary computation, in particular, a Genetic Algorithm (GA) has been used to efficiently explore a wider range of WRF physics schemes and identify optimal configurations to simulate hail events. Unlike traditional methods that test fixed or limited sets of configurations, GA applies evolutionary principles to iteratively refine WRF setups across multiple generations. GA significantly expands the solution space and systematically improves the simulation accuracy through continuous improvements [23]. In the proposed GA, the WRF configuration options are encoded as "genes" and each "individual" represents a specific configuration of the model. While GAs have successfully optimized WRF parameterizations for applications such as wind and solar energy prediction [28] and sea breeze prediction [31], their use in hail event simulations remains relatively unexplored.

This paper is organized as follows. In Sect. 2, the models and observational data used are described; Sect. 3 exposes the methodology for calibrating the physics parametrizations of WRF model applying the Genetic Algorithm; Sect. 4 discusses the results, and Sect. 5 summarizes the main conclusions and future work.

2   Models and Data

2.1   Observational Data and Study Case

In this study, we have used the European Severe Weather Database (ESWD), operated by the European Severe Storms Laboratory (ESSL) [11]. This database provides detailed information on severe storm events in Europe, including thunderstorms, hail, and tornadoes. The major hailstorms of 2022 in Europe were marked by several record-breaking events, including an overall total of 8,224 large hail reports: 1,334 very large hail reports ($\geq$ 5 cm), and 18 giant hail reports ($\geq$ 10 cm) [12]. These storms caused significant damage and economic losses. The storms from 3 June to 5 June affected France, Germany, Belgium and parts of Italy, with hailstones up to 11 cm in diameter [12]. In particular, on June 4th, it caused widespread hail damage across central France. The estimated insured losses from these storms in France reached EUR 650-850 million, mainly due to hail, although flooding and wind also contributed [13]. This work is focused on this particular extreme event, so, the area of interest is Central Europe. Since the WRF model is a regional model, it is essential to define the simulation area, which is known as the domain. To achieve high-resolution outputs, a nesting process is required, where a high-resolution inner domain is placed within a lower-resolution outer domain. Figure 1 shows the two domains used for the studied area, where $D01$ corresponds to domain 1 with a 9 km low-resolution domain, whereas $D02$ (domain 2) is an inner-nested domain with a 3 km resolution.

2.2   The WRF-HAILCAST Model

The Weather Research and Forecasting (WRF) model is a state of the art mesoscale numerical weather prediction system. In this work, we have used the WRF-HAILCAST model, which couples WRF with the HAILCAST module [1,27] providing a framework to simulate hailstone growth and trajectory within realistic three-dimensional atmospheric conditions rather than idealized profiles [20]. For simplicity, we will refer to it as the WRF model throughout the rest of this paper. Weather prediction involves physical processes such as evaporation, condensation, friction, radiation, and turbulent transfers of moisture, which cannot be directly resolved numerically. To emulate these processes, numerical weather prediction models like WRF use parametrizations that interact during simulations to represent atmospheric dynamics and thermodynamics. These physics schemes included in the model are summarized in Table 1.

For this study, we have selected a subset of the available schemes based on documented incompatibilities in the WRF User's Guide [21]. These schemes include: 27 microphysics schemes, 12 cumulus schemes, 6 longwave radiation schemes, 7 shortwave radiation schemes, 10 boundary layer schemes, 3 surface

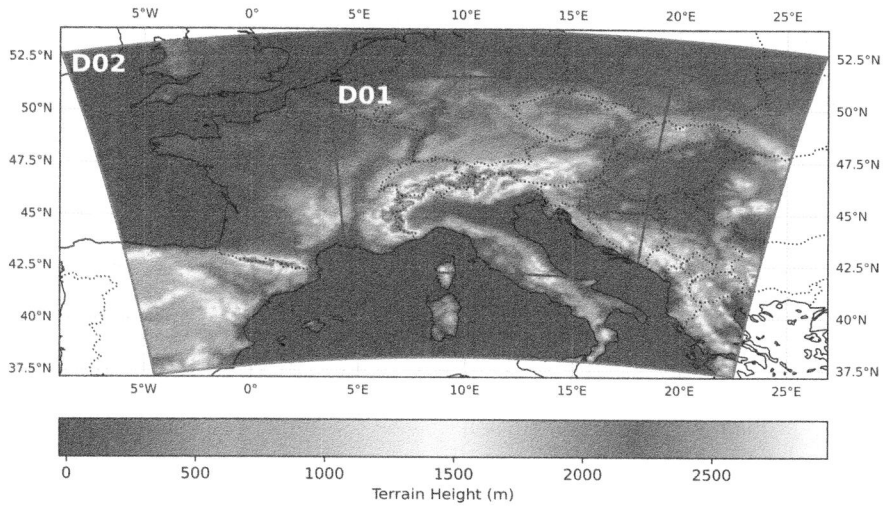

Fig. 1. WRF model domains. *D*01 9 km at resolution and *D*02 at 3 km resolution. Shading indicates topography height.

Table 1. Physics schemes included in the WRF model [21]

Physic schemes	WRF Parameterization	Description
Microphysics (MP)	mp_physics	Models cloud and precipitation processes, including rain, snow, ice, and graupel formation.
Cumulus (CU)	cu_physics	Represents the effects of convective clouds that are too small to be resolved by the model grid.
Longwave radiation (LW)	ra_lw_physics	Models the radiation emissions from atmospheric layers and surface emissivity.
Shortwave radiation (SW)	ra_sw_physics	Models solar radiation fluxes for both clear and cloudy conditions.
Planetary Boundary Layer (PBL)	bl_pbl_physics	Handles surface fluxes distribution and vertical mixing caused by boundary layer turbulence.
Surface layer (SL)	sf_sfclay_physics	Describes the exchange of energy and moisture between the surface and the atmosphere.
Land Surface Model (LSM)	sf_surface_physics	Models soil temperature, moisture, and snow water equivalent, influencing surface energy, and water fluxes.

layer schemes, and 6 land surface model schemes, resulting in a total of 2,449,440 possible combinations. Although the User's Guide provides detailed information about many known incompatibilities, some scheme combinations may still be incompatible but are not documented. Known incompatibilities are excluded from the initial population, as some of them are only suited for idealized tests or specific WRF compilation settings.

3   Methodology

Our methodology utilizes a GA to select a WRF model configuration, focusing on the physics schemes, to effectively simulate hail events. A Genetic Algorithm (GA) is a metaheuristic optimization method inspired by the principles of natural evolution [8]. This kind of optimization algorithm is well-suited for solving complex problems with large solution spaces by iteratively refining candidate solutions based on their performance. In this study, a GA is used to search for the WRF parameterization configuration that best reproduces the hail events in the study case. Each candidate solution, called *individual*, represents a unique combination of the seven WRF physics schemes listed in Table 1, which act as the *genes* of the individual (Fig. 2).

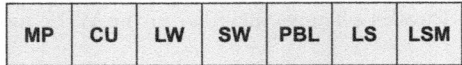

Fig. 2. Schematic representation of an individual, where each physics scheme act as "gene". Here MP, CU, LW, SW, PBL, LS, and LSM refer to microphysics, cumulus, longwave radiation, shortwave radiation, Planetary Boundary Layer, Land Surface Model, and Surface physics schemes, respectively

A set of individuals (candidate solution) defines a *population*. The GA evolves an initial population through genetic operators (selection, crossover and mutation) to identify the individual that most accurately simulates the hail events, based on a fitness score. More precisely, the GA starts by generating a random initial population of individuals. For each individual, the underlying WRF model is executed and the resulting output is compared to the observed data by evaluating a certain fitness function, which quantifies how well each individual's simulation matches the observed event. Based on these fitness scores, the next generation of individuals is created using the following operators:

- *Elitism*: The best-performing individuals from the current generation are carried over directly to the next generation without any changes. This helps preserve the most successful solutions. Elitism is implemented by ranking individuals based on their fitness scores and retaining a fixed percentage of the top performers, determined by the elitism rate. In this work, 6% of the individuals have been retained as elite.

- *Selection*: Individuals are chosen for reproduction (mating population) using a linear ranking selection method. First, the population is ranked by fitness, and selection probabilities are assigned according to each individual's rank. High-ranked individuals have a greater probability of being selected for reproduction.
- *Crossover*: New individuals (offspring) are generated by mixing the genes of selected parents. Each gene of the offspring is randomly inherited from one of the parents. The crossover process is governed by a predefined crossover rate. The crossover probability in this work is set to 0.75.
- *Mutation*: Mutation adds randomness by changing some of the genes in an individual. According to the mutation rate, a subset of genes is randomly selected, and each chosen gene is replaced with a new value. In this work we have set up the mutation probability for each gene at 0.1.

These steps are iterated over n generations. After each iteration, the population is evaluated and the best individuals are carried forward. The process continues until either convergence is achieved or a maximum number of generations is reached. Convergence is determined by analyzing the fitness scores of previous generations. If the difference between the maximum fitness scores of consecutive generations is less than a specified threshold (0.001 in this case), it indicates that the algorithm is no longer making significant improvements. When this occurs on a set number of past generations, the algorithm is considered to have converged. If convergence does not occur before reaching the maximum number of generations, the process will stop regardless. Table 2 summarizes the GA configuration parameters and Fig. 3 illustrates this entire GA process.

Table 2. Parameters used for the Genetic Algorithm

Parameter	
Maximum N^o of generations	35
Population size	100
Elite percentage	6%
Crossover probability	0.75
Mutation probability	0.1
Convergence threshold	0.001

3.1 Fitness Evaluation

Validating hail predictions is challenging because hail events are highly localized events, and there is a limited availability of observational data. While observational networks are essential for validation, they often fail to capture localized

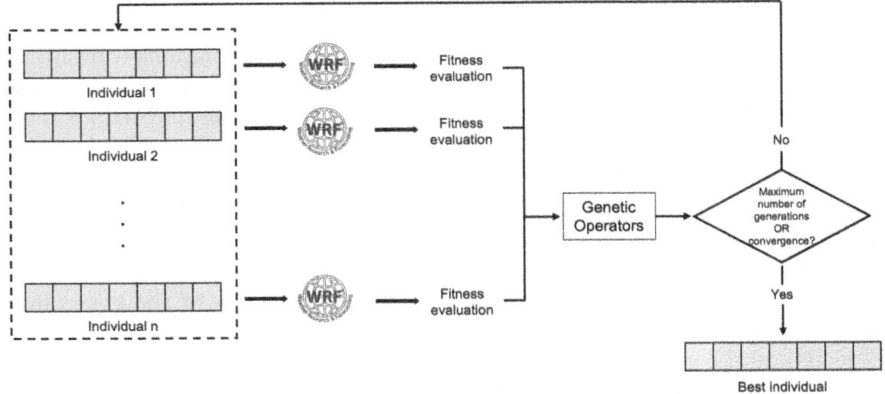

Fig. 3. Schematic illustration of the GA framework steps

hail occurrences, resulting in an over-representation of non-events. This sampling bias creates a significant class imbalance, with non-hail cases dominating the dataset. Furthermore, the rarity of hail and positional uncertainties in both forecasts and observations further complicate the validation, making traditional methods less effective. The choice of a fitness function is particularly important, as different functions can yield significantly different results, which directly influences the evaluation of hail prediction performance [7]. To evaluate each individual, the hail map is analyzed at grid cell level, where the WRF model indicates the presence or absence of hail. This assessment is a true-false event and is typically represented in a contingency table (see Table 3).

Table 3. Binary contingency table for whether or not hail is predicted.

		Observed	
		Yes	No
Predicted	Yes	Hits	False Alarms
	No	Missed	Correct Negatives

The comparison between the simulated and the real hail event presents four possible situations:

- Hits (H): hail both observed and predicted.
- False Alarms (FA): hail predicted but not observed.
- Missed (M): hail observed but not predicted.
- Correct Negatives (CN): hail neither observed nor predicted.

As previously mentioned, the class imbalance in this problem poses a challenge for standard evaluation metrics, such as POD and FAR [4], which can be dominated by correct negatives. When a metric is influenced by non-hail events, it may produce misleadingly high scores even if it fails to detect hail. To mitigate this issue, we use the *F2* score, a member of the *F-beta* family, which ignores correct negatives and emphasizes the detection of both observed and simulated hail events (Eq. 1).

$$F_2 = \frac{(1 + 2^2) \cdot H}{(1 + 2^2) \cdot H + 4 \cdot M + FA} \tag{1}$$

This metric gives more weight to recall, making the metric more sensitive to false negatives, that is, we prioritize the identification of hail events, even if it leads to an increase in false alarms. This aligns with our main objective of minimizing missed detection, that is, missing actual hail events. Missing a hail event can have serious consequences, including missed warnings, insufficient preparation, and potential damage. Therefore, we prioritize ensuring that the model identifies as many hail events as possible, even if it occasionally predicts hail where it did not occur.

High-resolution weather forecasts often struggle to precisely match observed small-scale events in space, time, or intensity. These discrepancies arise due to the natural variability of the phenomena and differences between model predictions and observational data. For example, small differences in storm location may not indicate a significant forecast error but can lead to poor performance metrics when comparing the model output directly with observations at individual grid points. To address this issue, we have applied a convolutional filter that smooths the comparison between forecasts and observations, preventing small spatial misalignment from overly penalizing the model's performance. To do that, first, we upscale the point-based observational data to match the model's 3 km grid resolution, ensuring that both datasets are comparable. Once we have both maps, the observations and the predictions, at the same resolution (3 km) we overlap them to obtain the cells corresponding to hits, misses, and false alarms. These cells will be assigned a weight of 1 when considered to evaluate the *F2* score (Eq. 1). Figure 4 illustrates this using a 6 × 6 grid map, where blue cells represent predicted hail events (False Alarms), yellow cells represent observations (Misses), and green cells highlight exact matches (Hits). The map on the left depicts this direct overlapped map. In this map, there is only one green cell corresponding to an exact hit, then, this hit will be weighted by 1 as well as the yellow and blue cells, which correspond to misses and false alarms, respectively.

Furthermore, to mitigate the impact of small spatial mismatches, the *near-hit* concept is introduced. Then, for each cell corresponding to a false alarm or a miss, the Moore neighborhood is considered (purple squares in Fig. 4). The Moore neighborhood is the set of cells orthogonally or diagonally adjacent to a given cell map [32].

To account for spatial mismatches, we classify a false alarm (FA) or miss (M) as a near-hit if its Moore neighborhood contains: (1) if the FAs Moore neighbourhood contains any observed event (Fig. 4(d)), or if the Ms Moore neighbourhood contains any predicted event (Fig. 4(c)). Such cases are treated as near-hits and assigned a reduced weight of 0.5 in the F2-score calculation (Eq. 1). Cases without neighbouring real events or predicted ones maintain their original classification as FAs or Ms with full weight (1.0), as illustrated in Figs. 4(a-b).

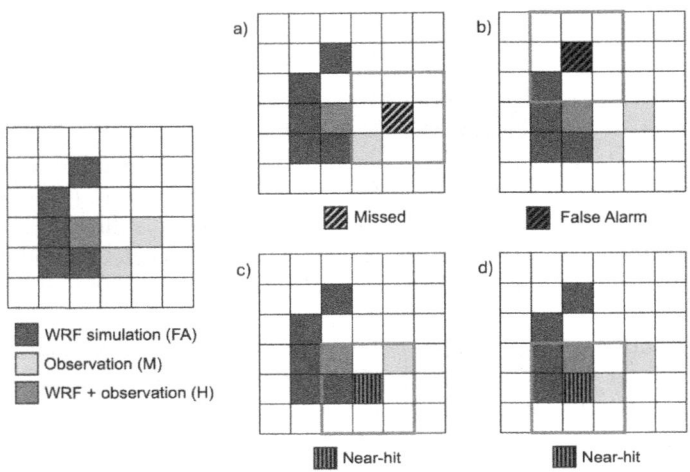

Fig. 4. Fitness evaluation using a 6 × 6 grid at 3 km resolution. Green cells are exact hits, blue cells represent WRF-predicted hail events (false alarms) and yellow cells correspond to observations (misses). Vertically striped cells are near-hits.(Color figure online)

This approach mitigates the impact of small spatial mismatches, ensuring that they are not overly penalized as false alarms or misses. By incorporating near-hits, the validation process emphasizes the model's ability to capture the event's general location. This reduces false alarms and ensures that the model accurately predicts both the occurrence and the general location of events, even with minor spatial deviations. By accounting for spatial uncertainty, this validation method offers a fairer and more robust evaluation of high-resolution forecasts, improving their reliability in assessing hail prediction performance.

4   Experimental Results

All WRF simulations described in this section were executed on the MareNostrum 5 General Purpose Partition (GPP) supercomputer [3]. MareNostrum 5 GPP consists of 6,408 nodes powered by Intel Sapphire Rapids processors, each configured with two 8480+ CPUs running at 2 GHz, providing 112 cores per node. Each standard node includes 256GB of DDR5 memory, with 216 nodes upgraded to 1TB of memory and 960GB of NVMe storage for high-speed data access. The system features NDR200 interconnects, with each link shared between two nodes, delivering 100Gb/s bandwidth per node. Additionally, 72 high-bandwidth memory (HBM) nodes use Intel Sapphire Rapids 03H-LC processors, each with 112 cores at 1.9GHz and 128GB of HBM memory, offering an impressive 2TB/s memory bandwidth per node. The entire machine achieves a peak performance of 45.9 PFlops and operates on a fat-tree network topology, ensuring high-speed communication and scalability.

Our executions have been carried out on 100 nodes per generation, and a runtime limit of three hours per individual is enforced. This setup allows for the simultaneous execution of multiple WRF configurations within a generation, ensuring efficient use of resources. If a job exceeds the three-hour runtime limit, it is flagged for review. This computational setup not only enables the execution of thousands of WRF simulations within a reasonable time frame (hours), but also guarantees the reproducibility and scalability of the simulations. Using the power of MareNostrum 5's advanced infrastructure, the GA is able to thoroughly explore the complex configuration space of the WRF physics options.

The setup of the WRF model consisted of two one-way nested domains: the outer domain with a horizontal grid resolution of 9 km (260 × 190 grid points) and the inner domain with a resolution of 3 km (361 × 340 grid points) (Fig. 1). Vertical resolution was maintained at 50 levels, with the model top set at 50 hPa. Boundary conditions were updated hourly using ERA5 reanalysis data with a horizontal resolution of 0.25°, 38 levels, and 1 h temporal resolution [15]. All simulations were initialized at 05 UTC, with 6-hour spin-up, and run for a total of 40 h to better capture the hail events and cover the development and decay of storms. Model outputs were generated hourly for both domains, with HAILCAST module activated only in the inner domain at a 60-second interval. The physics schemes changed between simulations, but certain aspects remained fixed as, for instance, the cumulus parameterization was only applied in the outer domain.

As it has been previously mentioned, although certain physics scheme combinations are known to be incompatible, there exist others that may still be incompatible but are not documented. In order not to prevent GA of exploring as much searching space as possible, when an individual is executed and it returns an incompatibility error, the fitness score assigned to the corresponding individual will be a low value (this point is explained later in this section). This way, they are considered for generating the next population but with a very low probability of being used in any genetic operator.

A total of 35 generations were run with a population size of 100 individuals, resulting in 3500 simulations. From these, 1873 unique WRF configurations were created, reflecting the extensive search space explored by the GA. This extensive search allowed the GA to identify combinations of physics parameterizations that might not have been considered through traditional sensitivity analysis methods. Although the 3500 simulations represent only a fraction of the space, they provide valuable insights into the algorithm's ability to identify optimal configurations and demonstrate its utility for practical applications. The best-performing configuration was identified in generation 25. The detailed breakdown of this configuration is shown in Table 4. We chose to stop at generation 35 because it allowed us to verify the stability and to ensure that the best-performing configuration remained across multiple generations.

Table 4. Best-performing WRF configuration identified by the genetic algorithm

WRF Parameterization	Option	Description
cu_physics	14	KIAPS SAS (KSAS) [14,18]
mp_physics	14	WRF Double-Moment 5-class (WDM5) [19]
ra_lw_physics	5	New Goddard [9,10]
ra_sw_physics	4	RRTMG [16]
bl_pbl_physics	9	UW [6]
sf_sfclay_physics	1	Revised MM5 [17]
sf_surface_physics	4	Noah-MP [22,30]

Figure 5 shows the results of the best-performing configuration, comparing observed hail events (red dots) and WRF-simulated hail occurrences (purple) within the 3 km domain. While some discrepancies are present due to model uncertainty and observational limitations, the overall consistency reinforces the GA's ability to optimize model performance effectively. The alignment of storm clusters highlights the effectiveness of the optimized configuration, although discrepancies arise from model uncertainty and observational limitations.

The progression of the fitness score across generations is shown in Fig. 6. Initially, the population consisted of random configurations, with fitness scores relatively low. The first notable improvement occurred in generation 3, where the fitness score reached 0.2273. This was followed by another increased to 0.2626 in generation 6. By generation 10 the fitness score stabilized at 0.2629. At generation 20 the fitness score improved slightly reaching 0.2635 and up to 0.2639 in generation 24. But the most significant jump occurred at generation 25, when the score reached 0.2915. This score remained steady through generation 35.

While the final fitness score of 0.2915 is not very high, it is important to consider the uncertainty of the model and the scarcity of the observational data available for validation. The observed jumps in the fitness score can be due to the chosen population size. Given the size of the search space – approximately 2.4 million possible combinations – and a relatively small population of 100 individuals, only a small fraction of the configurations is explored.

Despite these challenges, the algorithm's performance is still promising. The results from the spatial analysis (Fig. 5) further support the GA's effectiveness, as the final configuration was able to capture most of the storm clusters.

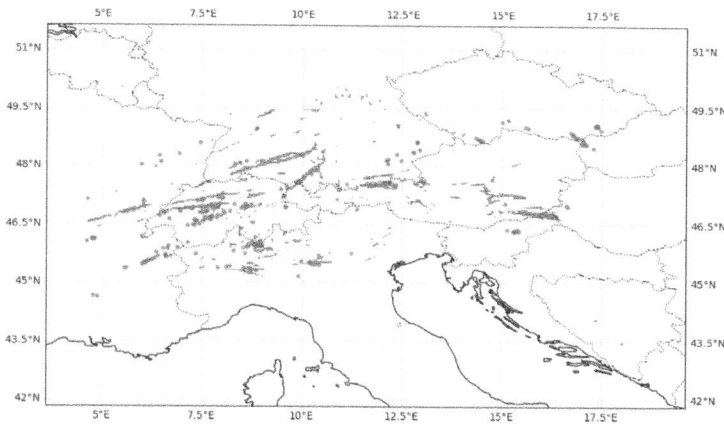

Fig. 5. Spatial comparison of observed hail events (red dots) and simulated hail occurrences (purple) using the best-performing WRF configuration. The map focuses on the inner domain, showcasing the alignment of storm clusters and the ability of the configuration to reproduce hail events. (Color figure online)

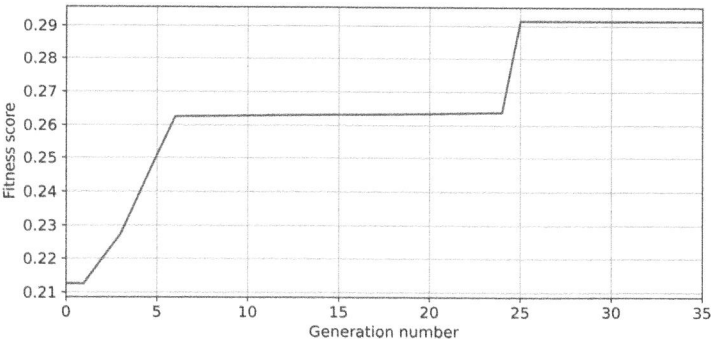

Fig. 6. Evolution of the best fitness scores across generations. Significant improvements were observed in generations 3, 6, and 25, with the score stabilizing at 0.2915 from generation 25 to 35

5   Conclusions and Further Work

In this study, a Genetic Algorithm was used to assess the physics parameterizations of the WRF model for hailstorm forecasting. Despite challenges like limited observational data, model uncertainty, and the use of fixed discrete parameter choices, which limits the fine-tuning of settings, the algorithm showed promising results in identifying effective WRF configurations for hail predictions. The proposed approach has been applied in a severe hail event that took place in Central Europe on June of 2022. The results captured the majority of the extreme storm clusters proving the ability of the system to assess discrete model configurations. Typically, there are small spatial mismatches when predicting the location of a hailstorm, thus, to mitigate this issue, the concept of near-hit has been introduced. The original 3 km map has been upscaled to a 9 km Moore neighborhood map, this way, the fitness score used in the GA process does not completely dismiss certain false alarms and misses when they are close to a exact hit but giving them partial weight in the fitness score.

Future work should consider expanding this flexibility to a larger radius, such as 25 km, to better account for the spatial characteristics of hailstorms, which often extend over 10 km-25 km due to storm motion and hailstone advection. This change would align with the needs of operational users, such as emergency managers and agricultural decision-makers, who require broader regional risk awareness for effective response. To assess the robustness of the methodology, future research should involve running the GA on multiple dates to evaluate its performance across diverse meteorological conditions. This would provide a broader understanding of the model's capabilities and its adaptability to different weather scenarios.

Acknowledgments. This work is supported by the Industrial Doctorate Plan of the Department of Research and Universities of the Generalitat de Catalunya under contract 2023DI00013. It also has been granted by the Spanish Ministry of Science and Innovation MCIN AEI/10.13039/501100011033 under contract PID2023-146193OB-I00 and by the Catalan government under grant 2021 SGR-00574. The simulations and data analysis presented in this study were supported by the Barcelona Supercomputing Center (BSC) and Red Española de Supercomputación (RES) under project Inno-2-2024-0002.

Disclosure of Interests. The authors have no competing interests to declare that are relevant to the content of this article.

References

1. Adams-Selin, R.D., Ziegler, C.L.: Forecasting hail using a one-dimensional hail growth model within wrf. Mon. Weather Rev. **144**(12), 4919–4939 (2016). https://doi.org/10.1175/MWR-D-16-0027.1
2. American meteorological society: hail. Glossary Meteorol. (2017). https://glossary.ametsoc.org/wiki/Hail Accessed 27 Nov 2024
3. Barcelona supercomputing center (BSC): Marenostrum 5 supercomputer (2025). https://www.bsc.es/marenostrum/marenostrum-5 Accessed 09 Jan 2025
4. Bhavyasree, Panda, S., Wasson, G., Mondal, U., Kumar, A., Sharma, D.: Assessing the performance of wrf model in simulating severe hailstorm events over assam and bihar, india. Modeling Earth Systems and Environment **10**, 6013–6034 (2024). https://doi.org/10.1007/s40808-024-02114-z
5. Bollareddy, R., C., V.S., JR, R., Balasubramaniam, V.: Numerical simulation of horizontal convective rolls over a tropical coastal site using wrf: sensitivity to land surface physics. Pure Appl. Geophys. **180**, 1–24 (2023). https://doi.org/10.1007/s00024-023-03361-4
6. Bretherton, C.S., Park, S.: A new moist turbulence parameterization in the community atmosphere model. J. Clim. **22**(12), 3422–3448 (2009). https://doi.org/10.1175/2008JCLI2556.1

7. Carrillo, C., Artes, T., Cortés, A., Margalef, T.: Error function impact in dynamic data-driven framework applied to forest fire spread prediction. Procedia Comput. Sci. **80**, 418–427 (2016). https://doi.org/10.1016/j.procs.2016.05.342

8. Chaudhry, I.A., Usman, M.: Integrated process planning and scheduling using genetic algorithms. Technical Gazette **24**, 1401–1409 (2017). https://doi.org/10.17559/TV-20151121212910

9. Chou, M.D., Suarez, M.J.: A solar radiation parameterization for atmospheric studies. NASA Tech. Memorandum NASA/TM-1999-104606/VOL15, NASA Goddard Space Flight Center (1999). https://doi.org/10.2172/19990060930

10. Chou, M.D., Suarez, M.J., Liang, X.Z., Yan, M.M.H., Cote, C.: A thermal infrared radiation parameterization for atmospheric studies. NASA Technical Memorandum NASA/TM-2001-104606/VOL19, NASA Goddard Space Flight Center (2001). https://doi.org/10.2172/20010072848

11. Dotzek, N., Groenemeijer, P., Feuerstein, B., Holzer, A.M.: Overview of essl's severe convective storms research using the european severe weather database eswd. Atmospheric Res. **93**(1), 575–586 (2009), 4th European Conference on Severe Storms

12. European storm forecast experiment (ESSL): Major hailstorms of 2022. European Storm Forecast Experiment (2022). https://www.essl.org/cms/major-hailstorms-of-2022/ Accessed 11 Dec 2024

13. Guy carpenter: severe convective storms in europe (2022). Guy Carpenter Insights (2022). https://www.guycarp.com/insights/2022/06/SCS_Europe_June2022.html Accessed 11 Dec 2024

14. Han, J., Pan, H.L.: Revision of convection and vertical diffusion schemes in the ncep global forecast system. Weather Forecast. **26**, 520–533 (2011). https://doi.org/10.1175/WAF-D-10-05038.1

15. Hersbach, H., et al.: The era5 global reanalysis. Quart. J. Royal Meteorol. Soc. **146** (2020). https://doi.org/10.1002/qj.3803

16. Iacono, M.J., Delamere, J.S., Mlawer, E.J., Shephard, M.W., Clough, S.A., Collins, W.D.: Radiative forcing by long-lived greenhouse gases: calculations with the aer radiative transfer models. J. Geophys. Res. Atmospheres **113**(D13) (2008). https://doi.org/10.1029/2008JD009944

17. Jiménez, P.A., Dudhia, J., González-Rouco, J.F., Navarro, J., Montávez, J.P., García-Bustamante, E.: A revised scheme for the wrf surface layer formulation. Mon. Weather Rev. **140**(3), 898–918 (2012). https://doi.org/10.1175/MWR-D-11-00056.1

18. Kwon, Y., Hong, S.Y.: A mass-flux cumulus parameterization scheme across gray-zone resolutions. Monthly Weather Rev. **145** (2016). https://doi.org/10.1175/MWR-D-16-0034.1

19. Lim, K.S., Hong, S.Y.: Development of an effective double-moment cloud microphysics scheme with prognostic cloud condensation nuclei (ccn) for weather and climate models. Monthly Weather Rev. **138**, 1587–1612 (2010). https://doi.org/10.1175/2009MWR2968.1

20. Malečić, B., et al.: Performance of hailcast and the lightning potential index in simulating hailstorms in croatia in a mesoscale model – sensitivity to the pbl and microphysics parameterization schemes. Atmos. Res. **272**, 106143 (2022). https://doi.org/10.1016/j.atmosres.2022.106143

21. National center for atmospheric research (NCAR): WRF Model User's Guide. Mesoscale and Microscale Meteorology Laboratory (2023). https://www2.mmm.ucar.edu/wrf/users/wrf_users_guide Accessed 11 Nov 2024

22. Niu, G.Y., et al.: The community noah land surface model with multiparameter-ization options (noah-mp): 1. model description and evaluation with local-scale measurements. J. Geophys. Res. Atmospheres **116**(D12) (2011). https://doi.org/10.1029/2010JD015139

23. Oana, L., Spataru, A.: Use of genetic algorithms in numerical weather prediction. pp. 456–461 (2016). https://doi.org/10.1109/SYNASC.2016.075

24. Púčik, T., et al.: Large hail incidence and its economic and societal impacts across europe. Mon. Weather Rev. **147**(11), 3901–3916 (2019). https://doi.org/10.1175/MWR-D-19-0204.1

25. Rajeevan, M., Kesarkar, A., Thampi, S.B., Rao, T.N., Radhakrishna, B., Rajasekhar, M.: Sensitivity of wrf cloud microphysics to simulations of a severe thunderstorm event over southeast india. Ann. Geophys. **28**(2), 603–619 (2010). https://doi.org/10.5194/angeo-28-603-2010

26. Shenoy, M., Raju, P., Prasad, V., Prasad, K.: Sensitivity of physical schemes on simulation of severe cyclones over bay of bengal using wrf-arw model. Theoretical Appl. Climatol. **149**, 1–15 (2022). https://doi.org/10.1007/s00704-022-04102-8

27. Skamarock, C., et al: A description of the advanced research wrf model version 4 (2019). https://api.semanticscholar.org/CorpusID:196211930

28. Sward, J., Ault, T., Zhang, K.: Genetic algorithm selection of the weather research and forecasting model physics to support wind and solar energy. SSRN Electron. J. (2022). https://doi.org/10.2139/ssrn.3999150

29. Xu, H., Zhang, D., Li, X.: The impacts of microphysics and terminal velocities of graupel/hail on the rainfall of typhoon fitow (2013) as seen from the wrf model simulations with several microphysics schemes. J. Geophys. Res. Atmospheres **126**(23) (2021). https://doi.org/10.1029/2020JD033940

30. Yang, Z.L., et al.: The community noah land surface model with multiparameter-ization options (noah-mp): 2. evaluation over global river basins. J. Geophys. Res. Atmospheres **116**(D12) (2011). https://doi.org/10.1029/2010JD015140

31. Yoon, J.W., Lim, S., Park, S.K.: Combinational optimization of the wrf physical parameterization schemes to improve numerical sea breeze prediction using micro-genetic algorithm. Appl. Sci. **11**(23) (2021). https://doi.org/10.3390/app112311221

32. Zaitsev, D.A.: A generalized neighborhood for cellular automata. Theoret. Comput. Sci. **666**, 21–35 (2017). https://doi.org/10.1016/j.tcs.2016.11.002

Domain Solutions Obtained by the FPIES for Potential 2D BVPs

Andrzej Kużelewski$^{(\boxtimes)}$ and Eugeniusz Zieniuk

Faculty of Computer Science, University of Bialystok,
Ciolkowskiego 1M, 15-245 Bialystok, Poland
{a.kuzelewski,e.zieniuk}@uwb.edu.pl

Abstract. The paper presents the fast parametric integral equations system (FPIES) for domain solutions of potential 2D boundary value problems (BVPs). The FPIES has been successfully applied in modelling 2D potential BVPs and finding solutions on the boundary. The combination of the modified fast multipole technique with the PIES reduced the numerical computation time and RAM usage in the fast PIES. Similar techniques are used to find solutions in the domain. The method is demonstrated with the solution in the domain of two BVPs.

Keywords: Fast PIES · BVPs · Fast Multipole Method

1 Introduction

The fast parametric integral equations system (FPIES) [1, 2] is a robust numerical tool for solving boundary value problems (BVPs). The method is an extension of the conventional PIES [3], allowing the solution of large-scale BVPs on a standard PC. The PIES does not require a mesh. Therefore, it can be considered as a mesh-free method. However, unlike other mesh-free methods (such as particle methods [4], Galerkin methods [5] or cloud methods [6]), which are mainly used in simulations related to, for example, plastic materials, fluid dynamics or crack simulations, the application of the PIES is comparable to widely used numerical element methods such as the FEM [7–9] and the BEM [10–12] or the still being developed the FEM-BEM hybrids [13], isogeometric analysis (IgA) [14] or the Virtual Element Method (VEM) [15].

The FPIES, as the successor of the PIES, offers the same remarkable advantages compared to the FEM and the BEM. At first, the lack of discretization of the problem's boundary and domain is due to the direct inclusion of parametric functions describing the boundary of the problem into the mathematical formalism of the FPIES. These functions are widely used in computer graphics and include, among others, Bézier and B-spline curves, Coons and Bézier surface patches. Therefore, a small number of control points is required to define the shape of the boundary in the FPIES. In addition, the use of parametric functions reduces the dimensionality of a solved problem by one. At last, despite the high accuracy of the solutions obtained by the FPIES, it is possible to improve the

M. H. Lees et al. (Eds.): ICCS 2025, LNCS 15904, pp. 121–133, 2025.
https://doi.org/10.1007/978-3-031-97629-2_9

accuracy with only a slight modification of the input data (precisely, the number of collocation points), unlike the BEM and the FEM, where the discretization process has to be repeated.

Previous studies on the conventional PIES have focused on various aspects of the method, including the distribution of collocation points [16], the use of NURBS curves [17], or application to problems described by other PDEs, such as Navier-Lame equations [18]. PIES use for nonlinear problems is also described in [19]. All these studies have confirmed its higher accuracy compared to BEM or FEM. However, time-consuming calculations and RAM utilization increase with the problem size's square. Therefore, similarly to the BEM, to compute the PIES matrices, we need $O(N^2)$ operations and another $O(N^3)$ operations to solve the obtained system using direct solvers (where N - the number of equations of algebraic equations system).

On the other hand, the FPIES uses the fast multipole method (FMM) [20–22], which eliminates the main disadvantage of the PIES, namely the generation of dense non-symmetric coefficient matrices and the Gaussian elimination used to solve the final system of algebraic equations. Thanks to the FMM and modified binary tree [23], the system of algebraic equations $A \cdot x = b$ in the IFPIES is generated implicitly and solved by iterative GMRES solver [24]. It means that only the result of the multiplication of the matrix A by the vector of unknowns x is stored in memory, in contrast to the conventional PIES, which has to store the dense matrix A and the vector b in RAM. The application of the FMM allows for a significant reduction in computational time to order $O(NlogN)$ and a decrease in utilization of RAM to $O(N)$. Previous studies on the FPIES have focused on efficient and accurate solutions to the problems described by single- and multi-connected domains [2,25], as well as some analysis on key parameters [26,27]. Also, a comparison to a competing method such as fast multipole BEM (FMBEM) is described in [28].

So far, the authors of this paper have used the FPIES to find solutions of BVPs on the boundary. Solutions in the domain can be obtained using the technique applied in the conventional PIES [3]. The main goal of this paper is to present the FPIES applied to find numerical solutions for 2D potential BVPs in the domain using the fast multipole technique. The efficiency and accuracy of the FPIES are tested on the potential problems described by linear and curvilinear boundary segments.

2   The Fast Parametric Integral Equations System

The FPIES for 2D potential problems [2] was obtained as the result of modification of the conventional PIES [3]. The way of obtaining the FPIES included the modification of the PIES kernels (to allow for the Taylor series approximation used by the FMM) and the modification of the tree used by the FMM (to include the way of defining the boundary in the PIES) is clearly presented in, among others, [23] and [25]. The following formula presents the final form of the FPIES for solving BVPs on the boundary [2]:

$$\frac{1}{2}u_l(\widehat{s}) = \sum_{j=1}^{n} \mathbb{R}\left\{ \frac{1}{2\pi} \sum_{l=0}^{N_T}(-1)^l \cdot \left[\sum_{k=0}^{N_T}\sum_{m=l}^{N_T} \frac{(k+m-1)! \cdot M_k(\tau_c')}{(\tau_{el}-\tau_c)^{k+m}} \right. \right.$$

$$\left. \left. \cdot \frac{(\tau_{el}'-\tau_{el})^{m-l}}{(m-l)!} \right] \cdot \frac{(\widehat{\tau}-\tau_{el}')^l}{l!} \right\} - \sum_{j=1}^{n} \mathbb{R}\left\{ \frac{1}{2\pi} \sum_{l=0}^{N_T}(-1)^l \cdot \right. \quad (1)$$

$$\left. \left[\sum_{k=1}^{N_T}\sum_{m=l}^{N_T} \frac{(k+m-1)! \cdot N_k(\tau_c')}{(\tau_{el}-\tau_c)^{k+m}} \cdot \frac{(\tau_{el}'-\tau_{el})^{m-l}}{(m-l)!} \right] \cdot \frac{(\widehat{\tau}-\tau_{el}')^l}{l!} \right\},$$

where moments $M_k(\tau_c')$ and $N_k(\tau_c')$ are computed twice only and have the form:

$$M_k(\tau_c') = \sum_{l=0}^{k} \frac{(\tau_c-\tau_c')^{k-l}}{(k-l)!}M_l(\tau_c), \quad M_l(\tau_c) = \int_{s_{j-1}}^{s_j} \frac{(\tau-\tau_c)^l}{l!}p_j(s)J_j(s)ds,$$

$$N_k(\tau_c') = \sum_{l=1}^{k} \frac{(\tau_c-\tau_c')^{k-l}}{(k-l)!}N_l(\tau_c), \quad N_l(\tau_c) = \int_{s_{j-1}}^{s_j} \frac{(\tau-\tau_c)^{l-1}}{(l-1)!}n^{(c)}u_j(s)J_j(s)ds,$$

$$(2)$$

N_T is the number of terms in Taylor series, $J_j(s)$ is the Jacobian, s_{j-1} correspond to the beginning of j-th segment, while s_j to its end, $u_j(s)$ and $p_j(s)$ are boundary functions (3), $n^{(c)}$ is the complex notation of normal vector to the curve, which creates segment j, $\mathbb{R}$ is the real part of complex number and superscript (c) means the complex variable.

Expressions $\widehat{\tau}, \tau$ are the complex version of parametric functions describing the boundary:

$$\widehat{\tau} = S_l^{(c)}(\widehat{s}) = S_l^{(1)}(\widehat{s}) + iS_l^{(2)}(\widehat{s}), \quad \tau = S_j^{(c)}(s) = S_j^{(1)}(s) + iS_j^{(2)}(s),$$

where $S_k^{(i)}(s_n)$ ($i = \{1,2\}$, $k = \{j,l\}$, $s_n = \{\widehat{s},s\}$) are parametric curves or lines, which define particular segments of the boundary of the problem (l or j). The points τ_c' and τ_{el}' are obtained during tracing the tree structure.

Boundary functions $u_j(s)$ and $p_j(s)$ in (1) are approximated by the following series:

$$u_j(s) = \sum_{k=0}^{N} u_j^{(k)} L_j^{(k)}(s), \quad p_j(s) = \sum_{k=0}^{N} p_j^{(k)} L_j^{(k)}(s), \quad (3)$$

where $u_j^{(k)}$ and $p_j^{(k)}$ are unknown or given values of boundary functions in defined points of the segment j, N - is the number of terms in approximating series (3), which approximated boundary functions on the segment j and $L_j^{(k)}(s)$ - the base functions (Lagrange polynomials) on segment j.

3   Numerical Solutions in the Domain

Solving the FPIES (1) results in solutions on the boundary only. To find solutions in the domain, a modification of the integral identity from the conventional PIES is required. The original identity uses the solutions of the PIES, and it has the following form [3]:

$$u(x) = \sum_{j=1}^{n} \int_{s_{j-1}}^{s_j} \left\{ \widehat{U}_j^*(x,s)p_j(s) - \widehat{P}_j^*(x,s)u_j(s) \right\} J_j(s)ds. \tag{4}$$

Integrands $\widehat{U}_j^*(x,s)$ and $\widehat{P}_j^*(x,s)$ are presented in the following form:

$$\widehat{U}_j^*(x,s) = \frac{1}{2\pi} \ln \frac{1}{\sqrt{r_1^2 + r_2^2}},$$

$$\widehat{P}_j^*(x,s) = \frac{1}{2\pi} \frac{r_1 n^{(1)}(s) + r_2 n^{(2)}(s)}{r_1^2 + r_2^2}, \tag{5}$$

where $r_1 = x^{(1)} - S_j^{(1)}(s)$ and $r_2 = x^{(2)} - S_j^{(2)}(s)$, $x^{(1)}$ and $x^{(2)}$ are the coordinates of the solution point in the domain, $n^{(1)}$ and $n^{(2)}$ are components of vector normal to the curve. The shape of the boundary is included in (5) by r_1 and r_2 which contain parametric curves.

At first, we should write data in complex number system:

$$\tau = S_j^{(1)}(s) + iS_j^{(2)}(s), \quad x^{(c)} = x^{(1)} + ix^{(2)}, \quad n^{(c)}(s) = n^{(1)}(s) + in^{(2)}(s).$$

Then

$$x^{(c)} - \tau = x^{(1)} - S_j^{(1)}(s) + i\left(x^{(2)} - S_j^{(2)}(s)\right) = \left|x^{(c)} - \tau\right| e^{i \, \mathrm{atan} \frac{x^{(2)} - S_j^{(2)}(s)}{x^{(1)} - S_j^{(1)}(s)}},$$

where $\left|x^{(c)} - \tau\right| = \sqrt{\left(x^{(1)} - S_j^{(1)}(s)\right)^2 + \left(x^{(2)} - S_j^{(2)}(s)\right)^2} = \sqrt{r_1^2 + r_2^2}$.

Therefore, the complex integrands $\widehat{U}_j^{*(c)}(x^{(c)}, \tau)$ and $\widehat{P}_j^{*(c)}(x^{(c)}, \tau)$ have the following form:

$$\widehat{U}_j^{*(c)}(x^{(c)}, \tau) = -\frac{1}{2\pi} \ln \left(x^{(c)} - \tau\right), \quad \widehat{P}_j^{*(c)}(x^{(c)}, \tau) = \frac{1}{2\pi} \frac{n^{(c)}(s)}{x^{(c)} - \tau}, \tag{6}$$

while

$$\widehat{U}_j^*(x,s) = \mathbb{R}\{\widehat{U}_j^{*(c)}(x^{(c)}, \tau)\} = \mathbb{R}\left\{ -\frac{1}{2\pi} \left[\ln \left|x^{(c)} - \tau\right| + i \, \mathrm{atan} \frac{x^{(2)} - S_j^{(2)}(s)}{x^{(1)} - S_j^{(1)}(s)} \right] \right\} =$$

$$= -\frac{1}{2\pi} \ln \left|x^{(c)} - \tau\right| = \frac{1}{2\pi} \ln \frac{1}{\sqrt{r_1^2 + r_2^2}},$$

$$\widehat{P}_j^*(x,s) = \mathbb{R}\{\widehat{P}_j^{*(c)}(x^{(c)}, \tau)\} = \mathbb{R}\left\{ \frac{1}{2\pi} \frac{n^{(c)}(s) \cdot (x^{(c)} - \tau)^*}{(x^{(c)} - \tau) \cdot (x^{(c)} - \tau)^*} \right\} =$$

$$= \frac{1}{2\pi} \frac{n^{(1)}(s)\left(x^{(1)} - S_j^{(1)}(s)\right) + n^{(2)}(s)\left(x^{(2)} - S_j^{(2)}(s)\right)}{\left(x^{(1)} - S_j^{(1)}(s)\right)^2 + \left(x^{(2)} - S_j^{(2)}(s)\right)^2} = \frac{1}{2\pi} \frac{r_1 n^{(1)}(s) + r_2 n^{(2)}(s)}{r_1^2 + r_2^2},$$

where $(x^{(c)} - \tau)^*$ is a complex conjugate to $(x^{(c)} - \tau)$.

Then, similarly to the FPIES, we can use the same fast multipole tree and Taylor series. Therefore, moments $M_k(\tau_c)$ and $N_k(\tau_c)$ are calculated previously. Substituting kernels $\widehat{U}_j^{*(c)}\left(x^{(c)}, \tau\right)$ and $\widehat{P}_j^{*(c)}\left(x^{(c)}, \tau\right)$ into (4), we obtain the following expression:

$$u(x) = \frac{1}{2\pi} \sum_{j=1}^{n} \mathbb{R}\left\{ \sum_{k=0}^{N_T} U_k(x^{(c)}, \tau_c) M_k(\tau_c) - \sum_{k=1}^{N_T} P_k(x^{(c)}, \tau_c) N_k(\tau_c) \right\}, \quad (7)$$

where:

$$U_k(x^{(c)}, \tau_c) = \begin{cases} -\ln\left(x^{(c)} - \tau_c\right) & \text{for } k = 0 \\ \dfrac{(k-1)!}{\left(x^{(c)} - \tau_c\right)^k} & \text{for } k \geq 1 \end{cases},$$

$$P_k(x^{(c)}, \tau_c) = \frac{(k-1)!}{\left(x^{(c)} - \tau_c\right)^k} \quad \text{for } k \geq 1.$$

To find solutions in the domain, only moments in the leaves are used.

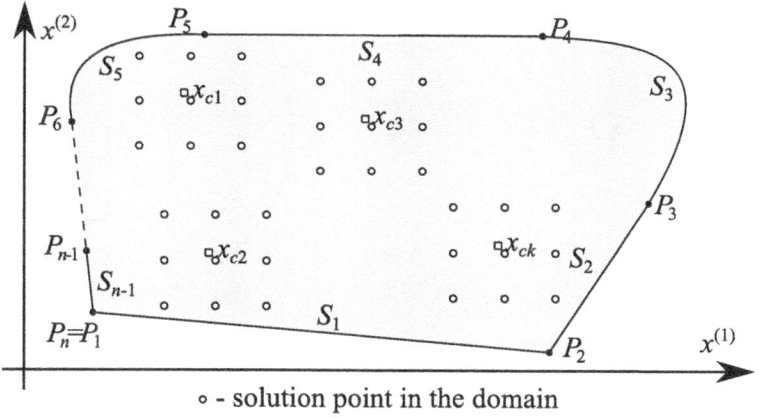

$\circ$ - solution point in the domain

Fig. 1. Solutions in the domain

To reduce the number of computations, we also expanded equation (7) using Taylor series about any point $x_c \in \{x_{c1}, x_{c2}, ..., x_{ck}\}$ close to the points of solutions (presented in Fig. 1). Therefore, the integral identity in the FPIES has the following form:

$$u(x) = \sum_{j=1}^{n} \mathbb{R} \sum_{l=0}^{N_T} \frac{(-1)^l}{2\pi} \left(\sum_{k=0}^{N_T} \frac{(k-1)! \cdot \left(M_k(\tau_c) - N_k(\tau_c)\right)}{(x_c - \tau_c)^k} \right) \frac{\left(x^{(c)} - x_c\right)^l}{l!}, \quad (8)$$

where $N_0(\tau_c) = 0$.

4   Numerical Results

All tests of the presented algorithm are performed on PC based on Intel Core i5-4590S with 32 GB RAM. The program is compiled by g++ 7.5.0 (with -O2 optimization) on 64-bit Ubuntu Linux operation system (kernel 6.8.0). conventional PIES was a bit modified and uses iterative solver GMRES to find the solution of the system of algebraic equations.

4.1   Current Flow Through the Square Plate

The first example is the current flow through the square plate presented in Fig. 2. Boundary conditions presented in the figure mean potential u (red electrodes) and flux p on the rest of the boundary.

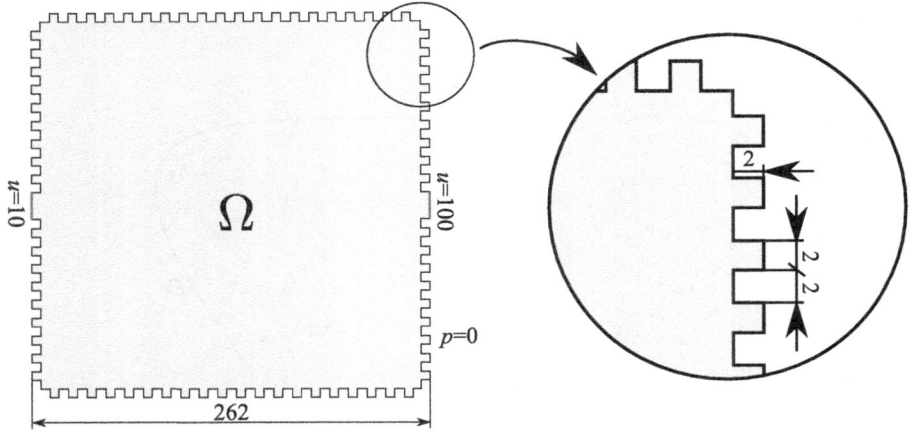

Fig. 2. The example of current flow through the square plate

The research focused on computation speed of solutions in the domain. The FPIES approximation of the modified PIES kernels uses 20 terms in the Taylor series, and the GMRES tolerance is 10^{-8}. The number of collocation points is the same for all segments (from 2 to 8). Therefore, we should solve systems from 2 040 to 8 160 algebraic equations. The same number of terms in Taylor series was also used to approximate the integral identity. The number of groups of solution points (points x_c) was equal to 289, while the number of all solution points in the domain was equal to 65 025 (uniformly distributed over the entire domain).

Table 1 presents the obtained time of solving the problem (CPU time - solving), the computation of solutions in the domain (CPU time - domain) and the RAM usage of the applications.

Table 1. CPU time and RAM utilization between the fast and conventional PIES

Number of col. pts.	CPU time - solving [s]		CPU time - domain [s]		RAM utilization [MB]	
	$FPIES$	$PIES$	$FPIES$	$PIES$	$FPIES$	$PIES$
2	0.86	7.79	9.82	270.89	6.73	70.14
3	1.61	16.42	13.26	369.18	8.47	148
4	2.88	30.92	19.43	559.84	10.85	260
5	4.42	51.12	24.44	733.08	13.61	403
6	6.94	78.12	31.65	926.5	16.89	578
7	8.80	112.66	35.59	1 174.21	20.68	786
8	12.40	154.06	44.14	1 462.57	24.73	1 024

As can be seen from Table 1, the conventional PIES is much slower than the FPIES. For the highest number of equations (8 collocation points), the FPIES required less than 1 minute and 24.73 MB of RAM to solve the problem and compute solutions in the domain, while the conventional PIES required almost 27 minutes and 1024 MB of RAM. Figure 3 presents the overview of the CPU time and RAM utilization between both applications.

We also calculated the mean square error (MSE) of the domain solutions between the FPIES and the conventional PIES to show the accuracy of the method.

Table 2. MSE of domain solutions between the fast and conventional PIES

Number of collocation points						
2	3	4	5	6	7	8
$1.97 \cdot 10^{-11}$	$2.18 \cdot 10^{-11}$	$2.18 \cdot 10^{-11}$	$2.12 \cdot 10^{-11}$	$1.78 \cdot 10^{-11}$	$2.17 \cdot 10^{-11}$	$1.95 \cdot 10^{-11}$

As can be seen from Table 2, solutions in the domain in the FPIES are as accurate as in the conventional PIES. The MSE for over 65 000 solution points does not exceed $3 \cdot 10^{-11}$. A graphical representation of solutions in the domain in the form of potential distribution on the plate is presented in Fig. 4.

4.2   The Perforated Plate

The second example is the perforated plate (square plate with many holes) shown in Fig. 5. The boundary conditions are given in this figure.

As in the first example, the research focused on the computation speed of solutions in the domain. The same number of terms in the Taylor series ($N_T = 20$) is used in the FPIES kernels, and the GMRES tolerance is also equal to 10^{-8}. The number of collocation points is the same on all segments (from 2 to 8). We should solve systems from 4 160 to 16 640 algebraic equations in this example.

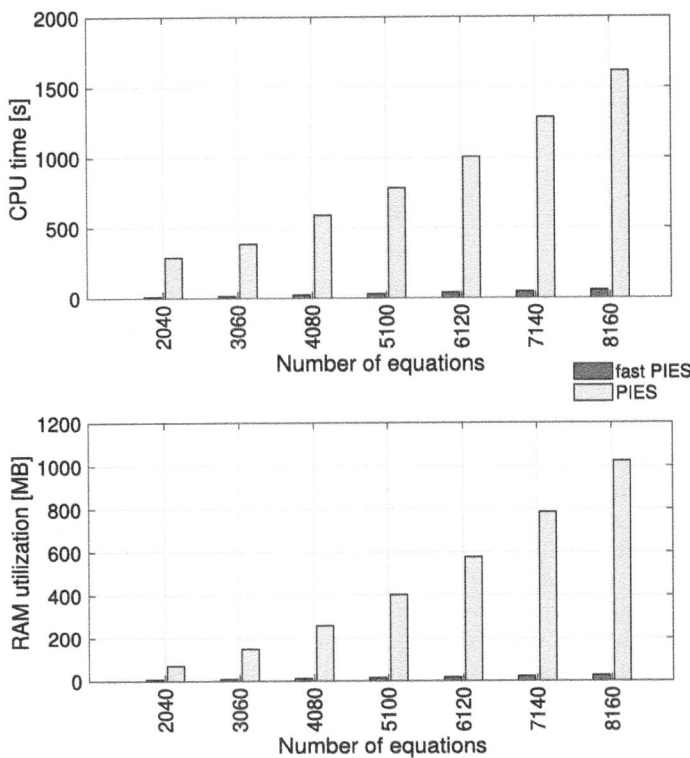

Fig. 3. CPU time and RAM utilization between the PIES and conventional PIES

The same number of terms in Taylor series was also used to approximate the integral identity. The number of groups of solution points (points x_c) was equal to 1 600, while the number of all solution points in the domain was close to 300 000 (uniformly distributed over the whole domain).

This research also confirms that the conventional PIES is much slower than the FPIES. As can be seen from Table 3, for the highest number of equations (8 collocation points), the FPIES required less than $4\frac{3}{4}$ minutes and 165 MB of RAM to solve the problem and compute solutions in the domain, while the conventional PIES used over $4\frac{3}{4}$ hours and 4 247 MB of RAM. Figure 6 gives an overview of the CPU time and RAM usage between the two applications.

The highest MSE value of domain solutions between the FPIES and conventional PIES is less than $5.0 \cdot 10^{-8}$. Therefore, the accuracy of the FPIES method is as accurate as the conventional PIES.

We also compared the FPIES with the FMBEM [28]. It has been proved previously (among others in [3]) that the application of the BEM requires meshes with a large number of elements to obtain accuracy similar to the PIES. Therefore, the boundary in the FMBEM is composed of 10 400 linear elements.

Fig. 4. Potential distribution on the plate

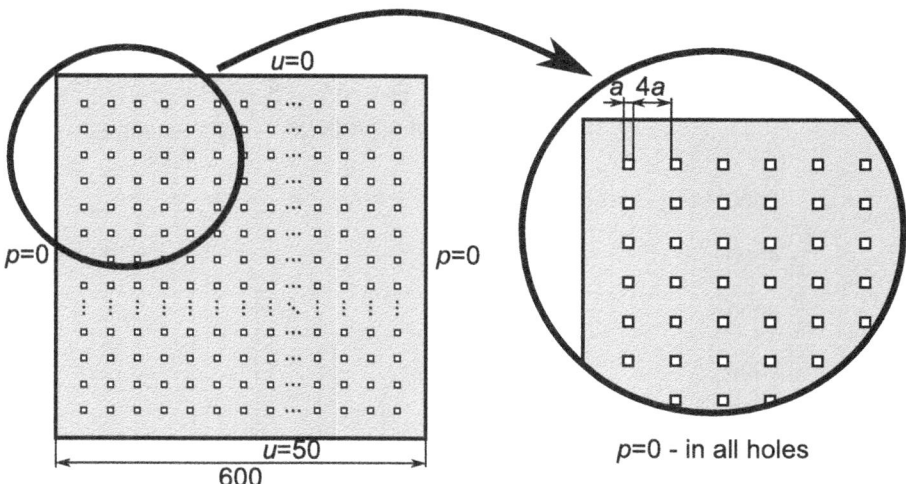

Fig. 5. The example of perforated plate

Solutions are compared with ones obtained from 4 collocation points FPIES. The GMRES tolerance (convergence criterion) equals 10^{-8}, and the number of terms in the Taylor series is set to 20, similar to previous examples. The accuracy

Table 3. CPU time and RAM utilization between the fast and conventional PIES

Number of col. pts.	CPU time - solving [s]		CPU time - domain [s]		RAM utilization [MB]	
	FPIES	*PIES*	*FPIES*	*PIES*	*FPIES*	*PIES*
2	10.01	31.76	35.18	2 855.86	19.01	272
3	18.59	75.34	52.42	4 167.13	32.77	604
4	30.79	142.77	73.78	6 380.91	50.76	1 068
5	47.28	226.65	94.62	8 295.64	73.72	1 665
6	68.11	343.78	116.39	10 481.7	98	2 393
7	92.80	498.10	138.73	13 025.4	129	3 254
8	121.85	679.37	161.31	16 566.5	165	4 247

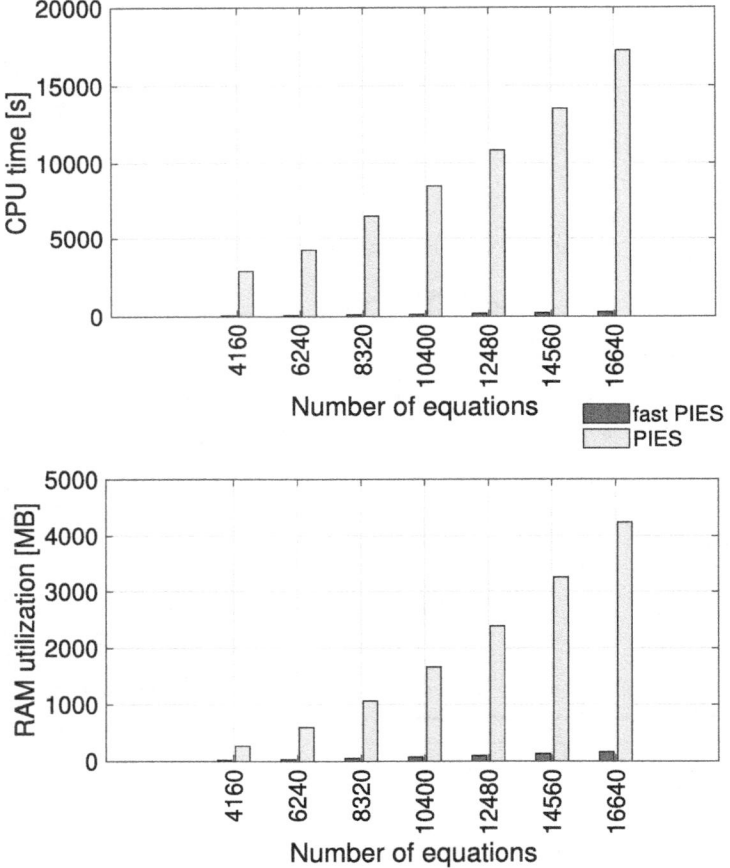

Fig. 6. CPU time and RAM utilization between the PIES and conventional PIES

of the solutions is calculated as the mean square error (MSE) between the results of the FPIES and the fast multipole BEM.

Table 4. Comparison of CPU time and RAM utilization between the FMBEM and the fast PIES

CPU time [s]		RAM utilization [MB]		No. of GMRES it.		MSE
FPIES	FMBEM	FPIES	FMBEM	FPIES	FMBEM	
104.57	217.38	50.76	121	17	77	$2.3904 \cdot 10^{-5}$

As can be seen from Table 4, both CPU time and RAM utilization are smaller in the FPIES. The MSE between the FPIES and the FMBEM is not as small as in the previous examples.

5   Conclusions

The paper presents the FPIES for domain solutions of potential 2D BVPs. The FPIES has previously been successfully applied to modelling 2D potential BVPs and finding solutions on the boundary. The fast multipole technique applied to the integral identity significantly reduces the CPU time for computing domain solutions. Presented examples confirm the high efficiency of the FPIES in solving complex engineering problems on a standard PC in a reasonable time. However, the real power of the FPIES is related to the size of the problems to be solved. The conventional PIES to solve the problems with a system of 16 640 equations (perforated plate, 8 collocation points) uses almost 4.25 GB RAM in $4\frac{3}{4}$ h, while the FPIES requires only 165 MB RAM in $4\frac{3}{4}$ min.

Obtained results strongly suggest that this line of research should be continued. The authors intend to extend the FPIES algorithm to problems modelled by other differential equations.

References

1. Kużelewski, A., Zieniuk, E.: The FMM accelerated PIES with the modified binary tree in solving potential problems for the domains with curvilinear boundaries. Numer. Algorithms **88**(3), 1025–1050 (2021). https://doi.org/10.1007/s11075-020-01066-6
2. Kużelewski, A., Zieniuk, E.: Solving of multi-connected curvilinear boundary value problems by the fast PIES. Comput. Methods Appl. Mech. Eng. **391**, 114618 (2022)
3. Zieniuk, E.: Hermite curves in the modification of integral equations for potential boundary-value problems. Eng. Comput. **20**(1–2), 112–128 (2003)
4. Seo, H.-D., Park, H.-J., Kim, J.-I., Lee, P.S.: The particle-attached element interpolation for density correction in smoothed particle hydrodynamics. Adv. Eng. Softw. **154**, 102972 (2021)

5. Jaśkowiec, J., Pluciński, P.: Discontinuous galerkin method in numerical simulation of two-dimensional thermoelasticity problem with single stabilization parameter. Adv. Eng. Softw. **122**, 62–80 (2018)
6. Duarte, C.A., Oden, J.T.: H-p clouds - an h-p meshless method. Numer. Methods Partial Differ. Eqn. **12**, 673–705 (1996)
7. Zienkiewicz, O.C.: The Finite Element Method. McGraw-Hill, London (1977)
8. Zienkiewicz, O.C., Taylor, R.L., Zhu, J.Z.: The Finite Element Method: Its Basis and Fundamentals, 7th edn. Butterworth-Heinemann, Oxford (2013)
9. Babuska, I., Banerjee, U., Osborn, J.E.: Generalized finite element methods: main ideas, results, and perspective. Int. J. Comput. Methods **1**(1), 67–103 (2004)
10. Brebbia, C.A., Telles, J., Wrobel, L.C.: Boundary element techniques, theory and applications in engineering. Springer-Verlag, New York (1984)
11. Banerjee, P.K., Butterfield, R.: Boundary Element Methods in Engineering Science. McGraw-Hill, London (1981)
12. Katsikadelis, J.T.: Boundary Elements Theory and Applications. Elsevier, Amsterdam (2002)
13. Nedjar, B.: A coupled BEM-FEM method for finite strain magneto-elastic boundary-value problems. Comput. Mech. **59**(5), 795–807 (2017)
14. Hughes, T., Cottrell, J.A., Bazilevs, Y.: Isogeometric analysis: CAD, finite elements, NURBS, exact geometry and mesh refinement. Comput. Methods Appl. Mech. Eng. **194**(39–41), 4135–4195 (2005)
15. L. Beirao Da Veiga, A. Russo, G.V.: The virtual element method with curved edges. ESAIM: Mathe. Model. Numeri. Anal. **53**(2), 375-404 (2019)
16. Zieniuk, E., Szerszeń, K., Bołtuć, A.: Genetic algorithms applied to optimal arrangement of collocation points in 3D potential boundary-value problems. In: Saeed, K., Pejaś, J. (eds.) Information Processing and Security Systems, pp. 113–122. Springer, Boston (2005)
17. Zieniuk, E., Kapturczak, M., Sawicki, D.: The NURBS curves in modelling the shape of the boundary in the parametric integral equations systems for solving the laplace equation. In: International Conference of Numerical Analysis and Applied Mathematics 2015, ICNAAM 2015, AIP Conference Proceedings 1738, 480100 (2016)
18. Zieniuk, E., Bołtuć, A.: Non-element method of solving 2D boundary problems defined on polygonal domains modeled by Navier equation. Int. J. Solids Struct. **43**(25–26), 7939–7958 (2006)
19. Zieniuk, E., Bołtuć, A., Szerszeń, K.: Shape identification in nonlinear boundary problems solved by PIES method. Acta Mechanica et Automatica **8**(1), 16–21 (2014)
20. Rokhlin, V.: Rapid solution of integral equations of classical potential theory. J. Comput. Phys. **60**(2), 187–207 (1985)
21. Greengard, L.F., Rokhlin, V.: A fast algorithm for particle simulations. J. Comput. Phys. **73**(2), 325–348 (1987)
22. Greengard, L.F.: The rapid evaluation of potential fields in particle systems. The MIT Press, Cambridge (1988)
23. Kużelewski, A., Zieniuk, E., Bołtuć, A., Szerszeń, K.: Modified Binary Tree in the Fast PIES for 2D Problems with Complex Shapes. In: Krzhizhanovskaya, V.V., Závodszky, G., Lees, M.H., Dongarra, J.J., Sloot, P., Brissos, S., Teixeira, J. (eds.) ICCS 2020. LNCS, vol. 12138, pp. 1–14. Springer, Cham (2020). https://doi.org/ 10.1007/978-3-030-50417-5_1
24. Saad, Y., Schultz, M.H.: A generalized minimal residual algorithm for solving nonsymmetric linear systems. SIAM J. Sci. Stat. Comput. **7**, 856–869 (1986)

25. Kużelewski, A., Zieniuk, E.: The fast parametric integral equations system in an acceleration of solving polygonal potential boundary value problems. Adv. Eng. Softw. **141**, 102770 (2020)
26. Kużelewski, A., Zieniuk, E.: Searching for the best tree parameters in the IFPIES. In: The 37th Annual European Simulation and Modelling Conference ESM2023, Modelling and Simulation 2023, EUROSIS-ETI Publications, Ghent, Belgium, pp. 48–52 (2023)
27. Kużelewski, A., Zieniuk, E.: Influence of selected IFPIES parameters on CPU time and RAM utilization. In: The 38th Annual European Simulation and Modelling Conference ESM2024, Modelling and Simulation 2024, EUROSIS-ETI Publications, Ghent, Belgium, pp. 288–293 (2024)
28. Liu, Y.J., Nishimura, N.: The fast multipole boundary element method for potential problems: a tutorial. Eng. Anal. Boundary Elem. **30**(5), 371–381 (2006)

Detecting and Understanding Hateful Contents in Memes Through Captioning and Visual Question-Answering

Ali Anaissi[1,3(✉)], Junaid Akram[1,3,4], Kunal Chaturvedi[2], and Ali Braytee[2(✉)]

[1] School of Computer Science, The University of Sydney, 2008 Camperdown, NSW,
Australia
{ali.anaissi,junaid.akram}@sydney.edu.au,
{ali.anaissi,junaid.akram}@uts.edu.au
[2] School of Computer Science, University of Technology Sydney, Ultimo, Australia
{Kunal.Chaturvedi,ali.braytee}@uts.edu.au
[3] University of Technology Sydney, TD School, Ultimo, Australia
[4] Faber Business School, Australian Catholic University, 2060 Peter, North Sydney,
NSW, Australia
junaid.akram@acu.edu.au

Abstract. Memes are widely used for humor and cultural commentary, but they are increasingly exploited to spread hateful content. Due to their multimodal nature, hateful memes often evade traditional text-only or image-only detection systems, particularly when they employ subtle or coded references. To address these challenges, we propose a multimodal hate detection framework that integrates key components: OCR to extract embedded text, captioning to describe visual content neutrally, sub-label classification for granular categorization of hateful content, RAG for contextually relevant retrieval, and VQA for iterative analysis of symbolic and contextual cues. This enables the framework to uncover latent signals that simpler pipelines fail to detect. Experimental results on the Facebook Hateful Memes dataset reveal that the proposed framework exceeds the performance of unimodal and conventional multimodal models in both accuracy and AUC-ROC.

Keywords: Hateful Memes · Multimodal Detection · Optical Character Recognition · Classification

1 Introduction

Memes have emerged as a widely used medium on social media platforms, combining images and text overlays to convey humor, satire, or cultural commentary. Despite their seemingly innocuous appearance, memes are increasingly exploited to propagate hateful or discriminatory content [3,27]. Due to their multimodal nature, such content often bypasses conventional text-only or image-only detection algorithms. When image elements and textual components interact in subtle ways, hateful content may remain hidden, allowing offending material to circulate unchecked [6,32]. Empirical evidence from the Hateful Memes Challenge has shown that unimodal approaches typically fail to adequately capture the range of possible hateful expressions embedded within memes [17,26]. Consequently, there is a critical demand for robust, integrated solutions that can parse both textual and visual cues to identify underlying animosity or prejudice.

Recent studies [11,13,23] have attempted to bridge the gap between language and vision representations, revealing that combined multimodal strategies can achieve promising results for specific domains such as misogynistic memes [34] or harmful COVID-19 memes [29]. While these approaches have shown promise, they exhibit several key limitations that hinder their ability to comprehensively detect nuanced hateful content. First, many existing methods [1,4,35] rely on fixed multimodal representations, where text and image features are extracted independently and fused statically. This rigid approach fails to capture the dynamic interplay between textual and visual cues, making it difficult to detect contextually embedded hate signals, such as sarcasm, coded symbols, or ambiguous imagery [18]. Second, these methods typically lack real-time adaptive reasoning, instead relying on predefined classification heuristics [19,33]. As a result, they struggle with detecting veiled or evolving hate speech that requires contextual reasoning beyond surface-level analysis. Third, existing models often categorize hateful content using coarse-grained labels, such as simply hateful or non-hateful, without distinguishing between different forms of hate speech. This lack of specificity reduces interpretability and makes it harder to apply targeted moderation strategies. Such limitations highlight the need for a more systematic approach that incorporates iterative questioning, refined retrieval, and text-image fusion at a granular level.

To address these challenges, this paper introduces a framework that integrates optical character recognition (OCR), caption generation, retrieval- augmented classification, and a visual question answering (VQA) module. OCR reliably extracts overlaid text, while captioning supplies a neutral description of the visual scene. We enhance classification by leveraging a sub-labeling strategy, segmenting hateful content according to attributes such as race, religion, or others. This fine-grained division increases precision in retrieval-augmented steps, ensuring that exemplars align more closely with the observed meme. Additionally, the VQA system formulates targeted queries about potentially harmful symbols, background contexts, or linguistic cues that might escape notice in single-round analyses. By integrating these components, we aim to offer a system robust enough to detect concealed instances of hate speech.

The paper makes the following contributions:

- A multimodal approach that integrates OCR for textual extraction, neutral captioning for visual context, a sub-label retrieval, and a multi-turn VQA, to detect both explicit and implicit hateful cues in memes is proposed.
- A sub-label classification framework that partitions hateful content into race-based, gender-based, and other sub-dimensions is introduced to improve the accuracy of retrieval-augmented generation (RAG).

2   Related Works

Over the past few years, research on hateful meme detection has evolved considerably, emphasizing the need for integrated analysis of both textual and visual modalities [15,30]. Early attempts often separated images from text, applying standard classifiers to each modality in isolation. However, the limitation of such methods became apparent when memes contained subtle or implicit hateful references that only emerged through interaction between visual features and overlaid text. Consequently, various studies started to explore multimodal fusion. Kiela et al. [16] introduced the Hateful Memes Challenge, releasing a dataset that paired each image with short textual content to highlight the complexities of meme-based hate. Badjatiya et al. [5] and Davidson et al. [7] initially concentrated on textual classification, adopting lexicon-based approaches or neural architectures like CNNs and LSTMs, but these did not fully capture the compound nature of memes. Meanwhile, image-based methods such as Gmez et al. [10] and Howard et al. [14] attempted to detect hateful symbols or cues through CNNs and other vision models, yet struggled when the hatred was expressed solely via text.

Subsequent efforts introduced hybrid or multimodal models to process images and text jointly. Transformative architectures such as ViLBERT [24] and Visual BERT [21] harness cross-attention mechanisms to align textual and visual embeddings, thereby improving classification accuracy. In parallel, the Facebook Hateful Memes dataset [16] further prompted researchers to refine their multimodal pipelines, as it contained nuanced and challenging examples of encoded hate speech. Rizzi et al. [34] addressed misogynistic memes by proposing a fine-tuned VisualBERT that excelled at combining textual embeddings from OCR with high-level image features obtained from pretrained CNNs. Pramanick et al. [29] tackled COVID-19-related misinformation with a focus on harmful memes, showing that domain-specific training data could refine detection for medical or pandemic-oriented hate. Although these approaches outperformed unimodal baselines, they occasionally failed on memes whose meaning shifted dramatically depending on cultural or contextual details not captured by purely data-driven models.

Recent works have sought to incorporate advanced language models, retrieval techniques, and Visual Question Answering (VQA) to overcome the remaining challenges. Devlin et al. [8] illustrated the utility of contextual embeddings via BERT for language understanding, while Lewis et al. [20] introduced Retrieval-Augmented Generation (RAG) to infuse external knowledge into classification or generation tasks. Accordingly, sub-label methods emerged to partition hateful content into categories such as race or gender, facilitating more precise retrieval and classification [34]. Additionally, VQA-based systems proved beneficial for generating iterative queries about scene elements, as multi-turn dialogue can reveal latent meaning. By incorporating refined embedding models like CLIP [31] and advanced prompt engineering, researchers succeeded in capturing the interplay between textual overlays and visual symbolism at deeper levels [2,11]. Collectively, these investigations underscore the vital role of multimodality and contextual verification in tackling hateful memes, thereby guiding the develop-

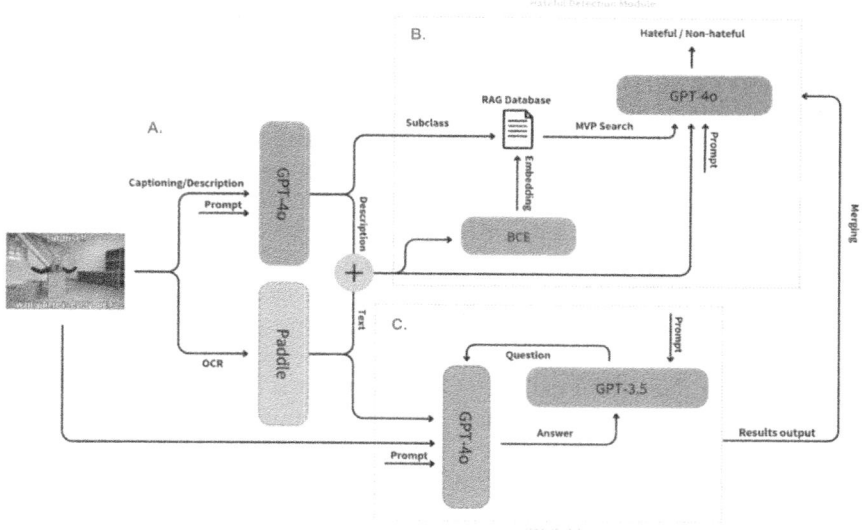

Fig. 1. The overall framework, RAG (sub_label + VQA) for detecting hateful content

ment of more robust pipelines that can identify concealed or culturally coded hatred.

3   Methods

In this section, we describe our proposed multimodal pipeline integrating OCR and captioning, VQA module, and a hateful detection module. The overall framework is shown in Fig. 1.

3.1   Captioning and OCR

A key challenge in detecting hateful content in memes arises from the interplay of textual and visual cues. As shown in Fig. 1A, we first extract all text and generate captions before passing the multimodal information to our detection modules. Specifically, we separate the input processing into two complementary procedures. First, Paddle OCR [9], an optical character recognition system, is used to extract textual messages in memes. To supplement OCR, we generate a caption describing the visual content of the meme using a large language model. The net effect is a more robust representation of the meme, combining the recognized text with a broad contextual description.

3.2   VQA Module

The Visual Question Answering module integrates OCR, multimodal analysis, and advanced language models to detect and analyze hateful content in memes.

As shown in Fig. 1C, the workflow begins with raw meme input, which includes both visual and textual data. Using the Paddle OCR module, textual information embedded within the image is extracted, enabling the identification of captions, phrases, or symbols that may carry hateful messages. This extracted text, along with the raw image, forms the foundation for multimodal analysis in the VQA pipeline. The second stage involves processing the inputs using GPT-based models for dynamic question generation and context-aware answering. Initially, GPT-4.0 generates a broad, context-sensitive question aimed at understanding the overall theme of the meme, with a focus on detecting hate signals such as stereotypes, offensive language, or harmful visual elements. The generated question is then processed by GPT-3.5, which provides detailed answers by integrating visual and textual cues. If hate-related elements are identified, such as racial slurs or stereotypical imagery, the system refines its analysis by generating follow-up questions through GPT-4.0, specifically targeting the hateful components.

To ensure coherence and accuracy, the system employs a multi-turn dialogue mechanism, where all questions and answers are stored in a contextual database. This enables the system to maintain continuity across interactions, eliminating redundancy and ensuring that every aspect of the meme is thoroughly examined. Once the analysis is complete, the structured Q&A pairs are fed into a specialized Hate Detection Module that works in conjunction with the Retrieval-Augmented Generation (RAG) pipeline. The RAG pipeline further contextualizes and validates the findings by cross-referencing against a knowledge base of hate-related symbols, phrases, and behaviors. The final output of the system includes a detailed summary of the detected hateful content, highlighting specific textual elements, visual cues, and their contextual implications. By integrating OCR, multimodal analysis, and dynamic reasoning, the VQA system provides a robust solution for detecting hateful content in memes. Its iterative question-answering logic, combined with adaptive refinement, ensures thorough exploration of nuanced hate signals, making it a powerful tool for content moderation in real-world scenarios.

3.3   Hateful Detection

In the hateful detection module, we aim to classify whether a meme is hateful or non-hateful by leveraging the OCR text, the generated caption, and outputs from the VQA module. Figure 1B illustrates the schematic of our hateful content detection module, which consists of multiple steps, including RAG, sub-labeling, and content explanation..

RAG Overview: The Retrieval-Augmented Generation (RAG) architecture [20] is utilized as a core component for hate detection by incorporating a vector database, embedding model, and ranking mechanism. The architecture processes each memes textual input (caption and OCR) by embedding it and querying a repository of labeled data, explanations, and examples. These retrieved chunks provide additional context in a prompt-like fashion to guide the model in producing more accurate and context-aligned inferences about hateful content.

Content Explanation for RAG: Another variant of the RAG architecture augmented the vector database with not only the caption and OCR text but also detailed explanations for why specific memes were labeled hateful or non-hateful. The assumption was that these explanations could help the model identify nuanced hate signals in new memes.

Sub-labeling for RAG: A specific implementation of RAG, known as sub-label classification, was applied. Instead of treating hatefulness as a single, broad category, the sub-labeling method divides hateful content into finer-grained categories, such as race, religion, and others. By embedding the memes caption and OCR text, the RAG system retrieves content related to the most relevant sub-label, providing contextual anchors that improve classification. For instance, a meme involving race-based hate speech retrieves examples and contextual references from the race sub-label category, making the detection more precise.

Our final framework, RAG (sub_label + VQA) integrated the outputs of the VQA module with the sub-labeling RAG method. The VQA results, which provide detailed contextual information by combining caption and OCR analysis with visual cues, were incorporated into the RAG pipeline as additional reference material to detect hateful memes.

4   Experiments and Results

4.1   Datasets

We utilized Facebook Hateful Memes (FHM) [26] dataset for our experiments. The dataset contains diverse meme examples with hateful vs. non-hateful labels. The data integrates text captions overlaid on images and is one of the primary resources provided by Facebook for the Hateful Memes Challenge. Next, we apply random transformations such as rotation, scaling, cropping, and mild color jitter. These transformations enrich the models exposure to diverse visual conditions, thus boosting resilience to typical noise or distortion in user-generated memes. In certain data splits, we note that hateful content is encoded through metaphors, coded language, or domain-specific references. We thus expand the dataset with carefully curated examples reflecting these nuances to reinforce the sub-labeling strategy in the RAG pipeline. This expansion aids in capturing cultural, linguistic, or other contextual factors that might not be evident from standard data subsets. Overall, these strategies enhance the systems capacity to tackle newly emerging hateful memes with novel textual or visual patterns.

4.2   Evaluation Metrics

For hateful content detection, we employ two primary metrics: Accuracy and AUC-ROC. For VQA, we adopt the VQAScore methodology [22]. We conducted five rounds of scoring to mitigate model variability. In each round, the VQA system generated answers to a collection of queries derived from the meme images. We then calculated VQAScore for each generated answer-image pair, and took the average over these five rounds. This approach ensures a more robust estimation of performance, reducing the influence of any single outlier run.

Table 1. Performance comparison across various models, including unimodal and multimodal methods. Acc. denotes Accuracy; AUROC stands for Area Under the Receiver Operating Characteristic.

Type	Model	Acc. (%)	AUROC (%)
	Human annotators	84.70	82.65
Unimodal	Image-grid [12]	52.00	52.63
	Image-region [16]	52.13	55.92
	Text BERT [8]	59.20	65.08
Multimodal	Late fusion [16]	59.66	64.75
	Concat BERT [16]	59.13	65.79
	MMBT-grid [16]	60.06	67.92
	MMBT-region [16]	60.23	70.73
	ViLBERT [25]	62.30	70.45
	Visual BERT [21]	63.20	71.33
	ViLBERT CC [16]	61.10	70.03
	Visual BERT COCO [16]	64.73	71.41
	GPT-4o mini [28]	69.50	75.02
Our Method	RAG (explanation)	59.20	63.01
	RAG (sub_label)	72.00	76.52
	RAG (sub_label + VQA)	**73.50**	**78.35**

4.3   Results

Quantitative Analysis. Table 1 summarizes the performance of various models and methods in detecting hateful memes. The table also includes comparisons with human annotations as an upper bound, as well as benchmark approaches from the challenge. Human annotations remain the most accurate, with 84.70% accuracy and 82.65% AUROC. Among unimodal methods, vision-only models such as Image-grid and Image-region perform poorly around 52% accuracy, highlighting the insufficiency of visual cues alone for detecting nuanced hateful content. Text BERT outperforms these with 59.20% accuracy and 65.08% AUROC, underscoring the greater informativeness of textual features in this domain. Multimodal baselines, which integrate image and text modalities, demonstrate marked improvements. Simple fusion techniques like Late Fusion and Concat BERT offer modest gains (5960% accuracy). More sophisticated architectures such as MMBT-Region, ViLBERT, and Visual BERT COCO further improve performance, reaching up to 64.73% accuracy and 71.41% AUROC. The strongest baseline, GPT-4o mini, achieves 69.50% accuracy and 75.02% AUROC, setting a high bar for general-purpose large multimodal models.

Our proposed method significantly outperforms all baselines. While the RAG (explanation) variant performs comparably to Text BERT with 59.20% accuracy, incorporating fine-grained sub-labels in RAG leads to a substantial boost with accuracy of 72.00% and AUROC of 76.52%. The highest gains are observed when this sub-label retrieval is further combined with VQA, yielding the best overall results of 73.50% accuracy and 78.35% AUROC. These findings confirm the synergy between sub_label-based retrieval and the contextual enhancements provided by the VQA module. Rather than only relying on raw text or naive retrieval from explanation templates, the sub_label approach retrieves precisely relevant hateful exemplars, while the VQA module helps uncover implicit cues that might not be evident through OCR captioning alone.

Qualitative Observations. Figure 2 offers an illustrative example, showing system outputs for both a positively identified hateful meme and a non-hateful instance. The figure includes how OCR extracts textual content, how the captioning module describes the image, and how the multi-round VQA interacts with the meme to highlight potentially hateful elements.

In the hateful example, OCR precisely captured key terms from the overlaid text, and the captioning module accurately noted contextual objects and background. The VQA dialog then focused on potentially discriminatory language, confirming hateful cues and retrieving relevant sub_label data through the RAG sub_label pipeline. The final classification was correct and accompanied by a short textual explanation consistent with the known ground truth.

In the non-hateful case, the system again accurately recognized textual and visual details but found no hateful signals. The RAG retrieval was less relevant, returning only examples bearing minimal resemblance to hateful content. As a result, the classification was non-hateful, aligning with the ground-truth label. This outcome underlines the pipelines capacity to remain conservative when the textual and visual signals do not suggest hateful references.

4.4 Discussion

These results indicate that combining large language models with sub_label-based retrieval and VQA modules yields significant gains over unimodal or simpler multimodal baselines. Several observations are worth highlighting:

Performance Gaps and Remaining Challenges. Although RAG (sub_label) + VQA achieves 73.50% accuracy, it still lags behind human annotators. This gap underscores the complexity and ambiguity of hateful memes, which often rely on cultural references, double meanings, or evolving slang not always

Image		
Text	SCREW FEMINISM MYPLACEISINTHE KITCHEN!	MARRIAGE IS... ALWAYS HAVING SOMEONE TO PITY-LAUGH AT YOUR JOKES RD
Description	This meme features a retro-styled image of a woman dressed in 1950s housewife attire, with a cheerful expression. It includes the text: "SCREW FEMINISM, MY PLACE IS IN THE KITCHEN!" The overall theme suggests a traditional view of gender roles, particularly emphasizing the idea that a woman's place is at home, specifically in the kitchen, which is juxtaposed against the concept of feminism.	The meme features a cheerful couple sitting together, each holding a cup. The text overlays the image and reads, "Marriage is... ALWAYS HAVING SOMEONE TO PITY-LAUGH AT YOUR JOKES." This combines a lighthearted view of marriage with a humorous take on the dynamics of relationship humor, suggesting that partners often support each other despite the quality of the jokes made. The tone is playful and aims to provoke a smile or a laugh.
VQA Dialogue	**Question:** Can you describe the image in detail, focusing on the people, their actions, emotions, notable objects, and the overall setting? **Answer:** A smiling woman in a retro kitchen wears a red dress with a white apron, expressing defiance toward feminism. **Question:** What specific elements in the retro kitchen contribute to the overall nostalgic feel of the setting? **Answer:** The retro kitchen features pastel colors, floral curtains, and vintage utensils, enhancing the nostalgic atmosphere.	**Question:** Can you describe the image in detail, focusing on the people, their actions, emotions, notable objects, and the overall setting? **Answer:** A smiling couple in casual attire share coffee and laughter on a porch, highlighting warmth and companionship. **Question:** What do you think might be the significance of the mugs they are holding in the image? **Answer:** The couple's body language is relaxed and affectionate, with genuine smiles and eye contact, as they enjoy coffee together.
Similar cases (The first one of top5)	that's it, back to the kitchen! where's my sammich? women's day is over! The meme features a woman on the ground in a casual pose, wearing shorts and a sweater, with a man standing behind her. The text overlay includes phrases that suggest traditional gender roles and a dismissal of a specific occasion, Women's Day. The tone appears sarcastic, using humor related to stereotypes about women's roles. The combination of the imagery and text creates a juxtaposition of casualness and a controversial statement regarding gender expectations. (gender)	when you see a black guy in your store but he isn't trying to steal anything impossible This meme features an image of a smiling woman wearing a purple headscarf and floral clothing. The text overlay reads: "i married my brother to commit immigration fraud." The meme appears to imply a controversial claim about the subject, suggesting inappropriate or illegal behavior related to marriage and immigration. The tone is informal and provocative, characteristic of many internet memes that aim to elicit strong reactions. (race)
Ground Truth	Hateful	Non-hateful
Prediction	Hateful	Non-hateful
Explanation	The statement is hateful because it explicitly denigrates feminism and promotes a traditional and restrictive view of women's roles. it perpetuates the stereotype that a woman's place is solely in the kitchen	The statement describes a meme that presents a humorous and affectionate take on marriage. the description emphasizes companionship

Fig. 2. Outputs for a hateful example (left) and a non-hateful example (right). The pipeline includes accurate OCR detection, objective captioning, multi-turn VQA addressing targeted hate cues, and final classification via RAG.

captured by static training data. Further refinement of sub-label categories, addition of external knowledge sources, and extended data augmentation strategies may narrow this human-machine divide.

Effectiveness of OCR With the rise of multimodal large language models, one may question the need for a dedicated OCR module. However, we find that incorporating explicit OCR (PaddleOCR) remains valuable, especially when dealing with stylized, distorted, or meme-specific fonts that challenge even state-of-the-art vision-language models. Explicit OCR ensures consistent and controllable extraction of embedded text, which downstream modules such as VQA and RAG rely on for accurate reasoning. Furthermore, separating text extraction from high-level reasoning supports interpretability, and modular debugging.

Utility of VQA Dialogue. Notably, RAG showed visible improvements after including VQA-derived context. The VQA system probes the meme with targeted questions, clarifying ambiguous cues and capturing nuanced correlations between text and visuals. This synergy is crucial in uncovering content that is hateful only when certain textual or symbolic aspects align with specific contexts or objects in the image. The average VQAScore of 75.04 also suggests that the system reliably produces answers consistent with the underlying image content, thereby strengthening the subsequent classification.

Explanation-based RAG Limitations. The RAG (explanation) configuration performed poorly for reasons related to noise in the textual explanations and potential misalignment with new memes. The assumption that labeled explanations from certain memes would be directly transferable or consistently interpreted appears flawed. In contrast, sub_label retrieval provides a more targeted anchor (e.g., detecting racial hate or religious hate specifically), improving retrieval precision.

Implications for Real-World Applications. Content moderation platforms or social media sites that must identify hateful memes in real time can benefit from adopting a pipeline that integrates a carefully designed retrieval mechanism, a multi-turn VQA system, and robust text-image analysis. However, real-time constraints require optimization to reduce computational overhead; our solution underscores the effectiveness of multi-step synergy but also reveals potential latency in large-scale deployments. In particular, the use of large language models for multi-turn VQA and the dependency on retrieval-augmented generation (RAG) with sub-label classification introduce substantial memory and processing demands. These may hinder responsiveness in high-throughput or latency-sensitive settings. Future engineering efforts would thus focus on accelerating sub_label lookups and streamlining the VQA query-response phase.

The quantitative results, shown in Table 1, and the qualitative analysis demonstrate that our integrated system effectively detects hateful memes and outperforms several multimodal baselines. Although there remains a gap relative to human-level comprehension of subtle, context-dependent hate, the results confirm that careful synergy among OCR, captioning, VQA, and specialized retrieval strategies can significantly improve classification performance.

5   Conclusion

The proposed framework demonstrates a robust approach to addressing hateful content detection within memes by integrating multiple modules for text extraction, captioning, retrieval, and visual question answering. The integrated pipeline achieves significant accuracy and AUC-ROC compared to existing methods on established benchmarks. These results underline the importance of uniting refined language strategies with methods that analyze images more deeply. Future work could extend the framework by incorporating culturally nuanced knowledge graphs, refining VQA prompts to reduce false positives, or integrating dynamic feedback loops for real-time detection.

Acknowledgment. We acknowledge the contributions of Bonnie Zhong, Haodi Yang, Yinuo Wang, Lu Tang, Yiqi Zhao, Yuanhao Huo to this project.

References

1. Aamir, M., Raut, R., Jhaveri, R.H., Akram, A.: AI-generated content-as-a-service in iomt-based smart homes: personalizing patient care with human digital twins. IEEE Transa. Consumer Electron. (2024)
2. Agarwal, S., Sharma, S., Nakov, P., Chakraborty, T.: MemeMQA: multimodal question answering for memes via rationale-based inferencing. arXiv preprint arXiv:2405.11215 (2024)
3. Akram, J., Tahir, A.: Lexicon and heuristics based approach for identification of emotion in text. In: 2018 International Conference on Frontiers of Information Technology (FIT). pp. 293–297. IEEE (2018)
4. Anaissi, A., Braytee, A., Akram, J.: Fine-tuning llms for reliable medical question-answering services. In: 2024 IEEE International Conference on Data Mining Workshops (ICDMW). IEEE (2024)
5. Badjatiya, P., Gupta, S., Gupta, M., Varma, V.: Deep learning for hate speech detection in tweets. In: Proceedings of the 26th International Conference on World Wide Web Companion (WWW '17 Companion) (2017)
6. Blaier, E., Malkiel, I., Wolf, L.: Caption enriched samples for improving hateful memes detection. arXiv preprint arXiv:2109.10649 (2021)
7. Davidson, T., Warmsley, D., Macy, M., Weber, I.: Automated hate speech detection and the problem of offensive language. In: Proceedings of the International AAAI Conference on Web and Social Media. vol. 11 (2017)
8. Devlin, J., Chang, M.W., Lee, K., Toutanova, K.: Bert: pre-training of deep bidirectional transformers for language understanding (2019)
9. Du, Y., et al.: Pp-ocr: a practical ultra lightweight ocr system (2020)
10. Gómez, R., Gibert, J., Gómez, L., Karatzas, D.: Exploring hate speech detection in multimodal publications. arXiv preprint arXiv:2005.04982 (2020)
11. Hamza, A., et al.: Multimodal religiously hateful social media memes classification based on textual and image data. ACM Transactions on Asian and Low-Resource Language Information Processing (2023)
12. He, K., Zhang, X., Ren, S., Sun, J.: Deep residual learning for image recognition (2015)

13. Hee, M.S., Lee, R.K.W., Chong, W.H.: On explaining multimodal hateful meme detection models. In: Proceedings of the ACM Web Conference 2022. p. 3651–3655. WWW '22, Association for Computing Machinery, New York, NY, USA (2022)
14. Howard, A., et al.: Searching for MobileNetV3. In: IEEE International Conference on Computer Vision (ICCV). pp. 1314–1324 (2019)
15. Khan, M., Saad, M.M., Tariq, M.A., Kim, D.: Spice-it: smart covid-19 pandemic controlled eradication over ndn-iot. Inform. Fusion **74**, 50–64 (2021)
16. Kiela, D., Firooz, H., Mohan, A., Goswami, V., Singh, A., Ringshia, P., Testuggine, D.: The hateful memes challenge: detecting hate speech in multimodal memes. arXiv preprint arXiv:2005.04790 (2021)
17. Kirk, H.R., et al.: Memes in the wild: assessing the generalizability of the hateful memes challenge dataset. arXiv preprint arXiv:2107.04313 (2021)
18. Kougia, V., Pavlopoulos, J.: Multimodal or text? retrieval or BERT? benchmarking classifiers for the shared task on hateful memes. In: Proceedings of the 2021 Workshop on Online Abuse and Harms (WOAH 2021) (2021)
19. Kovács, G., Alonso, P., Saini, R.: Challenges of hate speech detection in social media: data scarcity, and leveraging external resources. SN Comput. Sci. **2**(2), 95 (2021)
20. Lewis, P., et al.: Retrieval-augmented generation for knowledge-intensive NLP tasks. Neural Inform. Process. Syst. (NeurIPS) paper (2020)
21. Li, L.H., Yatskar, M., Yin, D., Hsieh, C.J., Chang, K.W.: Visualbert: a simple and performant baseline for vision and language (2019)
22. Lin, Z., et al.: Evaluating text-to-visual generation with image-to-text generation. arXiv preprint arXiv:2404.01291 (2024)
23. Liu, Z., Braytee, A., Anaissi, A., Zhang, G., Qin, L.: Ensemble pretrained models for multimodal sentiment analysis using textual and video data fusion. In: Companion Proceedings of the ACM Web Conference 2024. pp. 1841–1848 (2024)
24. Lu, J., Batra, D., Parikh, D., Lee, S.: ViLBERT: pretraining task-agnostic visiolinguistic representations for vision-and-language tasks. Neural Inform. Process. Syst. (NeurIPS) paper (2019)
25. Lu, J., Batra, D., Parikh, D., Lee, S.: Vilbert: pretraining task-agnostic visiolinguistic representations for vision-and-language tasks (2019)
26. Meta AI Research: hateful memes challenge and dataset. https://ai.meta.com/blog/hateful-memes-challenge-and-data-set/ Accessed 23 Nov 2024
27. Munzni, S., Dixit, S., Bhat, A.: Classification of hateful memes by multimodal analysis using CLIP. In: Proceedings of ConIT 2024. pp. 1–5 (2024)
28. OpenAI: Gpt-4o: Openai's most advanced multimodal model. https://openai.com/index/gpt-4o (2024) Accessed 03 Oct 2024
29. Pramanick, S., et al.: Detecting harmful memes and their targets. In: Zong, C., Xia, F., Li, W., Navigli, R. (eds.) Findings of the Association for Computational Linguistics: ACL-IJCNLP 2021, pp. 2783–2796. Association for Computational Linguistics, Online (2021)
30. Qian, C., Shi, X., Yao, S., Liu, Y., Zhou, F., Zhang, Z.: Optimized biomedical question-answering services with llm and multi-bert integration. In: 2024 IEEE International Conference on Data Mining Workshops (ICDMW). IEEE (2024)
31. Radford, A., et al.: Learning transferable visual models from natural language supervision (2021)
32. Rathore, R.S., Jhaveri, R.H., Akram, A.: Galtrust: generative adversarial learning-based framework for trust management in spatial crowdsourcing drone services. IEEE Transa. Consumer Electron. (2024)

33. Rehman, A.U., Rehman, Z., Ali, W., Shah, M.A., Salman, M.: Statistical topic modeling for urdu text articles. In: 2018 24th International Conference on Automation and Computing (ICAC). pp. 1–6. IEEE (2018)
34. Rizzi, G., Gasparini, F., Saibene, A., Rosso, P., Fersini, E.: Recognizing misogynous memes: biased models and tricky archetypes. Inf. Process. Manage. **60**(5), 103474–103474 (2023)
35. Zhang, Z., Robinson, D., Tepper, J.: Detecting hate speech on twitter using a convolution-gru based deep neural network. In: Gangemi, A., Navigli, R., Vidal, M.E., Hitzler, P., Troncy, R., Hollink, L., Tordai, A., Alam, M. (eds.) The Semantic Web, pp. 745–760. Springer International Publishing, Cham (2018)

A GPU-Accelerated Interior Point Method with Applications in Radiation Therapy Optimization

Felix Liu[1]([envelope])[ORCID], Albin Fredriksson[1], and Stefano Markidis[2]

[1] RaySearch Laboratories, Stockholm, Sweden
felix.liu@raysearchlabs.com
[2] KTH Royal Institute of Technology, Stockholm, Sweden

Abstract. Optimization plays a central role in modern radiation therapy, where it is used to determine optimal treatment machine parameters in order to deliver precise doses adapted to each patient case. In general, solving the optimization problems that arise can present a computational bottleneck in the treatment planning process, as they can be large in terms of both variables and constraints. As high precision is often sought, second-order optimization algorithms, such as sequential quadratic programming (SQP) and/or interior point methods (IPM) are commonly used. Existing implementations of these algorithms often use direct linear solvers internally, and are typically intended to run on CPUs. Utilizing iterative linear solvers instead is an active research topic in the optimization community, and one which carries the potential to enable efficient GPU acceleration for these types of optimization problems. Numerical stability issues make this a difficult problem for optimization solvers targeting problems from a wide range of application areas, however. In this paper, we develop and implement a GPU-accelerated interior point method for optimization problems from radiation therapy using iterative linear algebra. We utilize a so called doubly augmented formulation of the Karush-Kuhn-Tucker linear systems, together with a Jacobi-preconditioned conjugate gradient solver, which is able to find sufficiently accurate search directions while running on GPU. By evaluating our solver on real optimization problems from a commercial treatment planning system for radiation therapy, we show that our method can accelerate the aggregated time-to-solution by 1.4 and 4.4 times, respectively, for two patient cases.

1 Introduction

Optimization problems arise in a number of applications areas, including machine learning [3], operations research [23], radiation therapy planning [5,31] and many more. In many applications, the problems are large and challenging in terms of the number of variables and constraints, and the computational performance of the optimization solver is key. The focus in this work will be on interior point methods (IPMs) [22] for optimization, which are well known for their polynomial time complexity and good practical performance.

The specific application we have in mind for this work is optimization of radiation therapy treatment plans. In treatment planning, optimization is used to find control parameters for the treatment machine to deliver a precise dose that is concentrated to the tumor volume, thus achieving the desired killing of tumor cells while sparing surrounding healthy tissue as much as possible. The computations required in this process can be time-consuming, which makes computational speed a crucial factor. GPU computing is already widely used in radiation treatment planning [13], such as for dose calculation [8,17] and various image processing workloads [12]. With the advent of deep neural networks and machine learning, GPUs have also found uses in neural network based automatic segmentation algorithms [25].

One part of the computational workflow that has not yet benefited from GPU acceleration is the optimization algorithm itself. There may be many reasons for this, not least that other computational components, such as dose calculation [8], previously dwarfed the time required for optimization, a situation that is inevitably changing as large performance gains are realized in those areas. Another reason may be algorithmic in nature, in that current algorithms used for precise optimization may be inherently challenging to parallelize to the degree required to use GPUs efficiently.

An active research topic in the literature on IPMs is utilizing iterative linear solvers (e.g. Krylov subspace methods) as the method for solving linear systems internally. Traditionally, IPMs often rely on direct linear solvers, motivated in large part by numerical stability issues and inherent ill-conditioning of the linear systems involved. The move to iterative linear solvers, while challenging in terms of stability and preconditioning, may be crucial for the performance of IPMs on large-scale problems [10]. Another potential benefit of moving to iterative linear solvers, and one of the primary motivations for the method presented in this paper, is better suitability for massively parallel computing hardware such as GPUs,

In this paper, we present a GPU-accelerated IPM implementation for quadratic optimization problems, based on previous work on using Krylov subspace solvers to solve linear systems [16]. Furthermore, while our focus lies on interior point methods for quadratic optimization problems, we mainly consider the case where the quadratic problems are solved as part of a sequential quadratic programming (SQP) algorithm for general nonlinear optimization problems. We show that our solver can outperform existing, clinically used CPU-based solvers on real problems. To the best of our knowledge, this is the first of its kind in GPU accelerated IPMs for radiation therapy optimization.

2   Related Work

GPU accelerated optimization algorithms are already widely used in many contexts, especially for problems where first-order gradient-based algorithms (which do not require Hessian information) are used. Prominent examples include algorithms based on gradient descent for training deep neural networks and similar. For second-order methods with Hessian information such as IPMs and SQP, GPU accelerated solvers do not appear to be as widespread.

For linear programming, GPU accelerated IPMs have been studied previously by Smith et al. and Gade-Nielsen [9,27], which are both based on a matrix-free method proposed by Gondzio [11]. Notably, Gondzio's matrix-free method also uses a preconditioned conjugate gradient method, with regularization in the IPM itself as well as a custom preconditioner. GPU accelerated IPMs have also been studied for other types of optimization problems such as quadratic programming for training support vector machines [14], as well as more general nonlinear optimization [6]. The paper by Cao et al. [6] is similar to ours in that it uses a preconditioned conjugate gradient method with Jacobi preconditioning as well. However, they consider mainly equality constrained optimization problems and use a different formulation of the Karush-Kuhn-Tucker (KKT) system.

GPU acceleration for IPMs using direct linear solvers has also been studied previously, see [19], where the KKT-system is condensed into a dense form, which is more amenable to GPU accelerated factorization. In [30], a refactorization approach is considered, where pivots from previous factorizations are reused, since the sparsity pattern of matrices often remains the same between IPM iterations. Approaches combining inexact factorization and iterative refinement have also been considered [29], as well as hybrid methods combining factorization and iterative methods to solve KKT systems [24], with promising performance results demonstrated on optimal power flow problems, another application area where large-scale optimization problems are frequently encountered. An example of a software package for IPM with support for GPU acceleration is HiOp [20], which has also been used for optimal power flow problems [21].

For first-order optimization methods for quadratic optimization problems, GPU acceleration has been explored in for example the alternating direction method of multipliers (ADMM) based solver OSQP [28]. The GPU porting of OSQP is described in [26]. As a general rule, first-order methods trade achievable accuracy in favor of computational speed, which may be a very worthwhile trade-off for many applications, but may not be the most suited for radiation therapy where a high degree of accuracy is sought.

3   Background

The optimization algorithm used in this work is based on the method described in [16]. Our contribution in this work is to port the algorithm to GPU accelerators, address related challenges in performance optimization, and evaluate the performance compared to existing solvers on a set of realistic problems, in order to evaluate the performance compared to state-of-the-art. We give a brief overview of the optimization method used for completeness, but refer to [16] for more details.

3.1   Interior Point Methods

IPMs are commonly used for many types of constrained continuous optimization problems, including linear, quadratic, nonlinear and semidefinite programming. Our interest in this paper is in IPMs for quadratic programming (i.e. optimization problems with quadratic objective function and linear constraints). Generally those problems are of the form

$$\text{min.} \quad \frac{1}{2}x^T H x + p^T x$$
$$\text{s.t.} \quad l \le Ax \le u, \tag{1}$$

where H is the $n \times n$ Hessian of the objective function, $p \in \mathbb{R}^n$ are linear coefficients of the objective function and A is an $m \times n$ matrix with coefficients for the linear inequality constraints. A common trick in optimization solvers is to introduce slack variables s_l, s_u for inequality constraints, thereby transforming them into equality constraints and a simple positivity constraint for the slack variables instead. The positivity constraints for the slack variables are handled by replacing them with a logarithmic barrier term in the objective, yielding a problem of the form

$$\text{min.} \quad \frac{1}{2}x^T H x + p^T x - -\mu \sum_i \log((s_l)_i) - \mu \sum_i \log((s_u)_i)$$
$$\text{s.t.} \quad Ax - s_l - l = 0$$
$$-Ax - s_u + u = 0. \tag{2}$$

With $s_l, s_u \ge 0$ handled implicitly. The intuition is that the logarithmic terms in the objective tend towards infinity as the boundary of the feasible region is approached from within, or in this case, when s_l, s_u become close to zero. μ is known as the *barrier parameter*, and its value can be chosen by the solver. IPMs proceed by solving the barrier problem while successively decreasing the value of the barrier parameter μ towards 0.

It can be shown that there exists *Lagrange multipliers* λ such that solutions x to the barrier problem (2) satisfy the following system of equations:

$$r_H := Hx + p - A^T \lambda_l + A^T \lambda_u = 0$$
$$r_l := Ax - s_l - l = 0$$
$$r_u := -Ax + s_u + u = 0$$
$$r_{c_1} := (\lambda_l)_i (s_l)_i - \mu = 0, \quad i \in \{1, ..., m_l\},$$
$$r_{c_2} := (\lambda_u)_i (s_u)_i - \mu = 0, \quad i \in \{1, ..., m_u\}, \tag{3}$$

where λ_l denotes the multipliers for the lower bounds and λ_u for the upper bounds. The slack variables s are subscripted in the same way.

A popular approach is a so called primal-dual [22] approach, which is based on solving the system of equations (3) directly using Newton's method. Newton's method applied to (3) gives a linear system to solve of the form

$$
\begin{pmatrix}
H & -A^T & A^T & & \\
A & & & -I & \\
-A & & & & -I \\
& S_l & & \Lambda_l & \\
& & S_u & & \Lambda_u
\end{pmatrix}
\begin{pmatrix}
\Delta x \\
\Delta \lambda_l \\
\Delta \lambda_u \\
\Delta s_l \\
\Delta s_u
\end{pmatrix}
= -
\begin{pmatrix}
r_H \\
r_l \\
r_u \\
r_{c_1} \\
r_{c_2}
\end{pmatrix},
\tag{4}
$$

where Λ, S are diagonal matrices with the Lagrange multipliers and slack variables on the diagonal, respectively, and e is an appropriately sized vector of ones. Newton's method does not take into account the implicit condition that the slack variables and Lagrange multipliers remain positive throughout. This is accounted for by some line search method instead, where the search direction is scaled by some step length α such that the slacks and multipliers remain positive.

3.2   Sequential Quadratic Programming

Sequential quadratic programming (SQP) [4] is an optimization algorithm for solving nonlinear optimization problems with constraints. The basic idea is to solve, in each SQP iteration, a quadratic subproblem consisting of a quadratic approximation of the objective function or Lagrangian and linear approximation of the constraints. To give a concrete example, consider a problem of the form:

$$
\begin{aligned}
\text{min.} \quad & f(x) \\
\text{subject to} \quad & g(x) \le 0,
\end{aligned}
\tag{5}
$$

where $f : \mathbb{R}^n \to \mathbb{R}$ is the objective function and $g : \mathbb{R}^n \to \mathbb{R}^m$ are the constraints. We assume both $f(x)$ and $g(x)$ to be three times continuously differentiable. We define the Lagrangian of the problem as

$$
\mathcal{L}(x, \lambda) = f(x) - \lambda^T g(x).
\tag{6}
$$

In SQP, we find search directions to iteratively solve problem (5) from the quadratic sub-problem

$$
\begin{aligned}
\text{min.}_d \quad & d^T \nabla_{xx}^2 \mathcal{L}(x, \lambda) d + d^T \nabla f(x) \\
\text{subject to} \quad & d^T \nabla g(x) + g(x) \le 0,
\end{aligned}
\tag{7}
$$

where d is the search direction for the current iteration. These QPs solved in an SQP solver will often be referred to as *QP subproblems* in the remainder of the paper. For many practical problems, the Hessian of the Lagrangian may be too expensive to compute exactly. In such cases, it is common to use quasi-Newton type approximations of the Hessian instead. This is the approach used in the treatment planning optimization problems considered later in this paper, where a Broydon-Fletcher-Goldfarb-Shanno (BFGS) [15] type quasi-Newton approximation for the Hessian is used. The solution of the QP subproblems is a major computational burden in SQP solvers. Thus, SQP solvers both rely on an efficient solver fo quadratic programs internally, and also provide a way to extend a solver for QPs to the nonlinear setting. In this paper, we use our GPU accelerated IPM solver for the QP-subproblems.

4   Implementation

Algorithm 1. Interior Point Method

1: **for** $i \leftarrow 1$ to N **do**
2:     Solve (9) using preconditioned conjugate gradient (PCG) (GPU)
3:     Assemble full search direction from solution to (9) (CPU)
4:     Compute maximum step length α_x, α_λ (CPU)
5:     $x \leftarrow x + \alpha_x \Delta x$ (CPU)
6:     $\lambda \leftarrow \lambda + \alpha_\lambda \Delta\lambda$ (CPU)
7:     $s \leftarrow s + \alpha_x \Delta s$ (CPU)
8:     Update diagonal D in KKT system (CPU / GPU)
9:     Compute residuals r (CPU)
10:    **if** $||r|| < \mu$ **then**
11:        **if** $\mu \leq \mu_{tol}$ **then**
12:            Return solution
13:        **end if**
14:        $\mu \leftarrow \mu/10$
15:    **end if**
16: **end for**

Solving the system (4) is the computational core of our method. As is common in practical implementations, we reduce the size of the system through block-row elimination for efficiency reasons. Furthermore, it is common that our optimization problems will include bound constraints on the variables (of the form $a \leq x \leq b$). In the more general formulation (1), these are handled implicitly in the linear constraints. For computational efficiency however, it is beneficial to separate the rows of the constraint matrix A corresponding to such bound constraints. The result of these reductions gives us a system to solve of the form

$$\begin{pmatrix} Q & -B^T \\ B & D \end{pmatrix} \begin{pmatrix} \Delta x \\ \Delta\lambda_A \end{pmatrix} = \begin{pmatrix} r_1 \\ r_2 \end{pmatrix}, \tag{8}$$

where

$$Q = H + S_{l_x}^{-1}\Lambda_{l_x} + S_{u_x}^{-1}\Lambda_{u_x}, \quad B = \begin{pmatrix} A \\ -A \end{pmatrix},$$

$$D = \begin{pmatrix} \Lambda_{l_A}^{-1}S_{l_A} & \\ & \Lambda_{u_A}^{-1}S_{u_A} \end{pmatrix}, \quad \Delta\lambda_A = \begin{pmatrix} \Delta\lambda_{l_A} \\ \Delta\lambda_{u_A} \end{pmatrix}.$$

S denotes diagonal matrices with the slack variables on the diagonal, and Λ denotes diagonal matrices with the Lagrange multipliers on the diagonal. They are subscripted based on the type of constraint they correspond to: l_x and u_x for lower and upper bounds on the variables, respectively, and l_A and u_A for lower and upper bounds on the linear constraints, respectively. A more detailed derivation of the block reductions leading to the formulation above can be found in [16].

To symmetrize system (8), we consider a *doubly augmented* formulation due to Forsgren and Gill [7]

$$\begin{pmatrix} Q + 2B^T D^{-1} B & B^T \\ B & D \end{pmatrix} \begin{pmatrix} \Delta x \\ \Delta \lambda_A \end{pmatrix} = \begin{pmatrix} r_1 + 2B^T D^{-1} r_2 \\ r_2 \end{pmatrix}. \tag{9}$$

A high-level algorithmic overview of our method is shown in Algorithm 1 (adapted from [16]). The doubly augmented matrix in (9) is positive definite when Q is, which enables the use of a conjugate gradient (CG) solver. Furthermore, the ill-conditioning of the system arises in part due to the poor scaling of the diagonal block D, making Jacobi preconditioning a natural choice, which has the advantage of being very cheap to apply.

The main part of the computation that we have ported to GPU is the solution of the doubly augmented linear system (9) on line 2, while the remainder of the algorithm is run on CPU. The data transfer required in each iteration is not large, as we keep the doubly augmented KKT system on the GPU throughout the optimization, only updating the diagonal D, and the diagonal term of the Hessian block block each iteration. More concretely, the data transfer between CPU and GPU in each iteration consists of:

– The residuals which form the basis of the RHS of (9)
– The solution $(\Delta x, \Delta \lambda_A)$ from the PCG solver
– The diagonal matrix D
– The diagonal terms $S_{l_x}^{-1} \Lambda_{l_x} + S_{u_x}^{-1} \Lambda_{u_x}$ of the Hessian block.

4.1   GPU Acceleration

The most time-consuming part in the optimization algorithm is the preconditioned CG solver used to solve the KKT system at each iteration, which makes it a natural target for GPU acceleration. There are essentially three components to this, computing the matrix-vector products with the KKT-matrix on the GPU, the Jacobi preconditioner, and then general performance considerations for CG on GPUs.

Doubly Augmented KKT Matrix-Vector Multiplications. Multiplications with the doubly augmented matrix are relatively straightforward to implement efficiently on GPU, and we always work with the matrix in unassembled form by computing the products with different sub-blocks of the matrix separately. In the doubly augmented matrix, the Hessian H is stored on the GPU exclusively, as are the diagonal block D and the diagonal terms in Q. The constraint matrix B is stored on both CPU and GPU. Since the constraint matrix remains constant throughout the optimization, the copy to GPU is done when the problem is initialized and no further data transfer between CPU and GPU is needed for the constraint matrix in the solver. We store the sub-blocks of the B matrix in CSR format, and we also pre-compute and store their transposes. This enables us to use transpose-free SpMV kernels for all of our sparse matrix-vector products, which improves performance.

The C++ code for our solver is written to allow execution on both CPU and GPU for the solver. As an example of design choices to accommodate this, the class representing the doubly augmented matrix is templated on the dense vector type used. This is to allow both CPU and GPU execution. For the GPU accelerated case, we provide the template argument `CudaDenseVector` (representing a GPU dense vector implemented in CUDA) and for the CPU case, we use the corresponding CPU class `DenseVector`. The matrix-free approach enabled by the use of Krylov solvers is also seen in the code, as the multiplication is done by accumulating results from each separate block and term of the matrix separately. No fully assembled representation of the entire doubly augmented matrix is ever required, we only store the components of the full matrix (e.g. Hessian H and constraint matrix). The code makes extensive use of C++20's `std::span`, to extract views over contiguous blocks of arrays. This is especially useful in our application, as the doubly augmented matrix is naturally divided into blocks, and performing matrix-vector multiplications (or other operations) using it naturally means needing to perform calculations on sub-sections of the vector. `std::span` enables us to perform operations on chunks of the vector easily, without the use of costly additional copies. Being essentially a thin wrapper around a raw pointer and size, `std::span` is also usable for both CPU and GPU arrays, which we use to simplify code and interfaces for code meant for both CPU and GPU calculations.

Computing the Jacobi Preconditioner. For the Jacobi preconditioner, one needs to compute the diagonal elements of the KKT-matrix. For the bottom right block, consisting of the D matrix, this is trivial, since the matrix is already diagonal, so one can simply extract the elements directly. The top left block is different, since it is not explicitly formed in the solver.

The computation of the diagonal of the Hessian is relatively straightforward, and this calculation only needs to be performed once, since the Hessian does not change throughout the optimization. For the remainder of the optimization, we cache the pre-computed diagonal of the Hessian and provide that directly whenever required.

For the diagonal of the $2A^T D^{-1} A$ term in the top left block, the situation is more complicated. First, this term changes each IPM iteration, since the diagonal matrix D does, and secondly, computing the diagonal in parallel on the GPU is less straightforward. We address this by keeping an extra copy of the constraint matrix on the CPU, which is used to compute $\text{diag}(A^T D^{-1} A)$. This presents some overhead in data transfer between CPU and GPU, but since the diagonal only needs to be recomputed once per IPM iteration (of which there are typically fewer than 100), this trade-off was found to be acceptable.

5   Experimental Setup

In the following we describe the experimental setup, in terms of the hardware and software used as well as the source of the test problems.

5.1   Hardware and Software Environment

The following test systems were used to conduct the performance evaluations in this work.

– **Bluedog** is a local workstation equipped with an AMD Ryzen 9 7900X CPU, and 64 GB of DDR5 RAM @ 5200 MT/s. The GPU is an Nvidia GeForce RTX 4080 with 16 GB of GDDR6X RAM.
– **RS_WKS** is a local Windows workstation with RayStation version 2024A installed. The system is equipped with an Intel Core i9-7940x CPU and 64 GBs of DDR4 RAM @ 2666 MT/s.
– **NJ** is a local server at KTH equipped with an AMD EPYC 7302p 16 core CPU. The GPU is an Nvidia A100 with 40GB of HBM2 memory.

We evaluate the performance of our method in multiple ways. We measure the impact of GPU acceleration by evaluating the performance improvement compared to the CPU version of the solver. The CPU version of the solver is capable of parallel execution through the use of multithreaded BLAS. We use Open-BLAS 0.3.24 for the CPU version of the solver, with the default value used for the number of threads. Our SpMV implementation on the CPU is parallelized using OpenMP. These computational kernels (BLAS and SpMV) should occupy the majority of computational time in the solver. The GPU version of the solver utilizes the cuBLAS and cuSPARSE library heavily for dense and sparse linear algebra kernels. We also analyze the performance of our solver on a range of different GPUs, to see how the performance varies across GPUs for our problem case. Finally, to give an idea of how the GPU accelerated solver may improve solution times for radiation treatment planning in practice, we compare our optimization solver to the one implemented in RayStation, which is a commercial treatment planning system used in clinics around the world. RayStation's QP solver is capable of multi-threaded execution in many cases, but the degree of parallelization varies substantially between cases.

5.2   Test Problems from Radiation Treatment Planning

The optimization problems we use are quadratic programming subproblems exported directly from the RayStation SQP solver. These are the problems the SQP method solves to find search directions in each iteration, and represent the main computational burden. We consider two cases, one for cancer in the

Table 1. Dimensions of the optimization problems used in the performance analysis in terms of number of variables, linear constraints and bound constraints. The proton case is shown before and after spot filtering, which occurs after 100 SQP iterations.

Problem	Vars.	Lin. cons.	Bound cons.
Proton H&N	77373	0	77373
Proton H&N (after spot filtering)	33531	0	33531
VMAT H&N	13425	68618	13425

head and neck region treated using protons, and one head and neck case treated using photons with a treatment technique known as Volumetric Modulated Arc Therapy (VMAT) [18]. For the proton case, the SQP solver performs so-called spot filtering after 100 SQP iterations, which reduces the size of the optimization problem by eliminating variables that are close to zero. Spots, in this case, are intensities of the proton beam in discrete points along the scanning path which can be controlled to achieve the desired dose. Dimensions of the QP subproblems for our test problems are shown in Table 1. The total number of SQP iterations used were 200 and 33, for the proton and photon VMAT case, respectively. This also corresponds to the number of QP subproblems for the different cases.

We expect the QP subproblems to become more expensive to solve in later SQP iterations due to the quasi-Newton Hessian becoming larger for each iteration, since each iteration adds two terms to the BFGS Hessian approximation, which makes the (dense) matrix of update vectors two columns larger. Spot filtering resets the quasi-Newton Hessian approximation, which is another important factor in reducing computational cost after filtering.

5.3   Matrix Dimensions and Computational Cost

The size of the matrices in each QP subproblem varies slightly from iteration to iteration. The largest matrix components of the doubly augmented KKT linear system are the quasi-Newton Hessian $H = H_0 + UWU^T$, and the linear constraint matrix B. The remaining blocks and terms are diagonal matrices, and thus comparatively cheap to perform calculations with. In the quasi-Newton Hessian, H_0 is a square diagonal matrix (corresponding to the initial guess for the BFGS Hessian) with the dimension equal to the number of variables in the optimization problem, which can be found in Table 1. W is a diagonal $k \times k$ matrix, where k is the twice the SQP iteration that the current QP-subproblem corresponds to (since two new rank one updates are added to the BFGS Hessian each SQP iteration). Finally, U is a dense $n \times k$ matrix with the BFGS update vectors as columns.

6   Results

This section presents our results from benchmarking the CPU and GPU performance, as well as comparison with the RayStation solver on a set of realistic problems.

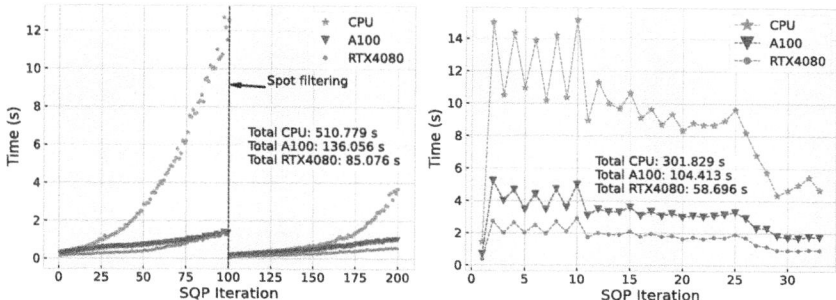

(a) Head and Neck proton arc problem. Spot filtering after 100 SQP iterations, where some variables that are close to zero are pruned from the problem.

(b) VMAT Head and Neck problem

Fig. 1. Comparison of performance for our solver on different GPUs and on the CPU. The CPU and RTX4080 benchmarks were run on Bluedog. The A100 benchmark was performed on NJ.

6.1   GPU and CPU Comparison

We begin by measuring the execution time of our solver on different GPUs and on the CPU to see the impact of GPU acceleration. Figure 1 shows some results from this comparison. We see that the GPU acceleration brings significant performance benefits to our solver, as we would expect, with an approximately 6× speedup of the total execution time when comparing the CPU baseline with the RTX4080 results for the proton head and neck case and approximately 5.1× for the VMAT head and neck case. The execution time of early SQP iteration for the VMAT case shows relatively large oscillations, which seem to be due to numerical differences between the problems causing slower convergence speed (i.e. more IPM or and/or Krylov iterations). The reason why the QP-subproblems alternate between being easier and more difficult is not fully understood. Interestingly, the solver performs better on the RTX4080 system (Bluedog) than on the A100 system (NJ), despite the peak throughput in both memory bandwidth and floating point operations being higher for the A100. Further profiling and analysis suggests that relatively small kernel sizes and larger latency for the A100 may be a large contributing factor to the performance deficit. Merging multiple smaller computational kernels into larger ones (and consequently relying less on cuBLAS for smaller kernels) may alleviate this issue [1]. Further performance engineering in that direction is left for future work.

6.2 Performance Comparison on Realistic Cases

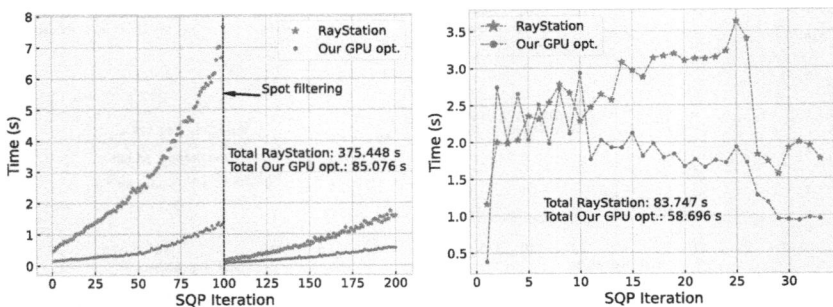

(a) Head and Neck proton arc problem. (b) VMAT Head and Neck problem.
Spot filtering occurs after 100 SQP iterations, marked with a dashed vertical line, where some variables that are close to zero are pruned from the problem.

Fig. 2. Comparisons of solution times for QP subproblems between RayStation's QP solver and our GPU accelerated solver. Only solution times for QP subproblems is measured, and does not include updating the quasi-Newton Hessian or other operations. Total solution time (on all subproblems) for the RayStation optimizer and our solver is shown in the text box.

Figure 2 shows the solution time comparison between our solver running on **Bluedog** (with the Nvidia RTX 4080 GPU) and RayStation running on **RS_WKS**. The times shown are solution times for QP subproblems in the SQP solver. The runs using our GPU accelerated optimizer are performed on exported QP subproblems from RayStation. The timing for the RayStation solver is isolated to only include the solution of the QP subproblems, in order to make a fair comparison. Other work in each SQP iteration is not included in the RayStation solution times. In total, we see that our optimization solver outperforms RayStation's optimizer by 4.4× for the proton problem and roughly 1.4× for the photon VMAT problem. For the proton case, the dashed vertical line in the plot shows the point where *spot filtering* occurs, which is an intermediate step in the SQP optimization where variables that are close to zero are pruned from the problem. The reason, we believe, for the relatively larger improvement for the proton case is twofold. First, the proton case is a bound constrained problem, and tended to require fewer CG iteration in each IPM iteration to converge. Secondly, the main computations in the VMAT case are sparse matrix-vector products, which may relatively speaking benefit less from GPU porting compared to the dense matrix-vector operations for the proton case. While a completely fair comparison between a CPU and GPU implementation is challenging—and is made more complicated by the fact that the algorithms used also differ (direct versus iterative linear solvers)—we emphasize that the RayStation optimization algorithm is originally developed for CPU only, and may not benefit from direct porting to GPU at all. The comparison above is rather intended to give an idea of the speedup obtainable in practice by shifting to the GPU based optimization solver instead.

7   Conclusions and Future Work

In this paper, we presented our GPU accelerated interior point method implementation which is tailored for solving optimization problems from treatment planning for radiation therapy. The move to GPU was enabled by a shift from using direct linear solvers to iterative linear solvers internally, based on the method proposed in [16]. Much research in the optimization literature has been conducted on the use of iterative linear solvers for IPMs, however, most optimization packages still use direct linear solvers. We showed in this paper that a GPU accelerated implementation based on iterative linear algebra can outperform existing approaches on real problems from radiation therapy optimization.

Some aspects of the IPM implementation used in this work are still rather crude and could be further investigated and improved. An example is the rather conservative method to update the barrier parameter μ, where we decrease its value only when the current barrier subproblem is solved somewhat accurately. Further work on improving the selection of barrier parameter values, which should likely take into account the inexactness of the search direction due to the iterative linear solver [2], and similar may push the performance even further. Additionally, our comparison of performance between GPU models suggests that there is still room for further performance optimizations in the implementation. For example, kernel launch overhead for small computational kernels could be addressed by merging smaller computational kernels into larger ones, thus relying less on cuBLAS for computations. However, it is encouraging that even with a relatively simple IPM implementation, the use of iterative linear solvers and GPUs is able to improve the overall time-to-solution compared to the clinically used solver in RayStation.

Overall, our GPU accelerated solver was able to improve total optimization times by 1.4× and 4.4× respectively, when compared to an optimization solver from a clinically used treatment planning system on realistic problems. Furthermore, enabling the use of more powerful computational hardware may provide future proofing in a field where demands for computational efficiency are ever increasing.

References

1. Anzt, H., Tomov, S., Luszczek, P., Sawyer, W., Dongarra, J.: Acceleration of GPU-based Krylov solvers via data transfer reduction. Int J. High Perform. Comput. Appl. **29**(3), 366–383 (2015)
2. Bellavia, S.: Inexact interior-point method. J. Optim. Theory Appl. **96**, 109–121 (1998)
3. Bennett, K.P., Parrado-Hernández, E.: The interplay of optimization and machine learning research. J. Mach. Learn. Res. **7**, 1265–1281 (2006)
4. Boggs, P.T., Tolle, J.W.: Sequential quadratic programming. Acta Numer **4**, 1–51 (1995)
5. Bortfeld, T., Thieke, C.: Optimization of treatment plans, inverse planning. In: New Technologies in Radiation Oncology, pp. 207–220. Springer (2006)
6. Cao, Y., Seth, A., Laird, C.D.: An augmented Lagrangian interior-point approach for large-scale NLP problems on graphics processing units. Comput. Chem. Eng. **85**, 76–83 (2016)
7. Forsgren, A., Gill, P.E., Griffin, J.D.: Iterative solution of augmented systems arising in interior methods. SIAM J. Optim. **18**(2), 666–690 (2007)
8. Fracchiolla, F., Engwall, E., Janson, M., Tamm, F., Lorentini, S., Fellin, F., Bertolini, M., Algranati, C., Righetto, R., Farace, P., et al.: Clinical validation of a GPU-based Monte Carlo dose engine of a commercial treatment planning system for pencil beam scanning proton therapy. Physica Med. **88**, 226–234 (2021)
9. Gade-Nielsen, N.F.: Interior point methods on GPU with application to model predictive control. Ph.D. thesis, Technical University of Denmark (2014)
10. Gondzio, J.: Interior point methods 25 years later. Eur. J. Oper. Res. **218**(3), 587–601 (2012)
11. Gondzio, J.: Matrix-free interior point method. Comput. Optim. Appl. **51**, 457–480 (2012)
12. Gu, X., Pan, H., Liang, Y., Castillo, R., Yang, D., Choi, D., Castillo, E., Majumdar, A., Guerrero, T., Jiang, S.B.: Implementation and evaluation of various demons deformable image registration algorithms on a GPU. Phys. Med. Biol. **55**(1), 207 (2009)
13. Jia, X., Ziegenhein, P., Jiang, S.B.: GPU-based high-performance computing for radiation therapy. Phys. Med. Biol. **59**(4), R151 (2014)
14. Li, T., Li, H., Liu, X., Zhang, S., Wang, K., Yang, Y.: GPU acceleration of interior point methods in large scale SVM training. In: 2013 12th IEEE International Conference on Trust, Security and Privacy in Computing and Communications, pp. 863–870. IEEE (2013)
15. Liu, D.C., Nocedal, J.: On the limited memory BFGS method for large scale optimization. Math. Program. **45**(1), 503–528 (1989)
16. Liu, F., Fredriksson, A., Markidis, S.: Krylov solvers for interior point methods with applications in radiation therapy and support vector machines. In: Computational Science – ICCS 2024, pp. 73–77. Springer (2024)
17. Liu, F., Jansson, N., Podobas, A., Fredriksson, A., Markidis, S.: Accelerating radiation therapy dose calculation with Nvidia GPUs. In: 2021 IEEE International Parallel and Distributed Processing Symposium Workshops (IPDPSW), pp. 449–458. IEEE (2021)
18. Otto, K.: Volumetric modulated arc therapy: IMRT in a single gantry arc. Med. Phys. **35**(1), 310–317 (2008)

19. Pacaud, F., Shin, S., Schanen, M., Maldonado, D.A., Anitescu, M.: Accelerating condensed interior-point methods on SIMD/GPU architectures. J. Optim. Theory Appl. 1–20 (2023)
20. Petra, C.G.: A memory-distributed quasi-Newton solver for nonlinear programming problems with a small number of general constraints. J. Parallel Distrib. Comput. **133**, 337–348 (2019)
21. Petra, C.G., Aravena, I.: Solving realistic security-constrained optimal power flow problems. arXiv preprint arXiv:2110.01669 (2021)
22. Potra, F.A., Wright, S.J.: Interior-point methods. J. Comput. Appl. Math. **124**(1–2), 281–302 (2000)
23. Rais, A., Viana, A.: Operations research in healthcare: a survey. Int. Trans. Oper. Res. **18**(1), 1–31 (2011)
24. Regev, S., Chiang, N.Y., Darve, E., Petra, C.G., Saunders, M.A., Świrydowicz, K., Peleš, S.: HyKKT: a hybrid direct-iterative method for solving KKT linear systems. Optim. Methods Softw. **38**(2), 332–355 (2023)
25. Samarasinghe, G., Jameson, M., Vinod, S., Field, M., Dowling, J., Sowmya, A., Holloway, L.: Deep learning for segmentation in radiation therapy planning: a review. J. Med. Imaging Radiat. Oncol. **65**(5), 578–595 (2021)
26. Schubiger, M., Banjac, G., Lygeros, J.: GPU acceleration of ADMM for large-scale quadratic programming. J. Parallel Distrib. Comput. **144**, 55–67 (2020)
27. Smith, E., Gondzio, J., Hall, J.: GPU acceleration of the matrix-free interior point method. In: Parallel Processing and Applied Mathematics: 9th International Conference, PPAM 2011, Torun, September 11-14, 2011 Part I 9. pp. 681–689. Springer (2012)
28. Stellato, B., Banjac, G., Goulart, P., Bemporad, A., Boyd, S.: OSQP: an operator splitting solver for quadratic programs. Math. Program. Comput. **12**(4), 637–672 (2020)
29. Świrydowicz, K., Koukpaizan, N., Alam, M., Regev, S., Saunders, M., Peleš, S.: Iterative methods in GPU-resident linear solvers for nonlinear constrained optimization. arXiv preprint arXiv:2401.13926 (2024)
30. Świrydowicz, K., Koukpaizan, N., Ribizel, T., Göbel, F., Abhyankar, S., Anzt, H., Peleš, S.: GPU-resident sparse direct linear solvers for alternating current optimal power flow analysis. Int. J. Electr. Power Energy Syst. **155**, 109517 (2024)
31. Wedenberg, M., Beltran, C., Mairani, A., Alber, M.: Advanced treatment planning. Med. Phys. **45**(11), e1011–e1023 (2018)

An Empirical Assessment of LLM-Based Approaches to Malicious Webpage Detection

Gracjan Mak[ID], Mateusz Gniewkowski[ID], Paweł Walkowiak[ID],
and Arkadiusz Janz[(✉)][ID]

Wrocław University of Science and Technology, Wrocław, Poland
{mateusz.gniewkowski,arkadiusz.janz}@pwr.edu.pl

Abstract. Large language models (LLMs) are increasingly influential in advancing NLP technology and solving complex tasks, yet their potential misuse in cybersecurity poses significant risks. This paper addresses the challenge of detecting malicious webpages using LLMs, an area with limited research. We evaluate LLMs by expanding zero-shot and few-shot query formulations, testing previously unassessed open-source and proprietary models, and assessing robustness under adversarial conditions. Additionally, we verify model performance using Chain of Thought reasoning and compare these explanations with traditional methods. Our work aims to enhance the application of LLMs in cybersecurity, guiding the development of more effective detection systems.

1 Introduction

The role of large language models (LLMs) in the development of NLP technology and the resolution of complex NLP tasks continues to expand, contributing to solving novel problems in various application areas. The increasing reasoning capabilities of LLMs significantly contributed to open-domain question answering, mathematical question answering, commonsense reasoning, and logical reasoning. The use of large and diverse pretraining data facilitates zero-shot classification across a wide range of application domains, extending beyond typical benchmark applications. As technology advances, the threat of LLM misuse also increases. Thus, a proper response should focus on designing methods against threats that emphasize equally sophisticated techniques and technology.

The methods for leveraging large-scale language models (LLMs) in complex reasoning have advanced significantly, with diverse inference strategies such as Tree of Thoughts [12] and other innovative approaches enhancing their application across various domains. Despite these advancements, LLMs remain susceptible to adversarial attacks, a critical factor that must be considered when designing methods for detecting malicious webpages. Malicious actors may exploit the known limitations of LLMs, necessitating robust detection strategies that account for these vulnerabilities.

Currently, the application of LLMs in the realm of cybersecurity, particularly in detecting cyber threats, is still in its early stages [7]. Moreover, there is a

M. H. Lees et al. (Eds.): ICCS 2025, LNCS 15904, pp. 162–176, 2025.
https://doi.org/10.1007/978-3-031-97629-2_12

limited body of work exploring both the diverse inference techniques and the susceptibility of LLMs to adversarial attacks within this context. In this paper, we aim to bridge this gap by conducting a comprehensive evaluation of LLMs for the detection of malicious webpages, considering both common prompting-based inference methods and the models' resilience to adversarial conditions. Through this work, we seek to advance the understanding and application of LLMs in cybersecurity, providing insights that could inform the development of more effective detection systems. Our contributions are as follows:

1. We introduce a novel dataset for researching malicious webpage detection in an adversarial setting, including attacks at 5%, 25%, and 50% of page content change, as well as prompt injection and compromised URLs.
2. We extend the evaluation of zero-shot and few-shot approaches by expanding the query formulations, providing a more comprehensive assessment of model capabilities.
3. We test open source and proprietary models that have not been previously evaluated for this task, broadening the scope of model applicability.
4. We incorporate an evaluation under adversarial attack conditions and present the results of this evaluation across different models, offering insights into their robustness.
5. Finally, we verify the performance of the models using Chain of Thought (CoT) reasoning.
6. We examine the results obtained with traditional explainability methods.

2   Related Work

Before the advent of Large Language Models (LLMs), malicious website detection relied heavily on rule-based systems, statistical analysis, and then on traditional machine learning. These approaches were tailored to specific features of malicious activity, often requiring extensive domain knowledge and feature engineering.

Supervised models, including Support Vector Machines (SVMs), Decision Trees, and Random Forests, used labeled datasets to identify malicious webpages based on extracted features. Commonly used features included URL length, entropy, keyword patterns, metadata, and network traffic behaviors. These models achieved higher detection rates and better adaptability compared to rule-based systems. However, they were highly dependent on manual feature engineering, which required significant expertise and often failed to generalize across diverse or evolving threats. For example, in [5], the authors utilize lexical features and compare a wide set of machine learning algorithms. Similar approaches can be found in [11] or in [4].

In the early stages of applying natural language processing (NLP) to cybersecurity, text classification tasks such as protocol analysis, website categorization, and source code identification played a critical role. These tasks relied on foundational techniques designed to process and interpret text in a structured manner.

For example, spam detection systems leveraged basic NLP techniques like bag-of-words models or Term Frequency-Inverse Document Frequency (TF-IDF). Later advances in NLP introduced more sophisticated methods, taking advantage of deep learning-based models such as doc2vec, FastText, and transformers such as BERT [2]. These approaches significantly improved the ability, among others, to analyze and classify web content by capturing richer semantic and contextual information. The works in this field include: [3, 8, 10].

In recent years, large language models (LLMs) such as GPT or Llama have revolutionized the field of cybersecurity by offering powerful tools for analyzing and processing textual data. Their ability to understand context, generate coherent text, and adapt to various tasks with minimal fine-tuning has made them invaluable across diverse cybersecurity applications. Zero-Shot Learning and Few-Shot Learning deserve special attention. In the context of malicious webpage detection, [7] proposed a dataset for benchmarking LLMs in this task. The authors conducted a study in which they tested diverse prompting strategies to detect malicious webpages using LLMs. However, the dataset is not publicly available, and the authors limited their prompting strategies to few-shot learning, without exploring Chain-of-Thought (CoT) reasoning, explainability, or adversarial attacks.

3   Methods

In this work, we focus on the webpage content and URL classification task. One can use large-scale language models that offer various approaches to classification. In this study, these classification methods will be analysed in terms of their effectiveness and robustness against adversarial attacks. This chapter outlines the key methods and their adaptations to the problem at hand, forming the foundation for the experimental part of the study. For the experimental setup, we introduced a pipeline which is presented in Fig. 1. For each of two Webpage datasets, we created their adversarial setting versions with gradually attacked content. Our experiments for five size diversified LLMs (Llama3.1-8B[1], Llama3.1-70B[2], GPT4o-mini[3], GPT4o[4] and Gemini2.0-flash-exp[5]), includes checking their ability to classify Webpage maliciousness given:

- The **URL** of the webpage, in this setting we test if the model remembers some of the websites and to what extent it bases prediction on the URL.
- The Webpage **Content**, which tests how well a model can analyse html files.
- The **URL and Content**, model receives all the information from datasets, this setting also shows if there is a bias basing the prediction towards URL or content. In this setting, we also included the website JavaScript sections.

---

[1] https://huggingface.co/meta-llama/Llama-3.1-8B-Instruct.

[2] https://huggingface.co/meta-llama/Llama-3.1-70B-Instruct.

[3] https://platform.openai.com/docs/models#gpt-4o-mini.

[4] https://platform.openai.com/docs/models#gpt-4o.

[5] https://ai.google.dev/gemini-api/docs/models/experimental-models.

In order to test the models in most diverse way, we prepared various Zero-and Few-Shot versions of prompts. The URL setting was tested with **Simple Zero-Shot**, **CoT Zero-Shot**, **Advanced Zero-Shot**, **Simple Few-Shot** and **Advanced Few-Shot** (see Sect. 3.1). In the Content setting only CoT Zero-Shot was used. Finally, the URL and content was tested with Simple and CoT Zero-Shots.

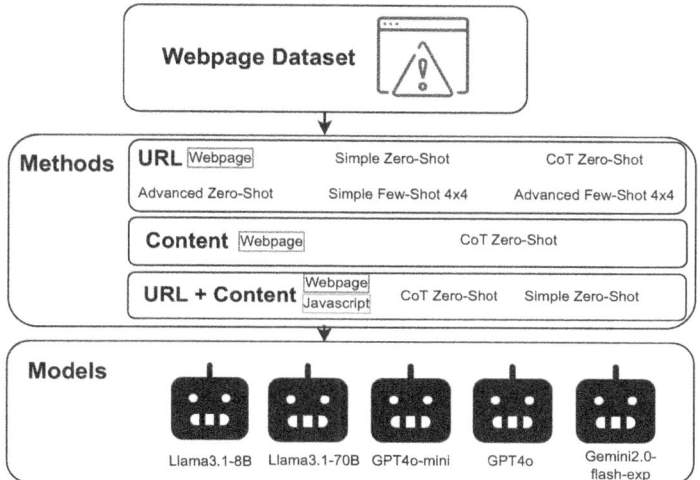

Fig. 1. Experimental pipeline for the Webpages dataset. Each dataset sample was evaluated using two primary methods: one based only on the URL and another incorporating both the URL and page content. Each method further includes sub-methods that leverage few-shot learning and chain-of-thought reasoning. Five different LLMs were tested to ensure diversity in model size and performance assessment.

3.1 Malicious Webpage Detection

The experiments on webpage classification involve inputting data, in the form of instructions containing URL, webpage content, or both, into large-scale language models. The models perform classification by executing the task specified in the prompt, predicting a label for the website as malicious or benign.

Zero-Shot Classification. In this setting, the model is provided with a detailed task description, including classification criteria, target classes, purpose, and clear instructions on the expected output format. Zero-Shot learning is a technique in natural language processing where a language model performs a task without prior specialized training, relying solely on its general knowledge from pretraining.

This approach is applied to webpage classification, where models process instructions and classify the site as malicious or benign. The technique supports both simple queries, where the model classifies with minimal context, and advanced queries, which provide additional definitions of malicious and safe websites. We consider two variants of Zero-shot setting: **(1) Simple Zero-shot**, where the model receives plain instructions about the task without any demonstrations of task execution, **(2) Advanced Zero-shot**, where the model is given with precise instructions about task execution and an explanation of expected URL and content violation techniques, **(3) Zero-shot CoT**, in which the model has to execute the task explaining every single step of its reasoning.

System prompt
You will be provided with a website URL and its content, and your task is to classify it as either malicious or benign.
Criteria for Classification:
    – Malicious: The URL leads to a website that hosts malware, phishing, scams
    – Benign: The URL leads to a safe website that does not pose any threat to users.
Purpose: Filtering harmful websites to protect users and systems.
Response Format: Your response should be :
(A) Malicious: If the URL is harmful.
(B) Benign: If the URL is safe.

User prompt
Please classify the following URL **http://chatgpt.trojanxx.xyz**, and its content: <content>

Response
(A) Malicious

Fig. 2. Zero-shot prompt for LLM-based approach to URL classification.

Few-Shot Classification. This technique involves providing the model with a small number of examples related to the task it is expected to perform, which helps improve the accuracy of its output. By examining these examples, the model can better grasp the context and requirements of the task. For this method, prompts can be created by extending the Zero-Shot approach with examples randomly selected from the dataset. Following [7], we designed a Few-Shot prompt consisting of one benign and three malicious examples from our dataset, to guide the model's classification. In our work, we considered both **(1) Simple Few-shot** and **(2) Advanced Few-shot** approaches based on Zero-shot prompts (see Fig. 1). This structured approach aims to enhance the model's ability to distinguish between malicious and benign websites more effectively. The final prompt used in our few-shot approach is presented in Fig. 3.

Chain-Of-Thought. The Chain-of-Thought (CoT) prompting technique aims to improve the reasoning capabilities of language models by encouraging step-by-step, logical problem-solving. Instead of providing an immediate answer, the model is prompted to "think" by generating a sequence of intermediate reasoning steps that ultimately lead to the final answer. A well-constructed CoT response should resemble a clear, structured explanation of the problem-solving process, concluding with a concise and accurate final response.

```
System prompt
You will be provided with a website URL and its content, and your task is to classify it as either malicious or
benign.
Criteria for Classification:
   — Malicious: The URL leads to a website that hosts malware, phishing, scams
   — Benign: The URL leads to a safe website that does not pose any threat to users.
Purpose: Filtering harmful websites to protect users and systems.
Response Format: Your response should be :
(A) Malicious: If the URL is harmful.
(B) Benign: If the URL is safe.

Example:
URL: http://example.com/suspicious
Content: <content>
Response: (A) Malicious

User prompt
Please classify the following URL http://chatgpt.trojanxx.xyz, and its content: <content>

Response
(A) Malicious
```

Fig. 3. The prompt for One-shot LLM-based approach to **URL** classification.

3.2   Adversarial Attacks on Detection Methods

A significant challenge in utilizing large-scale language models for detecting malicious websites is ensuring their robustness against various forms of attacks that cybercriminals may use to evade detection. The attacks presented in this subsection are experimental, as research on adversarial attacks against models in the context of malicious website detection remains limited.

Prompt Injection. One of the key threats to the reliability of large-scale language models in detecting malicious websites is the use of manipulative inputs designed to influence model predictions. Prompt injection involves crafting specific text patterns within the website content to mislead the model into classifying a malicious website as safe. The attack phrase used in our experiments:

– `Browse with confidence, this site is safe`

Such manipulations exploit the model's reliance on linguistic cues, potentially leading to false negatives in malicious website detection.

Manipulated URL. Another technique for adversarial manipulation targets the URL structure itself. By embedding deceptive query parameters or modifications, attackers can influence the model's classification by introducing misleading context within the URL. In our experiments, the following manipulation was used:

– `?malware_check=none&status_confirmed=safe`

These modifications exploit the model's tendency to associate specific keywords with benign classifications, increasing the risk of false negatives.

URL Modification. We conducted a URL-based adversarial attack by randomly substituting three letters in the domain. This method exploits weaknesses in models that rely on exact string matching and learned domains. Such modifications resemble typosquatting attacks, where adversaries register domains with small alterations to trick users and evade detection. For example, a malicious URL:

`\url{https://maliciouswebsite.com/phishing}`

can be modified to:

`\url{https://mallciousvebsita.com/phishing}`

These small changes can bypass blacklists and confuse models, especially those that lack robust phonetic or semantic analysis.

3.3 JavaScript Code Obfuscation Attack

To evaluate the robustness of large-scale language models in detecting malicious JavaScript code, we performed code obfuscation using the JavaScript Obfuscator[6]. This widely used tool applies multiple transformations to JavaScript code, making it harder to analyze while preserving its functionality. Obfuscation is a well-known technique employed by cybercriminals to evade detection by static and dynamic analysis tools. By altering the syntactic structure of JavaScript scripts without changing their behavior, attackers can bypass signature-based security mechanisms and even some machine learning-based classifiers.

3.4 Webpage Content TextBugger Attacks

To evaluate the impact of perturbing webpage content on malicious page detection, the page html body content was attacked using the adversarial example generation method – TextBugger [6]. The TextBugger introduces five disturbance techniques for generating adversarial example: insertion of spaces into the disturbed word, deletion of a random letter, swapping the order of two adjacent letters, substitution of letters with visually similar characters, and substitution word (replacing a word with its nearest equivalent using a pre-trained GloVe model). These methods are combined in the TextBugger approach, resulting in highly diverse adversarial examples due to the randomness introduced in word alterations and the capability to apply multiple disturbance methods concurrently. For the purpose of modifying HTML pages three of TextBuggers disturbance methods were used: letter deletion, adjacent letters swap and substitution of letters to its visually similar equivalent. The modification of the html pages consisted of changing the natural language texts in the body section where scripts and html tags have remained intact. In order to check model's ability for classifying modified pages, three levels of disruption were chosen, modification of 5%, 25% and 50% of words in page text.

---

[6] https://github.com/javascript-obfuscator/javascript-obfuscator.

4   Datasets

The literature offers numerous datasets for detecting malicious websites, but most of them typically do not include the content of the pages, focusing solely on their URL addresses. The popularity of URL-only datasets comes from their ease of creation and maintenance, as domain names are relatively static, persisting even after a website becomes inactive, which simplifies the process of dataset creation. In contrast, webpage content becomes inaccessible once the server hosting it is taken offline, making it imperative to capture such data while the malicious page is still active. A Comprehensive Dataset for Webpage Classification [1] comprises 1,069,715 webpages and is specifically created for developing algorithms to detect malicious websites. Each webpage in the dataset is annotated with multiple attributes, including URL, HTML code and source. The dataset was generated automatically from existing lists of malicious websites. However, such sites are often taken down quickly, leading to potential false positives, such as inactive domains that have been repurchased. In contrast, our dataset consists of manually verified, fresh malicious webpages, ensuring higher accuracy and relevance for real-world detection.

4.1   Adversarial Benchmark for Malicious Webpage Detection (ABW)

Given the limited availability of datasets, we decided to create our own dataset, smaller but more reliable. The approach focused on collecting data from various sources, ensuring a balanced set of malicious and benign websites.

For the malicious websites, we gathered URLs from sources such as Open-Phish[7], PhishTank[8], and email spam. These sources are widely recognized for their up-to-date lists of potentially dangerous sites. After collecting the URLs, each website was manually verified to confirm that it was malicious. For the benign websites, we selected a random sample of frequently visited and well-established pages from across the internet. These sites were chosen to represent a broad range of trustworthy, legitimate content. The dataset ultimately consists of 110 benign websites and 108 malicious websites. Each entry in the dataset contains the following attributes:

- url: the webpage's URL,
- label: the primary category indicating whether the webpage is safe or malicious,
- source: the origin of the URL (e.g., OpenPhish, PhishTank, etc.),
- html: the raw HTML content of the webpage, including HTML code and embedded JavaScript.

To evaluate the model's performance under adversarial attack, three versions of the adversarial dataset were created, with 5%, 25%, and 50% of content words in the page text modified.

---

[7] https://openphish.com/.

[8] https://phishtank.org/.

4.2 JavaScript Dataset

In addition to the webpage dataset, we also created our own dataset of JavaScript files. This dataset was designed to facilitate the analysis and classification of JavaScript code as either malicious or benign. For the malicious JavaScript files, we utilized samples from the publicly available repository JavaScript Malware Collection[9], which contains a wide variety of confirmed malicious scripts. For the benign JavaScript files, we randomly collected scripts by accessing legitimate and frequently visited websites, ensuring a diverse representation of safe JavaScript code. The dataset is structured with the following attributes for each record:

- `filename`: the name of the JavaScript file,
- `label`: a classification label indicating whether the script is benign or malicious,
- `source`: the origin of the script (e.g., *JavaScript Malware Collection* or a legitimate website),
- `javascript`: the raw content of the JavaScript file.

The resulting dataset consists of 100 benign JavaScript files and 100 malicious JavaScript files.

5 Experimental Setting

The experiments were conducted using the following large language models: Llama 3.1 (7B and 70B), GPT-4o-Mini, GPT-4o, and Gemini 2.0 Flash Exp. A default temperature was applied to maintain a balance between creativity and consistency in the model's responses. The models were evaluated on presented datasets consisting of URLs, webpage content, and JavaScript scripts. The experimental pipeline for webpage datasets is presented on Fig. 1. To ensure efficient processing, the HTML and JavaScript datasets were truncated to a maximum of 20 thousand tokens, preserving representative information while minimizing cost and processing time. For few-shot classification, four randomly selected examples from the dataset were added to the prompt to enhance the model's contextual understanding and classification accuracy.

6 Results and Discussion

Table 1 presents the classification results of various large language models for attack detection using the previously mentioned strategies. The evaluation was conducted using several publicly available models based on True Positive Rate (TPR), True Negative Rate (TNR), and the F1 score.

The Few-Shot Learning approach, in both its Simple and Advanced variants, demonstrates the best overall performance. Their main drawback is the need to provide several examples, which can make the model less robust to concept

---

[9] https://github.com/HynekPetrak/javascript-malware-collection.

Table 1. The performance of language models in the task of malicious webpage classification. We provide the results for particular variants of prompting-based detection methods, for both content and URL classification.

Method	Model	Metrics		
		TPR	TNR	F1
Simple Zero-Shot (URL)	Llama-3.1-8B-Instruct	0.22	0.38	0.23
	Llama-3.1-70B-Instruct	0.51	1.00	0.68
	gpt-4o-mini	0.82	0.97	0.89
	gpt-4o	0.23	0.99	0.37
	gemini-2.0-flash-exp	0.45	1.00	0.62
	AVG	0.45	0.87	0.56
Advanced Zero-Shot (URL)	Llama-3.1-8B-Instruct	0.44	0.94	0.58
	Llama-3.1-70B-Instruct	0.33	0.99	0.50
	gpt-4o-mini	0.81	1.00	0.89
	gpt-4o	0.54	1.00	0.70
	gemini-2.0-flash-exp	0.70	1.00	0.82
	AVG	0.56	0.99	0.70
Simple Few-Shot (URL)	Llama-3.1-8B-Instruct	0.74	0.47	0.65
	Llama-3.1-70B-Instruct	0.64	0.97	0.77
	gpt-4o-mini	0.73	1.00	0.85
	gpt-4o	0.89	0.99	0.93
	gemini-2.0-flash-exp	0.76	1.00	0.86
	AVG	0.75	0.89	0.81
Advanced Few-Shot (URL)	Llama-3.1-8B-Instruct	0.74	0.9	0.80
	Llama-3.1-70B-Instruct	0.66	0.98	0.78
	gpt-4o-mini	0.78	0.99	0.87
	gpt-4o	0.70	1.00	0.83
	gemini-2.0-flash-exp	0.84	0.99	0.91
	AVG	0.74	0.97	0.84
CoT Zero-Shot (URL)	Llama-3.1-8B-Instruct	0.31	0.81	0.42
	Llama-3.1-70B-Instruct	0.37	0.96	0.53
	gpt-4o-mini	0.75	1.00	0.86
	gpt-4o	0.50	0.98	0.65
	gemini-2.0-flash-exp	0.75	0.99	0.85
	AVG	0.54	0.95	0.66
Simple Zero-Shot (URL + Content)	Llama-3.1-8B-Instruct	0.27	0.11	0.24
	Llama-3.1-70B-Instruct	0.30	0.95	0.44
	gpt-4o-mini	0.80	0.95	0.87
	gpt-4o	0.94	0.95	0.95
	gemini-2.0-flash-exp	0.77	1.00	0.87
	AVG	0.62	0.79	0.64
CoT Zero-Shot (URL + Content)	Llama-3.1-8B-Instruct	0.23	0.16	0.22
	Llama-3.1-70B-Instruct	0.30	0.97	0.45
	gpt-4o-mini	0.88	1.00	0.93
	gpt-4o	0.90	1.00	0.95
	gemini-2.0-flash-exp	0.93	1.00	0.97
	AVG	0.65	0.83	0.70
CoT Zero-Shot (Content)	Llama-3.1-8B-Instruct	0.18	0.18	0.15
	Llama-3.1-70B-Instruct	0.13	0.21	0.64
	gpt-4o-mini	0.53	0.69	0.81
	gpt-4o	0.68	0.80	0.86
	gemini-2.0-flash-exp	0.69	0.81	0.87
	AVG	0.44	0.54	0.67

Table 2. The performance of language models in the task of malicious JavaScript files detection.

Method	Model	Metrics		
		TPR	TNR	F1
Simple Zero-Shot (Content)	Llama-3.1-8B-Instruct	0,56	0,66	0,59
	Llama-3.1-70B-Instruct	0,89	0,85	0,87
	gpt-4o-mini	1,00	0,69	0,87
	gpt-4o	1,00	0,96	0,98
	gemini-2.0-flash-exp	1,00	0,83	0,92
	AVG	0,89	0,80	0,85
CoT Zero-Shot (Content)	Llama-3.1-8B-Instruct	0,46	0,56	0,48
	Llama-3.1-70B-Instruct	0,92	0,84	0,88
	gpt-4o-mini	0,94	0,74	0,85
	gpt-4o	0,94	0,96	0,95
	gemini-2.0-flash-exp	0,85	0,91	0,88
	AVG	0,82	0,80	0,80

drift–deviations from the patterns present in the provided examples. However, in general, they should be used as the preferred approach because of their superior performance in most cases. The next best results were achieved by the Advanced Zero-Shot method (based solely on URL addresses) and CoT Zero-Shot Learning, but only when both URL addresses and page content were used. It is important to note that classification using LLM models can be expensive, so the benefits of employing more advanced methods seem marginal, considering that the best classification results were obtained with just URL addresses (shorter queries are less costly). We hope to demonstrate further that despite this limitation, CoT methods and the use of content may offer greater robustness. Similar conclusions can be drawn from Table 2. It also shows that the simple Zero-Shot Learning approach outperforms the CoT.

Table 3 presents the effectiveness of the attacks in the form of new metric values, together with the differences between these values and the corresponding ones in Table 1. Looking solely at the F1 or TPR values, the most effective attack method (or the most unstable method) turns out to be the simple prompt injection into the URL address (Simple Zero-Shot). Apart from this attack, all other examples show that any modifications to the URL or page content lead to an increase in TPR and a decrease in TNR. This means that the classification becomes overly sensitive and starts classifying everything as a threat (indicating that the methods are effectively detecting the attack). This is most noticeable in the case of methods that directly modify the URL addresses (Simple Zero-Shot [URL] and CoT Zero-Shot [URL]). This supports our previous observation that the model is most focused on the URL. In response to the question of whether using the content of the page is justified, we believe the answer is yes.

Table 3. The performance of large language models in the task of malicious webpage detection in adversarial setting.

Method & Attack	Model	Metrics		
		TPR	TNR	F1
Simple Zero-Shot (URL)	Llama-3.1-8B-Instruct	0.14/−0.08	0.38/−0.00	0.16/−0.07
Manipulated URL	Llama-3.1-70B-Instruct	0.68/+0.17	0.99/−0.01	0.80/+0.12
	gpt-4o-mini	0.50/−0.32	1.00/+0.03	0.66/−0.23
	gpt-4o	0.19/−0.04	0.94/−0.05	0.31/−0.06
	gemini-2.0-flash-exp	0.30/−0.15	1.00/−0.00	0.46/−0.16
	AVG	0.36/−0.09	0.86/−0.01	0.48/−0.08
Simple Zero-Shot (URL − Content)	Llama-3.1-8B-Instruct	0.21/−0.06	0.15/+0.04	0.20/−0.04
Manipulated URL Prompt injection	Llama-3.1-70B-Instruct	0.35/+0.05	0.96/+0.01	0.51/+0.07
	gpt-4o-mini	0.69/−0.11	0.97/+0.02	0.80/−0.07
	gpt-4o	0.92/−0.02	0.92/−0.03	0.92/−0.03
	gemini-2.0-flash-exp	0.79/+0.02	0.99/−0.01	0.88/+0.01
	AVG	0.59/−0.03	0.80/+0.01	0.66/+0.02
Simple Zero-Shot (URL) URL	Llama-3.1-8B-Instruct	0.23/+0.01	0.30/−0.08	0.23/−0.00
Modification	Llama-3.1-70B-Instruct	0.78/+0.27	0.67/−0.33	0.74/+0.06
	gpt-4o-mini	0.99/+0.17	0.31/−0.66	0.73/−0.16
	gpt-4o	0.24/+0.01	0.21/−0.78	0.23/−0.14
	gemini-2.0-flash-exp	0.69/+0.24	0.83/−0.17	0.73/+0.11
	AVG	0.59/+0.14	0.46/−0.40	0.53/−0.03
CoT Zero-Shot (URL) URL	Llama-3.1-8B-Instruct	0.66/+0.35	0.42/−0.39	0.58/+0.16
Modification	Llama-3.1-70B-Instruct	0.56/+0.19	0.64/−0.32	0.58/+0.05
	gpt-4o-mini	0.93/+0.18	0.28/−0.72	0.70/−0.16
	gpt-4o	0.78/+0.28	0.57/−0.41	0.70/+0.05
	gemini-2.0-flash-exp	0.94/+0.19	0.45/−0.54	0.75/−0.10
	AVG	0.77/+0.24	0.47/−0.48	0.66/−0.00
CoT Zero-Shot (URL - Content)	Llama-3.1-8B-Instruct	0.44/+0.21	0.07/−0.09	0.37/+0.15
Prompt injection	Llama-3.1-70B-Instruct	0.49/+0.19	0.86/−0.11	0.60/+0.15
	gpt-4o-mini	0.90/+0.02	0.94/−0.06	0.92/−0.01
	gpt-4o	0.93/+0.03	0.99/−0.01	0.96/+0.01
	gemini-2.0-flash-exp	0.95/+0.02	0.81/−0.19	0.88/−0.09
	AVG	0.74/+0.09	0.73/−0.09	0.75/+0.04
CoT Zero-Shot (Content) Prompt	Llama-3.1-8B-Instruct	0.28/+0.10	0.04/−0.11	0.24/+0.06
injection	Llama-3.1-70B-Instruct	0.30/+0.17	0.80/−0.07	0.40/+0.19
	gpt-4o-mini	0.69/+0.16	0.98/−0.01	0.80/+0.11
	gpt-4o	0.72/+0.04	0.94/−0.04	0.81/+0.01
	gemini-2.0-flash-exp	0.83/+0.14	0.82/−0.18	0.82/+0.01
	AVG	0.56/+0.12	0.72/−0.08	0.61/+0.08
CoT Zero-Shot (Content) 25% of	Llama-3.1-8B-Instruct	0.17/−0.01	0.13/−0.02	0.17/−0.01
page modified	Llama-3.1-70B-Instruct	0.26/+0.13	0.73/−0.14	0.40/+0.19
	gpt-4o-mini	0.54/+0.01	0.80/−0.19	0.68/−0.01
	gpt-4o	0.73/+0.05	0.83/−0.15	0.80/−0.00
	gemini-2.0-flash-exp	0.76/+0.07	0.86/−0.14	0.83/+0.02
	AVG	0.49/+0.05	0.67/−0.13	0.58/+0.04

Classification results are harder to alter when the content is included, as it provides additional context that helps the model make more stable and accurate predictions. This is proven by looking at CoT attacks that includes content.

In our study, our objective was to investigate why the model classifies websites in a particular way. We applied explainability techniques to highlight key

Table 4. The performance of large language models in the task of malicious JavaScript detection in adversarial setting.

Method	Model	Metrics		
		TPR	TNR	F1
Simple Zero-Shot (Content)	Llama-3.1-8B-Instruct	0,51/−0,05	0,28/−0,38	0,46/−0,13
	Llama-3.1-70B-Instruct	0,83/−0,06	0,63/−0,22	0,75/−0,12
	gpt-4o-mini	1.00/−0,00	0,06/−0,63	0,68/−0,19
	gpt-4o	1.00/−0,00	0,30/−0,66	0,74/−0,24
	gemini-2.0-flash-exp	1.00/−0,00	0,14/−0,69	0,70/−0,22
	AVG	0,87/−0,02	0,28/−0,52	0,67/−0,18
CoT Zero-Shot (Content)	Llama-3.1-8B-Instruct	0,50/+0,04	0,14/−0,42	0,42/−0,06
	Llama-3.1-70B-Instruct	0,85/−0,07	0,77/−0,07	0,82/−0,06
	gpt-4o-mini	0,91/−0,03	0,05/−0,69	0,64/−0,21
	gpt-4o	0,95/+0,01	0,42/−0,54	0,75/−0,20
	gemini-2.0-flash-exp	0,95/+0,10	0,35/−0,56	0,73/−0,15
	AVG	0,83/+0,01	0,35/−0,46	0,67/−0,14

elements that influence classification. The explainability was assessed for the classifications generated using an advanced few-shot prompting approach with the GPT-4o-mini model and SHAP values [9] obtained using logprobs for a specific tokens (related to classes). In Fig. 4 (https://brawllstar.ru), the substring "ll" was emphasized (red indicates correlation with malicious class), suggesting the model associates certain character sequences with specific website types. In this case, "ll" appears to be an unusual repetition within the word 'brawl', which may indicate a typographical deviation from the expected "brawlstars", potentially signaling a deceptive domain. Similarly, in Fig. 5 (https://vulkanbet-offers.com), the terms "-" and "offers" were highlighted. The presence of a hyphen in domain names is less common among legitimate brand domains, since major companies typically register names without hyphens for credibility and ease of recall. The emphasis of the model on these elements suggests an association between such patterns and potentially lower-reputation websites. Even when webpage content is present, the model still prioritizes the URL, but also considers semantic elements. In Fig. 6 (https://octoplazmatic.ru), despite the presence of textual data, the model focuses mainly on the URL, although phrases like "app with a focus" were marked as contributing to a benign classification. This suggests that while URL-based patterns are dominant, meaningful content can influence classification decisions. Notably, the model's strong reliance on URLs is a global issue; we observed similar behavior across numerous explained samples.

https://brawllstars.ru/

Fig. 4. Explainability of classification for https://brawllstar.ru without content.

https://vulkanbetjoffers.com/vp_vb_003/index.php?ref=vp_w58784c102647l10866p1283_1140 & click_id=102b5537166848674df82f3c67edfe& sub

Fig. 5. Explainability of classification for https://vulkanbet-offers.com/vp_vb_003/index.php? without content.

URL: https://octoplazmatic.ru/static/js/main.39648aa8.js Web Content: <!doctype html><html lang="en"><head><meta charset="utf-8"/><link rel="icon" href="/favicon.ico"/><meta name="viewport" content="width=device-width,initial-scale=1"/><meta name="theme-color" content="#000000"/><meta name="description" content="Telegram is a cloud-based mobile and desktop messaging app with a focus on security and speed."/><link rel="apple-touch-icon" href="/logo192.png"/><link rel="manifest" href="/manifest.json"/><title>Telegram Web</title><script defer="defer" src="/static/js/main.9f0967a0.js"></script><link href="/static/css/main.afe8ddbc.css" rel="stylesheet"></head><body><noscript>You need to enable JavaScript to run this app.</noscript><div id="root"></div></body></html>

Fig. 6. Explainability of classification for https://octoplazmatic.ru/static/js/main.39648aa8.js with content.

7   Conclusions

In this paper, we analyze the performance of various prompting-based approaches (Zero-Shot, Few-Shot, and CoT) in the tasks of malicious webpage detection and malicious JavaScript file classification. In both tasks we evaluated several language models, by introducing a novel dataset for these tasks. Our extensive evaluation revealed that Few-Shot learning approach delivers the best overall performance. We also explored classification based on solely the webpage URLs as well as joint classification using both URL and webpage content. Among Zero-Shot methods, the advanced prompt applied to URLs performed best, while incorporating page content through CoT reasoning improved results, especially when both URL and content were considered together. Furthermore, we examined the impact of different attack methods, observing that modifications to the URL, especially in Simple Zero-Shot, led to higher True Positive Rates (TPR), but also increased False Positives. Including content in the queried LLM stabilizes the classification, reducing false positives and improving accuracy. Finally, through explainability analysis, we explored how the model makes its decisions, finding that it primarily relies on URL patterns but also uses page content to refine its predictions. This highlights the value of combining both URL and content for more robust and accurate attack detection.

The research in this paper focuses on advanced computational techniques, particularly LLM-based inference techniques, for web-oriented classification. By applying LLMs to malicious webpage detection, this work might contribute to AI's role in scientific computing and has potential to enhance cybersecurity in web environments.

Acknowledgements. Financed by: (1) CLARIN ERIC (20242026), funded by the Polish Minister of Science (agreement no. 2024/WK/01); (2) CLARIN-PL, the European Regional Development Fund, FENG programme (FENG.02.04-IP.040004/24); (3) statutory funds of the Department of Artificial Intelligence, Wroclaw Tech; (4) the

EU project "DARIAH-PL", under investment A2.4.1 of the National Recovery and Resilience Plan.

Ethics Statement. This study applied large language models for the detection of malicious web pages and JavaScript files, using publicly available data sets in compliance with data protection regulations. We acknowledge the ethical implications of AI in cybersecurity, aiming to improve threat detection and minimize false positives for safer web environments. The research did not involve human participants, and no informed consent was required for the use of public data.

References

1. Al-Maamari, M., Istaiti, M., Zerhoudi, S., Dinzinger, M., Granitzer, M., Mitrovic, J.: A comprehensive dataset for webpage classification (part 1: Adult & malicious) (version 1) (2024). https://doi.org/10.5281/ZENODO.10775260
2. Devlin, J., Chang, M.W., Lee, K., Toutanova, K.: Bert: pre-training of deep bidirectional transformers for language understanding. In: Proceedings of the 2019 Conference of the North American Chapter of the Association for Computational Linguistics: Human Language Technologies, pp. 4171–4186 (2019)
3. Gniewkowski, M., Maciejewski, H., Surmacz, T., Walentynowicz, W.: Sec2vec: anomaly detection in http traffic and malicious urls. In: Proceedings of the 38th ACM/SIGAPP Symposium on Applied Computing, pp. 1154–1162 (2023)
4. Gupta, B.B., Yadav, K., Razzak, I., Psannis, K., Castiglione, A., Chang, X.: A novel approach for phishing urls detection using lexical based machine learning in a real-time environment. Comput. Commun. **175**, 47–57 (2021)
5. Johnson, C., Khadka, B., Basnet, R.B., Doleck, T.: Towards detecting and classifying malicious urls using deep learning. J. Wirel. Mob. Networks Ubiquitous Comput. Dependable Appl. **11**(4), 31–48 (2020)
6. Li, J., Ji, S., Du, T., Li, B., Wang, T.: Textbugger: generating adversarial text against real-world applications. In: 26th Annual Network and Distributed System Security Symposium, NDSS 2019, San Diego, February 24-27, 2019. The Internet Society (2019). https://www.ndss-symposium.org/ndss-paper/textbugger-generating-adversarial-text-against-real-world-applications/
7. Li, L., Gong, B.: Prompting large language models for malicious webpage detection. In: 2023 IEEE 4th International Conference on Pattern Recognition and Machine Learning (PRML). pp. 393–400 (2023). https://doi.org/10.1109/PRML59573.2023.10348229
8. Liu, Y., et al.: Roberta: a robustly optimized bert pretraining approach. arXiv preprint arXiv:1907.11692 (2019)
9. Lundberg, S.M., Lee, S.I.: A unified approach to interpreting model predictions. Adv. Neural Inf. Proc. Syst. textbf30 (2017)
10. Luo, C., Tan, Z., Min, G., Gan, J., Shi, W., Tian, Z.: A novel web attack detection system for internet of things via ensemble classification. IEEE Trans. Ind. Inform. **17**(8), 5810–5818 (2020)
11. Nguyen, H.T., et al.: Application of the generic feature selection measure in detection of web attacks. In: Computational Intelligence in Security for Information Systems, pp. 25–32. Springer (2011)
12. Yao, S.,: Tree of thoughts: deliberate problem solving with large language models. In: Proceedings of the 37th International Conference on Neural Information Processing Systems. NIPS '23, Curran Associates Inc., Red Hook, NY (2023)

Physics Informed Neural Network Code for 2D Transient Problems (PINN-2DT) Compatible with Google Colab

Paweł Maczuga[1], Maciej Sikora[1], Tomasz Służalec[1], Marcin Szubert[1], Łukasz Sztangret[1], Danuta Szeliga[1], Marcin Łoś[1], Witold Dzwinel[1], Keshav Pingali[2], and Maciej Paszyński[1]([⊠])

^{1} AGH University of Krakow, Kraków, Poland
`maciej.paszynski@agh.edu.pl`
^{2} Oden Institute, the University of Texas at Austin, Austin, USA

Abstract. We present an open-source Physics Informed Neural Network environment for simulations of transient phenomena on two-dimensional rectangular domains, with the following features: (1) it is compatible with Google Colab which allows automatic execution on cloud environment; (2) it supports 2D linear or non-linear time-dependent PDEs; (3) it provides simple interface for definition of the residual loss, boundary condition and initial loss, together with their weights; (4) it support Neumann and Dirichlet boundary conditions; (5) it allows for customizing the number of layers and neurons per layer, as well as for arbitrary activation function; (6) the learning rate and number of epochs are available as parameters; (7) it automatically differentiates PINN with respect to spatial and temporal variables; (8) it provides routines for plotting the convergence (with running average), initial conditions learnt, 2D and 3D snapshots from the simulation and movies (9) it includes a library of problems: (a) non-stationary heat transfer; (b) atmospheric simulations including thermal inversion; (c) tumor growth simulations; and (d) the Stokes problem.

Keywords: Physics Informed Neural Networks · Colab · Atmospheric simulations · Tumor growth simulations · Material science simulations

1 Introduction

The goal of this paper is to replace the functionality of the time-dependent solver we published using isogeometric analysis and fast alternating directions solver [5–7] with the Physics Informed Neural Network (PINN) python library that can be easily executed on Colab. The PINN proposed in 2019 by Prof. Karniadakis revolutionized the way in which neural networks find solutions to initial-value problems described using partial differential equations [1] This method treats the neural network as a function approximating the solution of the given partial

© The Author(s), under exclusive license to Springer Nature Switzerland AG 2025
M. H. Lees et al. (Eds.): ICCS 2025, LNCS 15904, pp. 177–191, 2025.
https://doi.org/10.1007/978-3-031-97629-2_13

differential equation $u(x) = PINN(x)$. After computing the necessary differential operators, the neural network and its appropriate differential operators are inserted into the partial differential equation. The residuum of the partial differential equation and the boundary-initial conditions are assumed as the loss function. The learning process involves sampling the loss function at different points by calculating the PDE residuum and the initial boundary conditions. The PINN methodology has had exponential growth in the number of papers and citations since its creation in 2019. It has multiple applications, from solid mechanics [15], geology [4], medical applications [11], and even the phase-field modeling of fracture [14]. Why use PINN solvers instead of classical or higher order finite element methods (e.g., isogeometric analysis) solvers? PINN/VPINN solvers have affordable computational costs. They can be easily implemented using pre-existing libraries and environments (like Pytorch and Google Colab). They are easily parallelizable, especially on GPU. They have great approximation capabilities, and they enable finding solutions to a family of problems. With the introduction of modern stochastic optimizers such as ADAM [3], they easily find high-quality minimizers of the loss functions employed.

In this paper, we present the PINN library with the following features

- It is implemented in Pythorch and compatible with Google Colab.
- It supports two-dimensional linear or non-linear problems defined on a rectangular domain.
- It is suitable for smooth problems without singularities resulting from large contrast material data.
- It enables the definition of the PDE residual loss function in the space-time domain.
- It supports the loss function for defining the initial condition.
- It provides loss functions for Neumann and Dirichlet boundary conditions.
- It allows for customization of the loss functions and their weights.
- It allows for defining an arbitrary number of layers of the neural network and an arbitrary number of neurons per layer.
- The learning rate, the kind of activation function, and a number of epochs are problem-specific parameters.
- It automatically performs differentiation of the PINN with respect to spatial and temporal variables.
- It provides tools for plotting the convergence of all the loss functions, together with the running average.
- It enables the plotting of the exact and learned initial conditions.
- It plots 2D or 3D snapshots from the simulations.
- It generates gifs with the simulation animation.

We illustrate our PINN-2DT code with four numerical examples. The first one concerns the model heat transfer problem. The second one is the simulation of the thermal inversion and the process of pollution removal by artificially generated shock waves, and the last one is the simulation of brain tumor growth.

There are the following available PINN libraries. First and most important is the DeepXDE library [12] by the team of Prof. Karniadakis. It is an extensive library with huge functionality, including ODEs, PDEs, complex geometries, different initial and boundary conditions, and forward and inverse problems. It supports several tensor libraries such as TensorFlow, PyTorch, JAX, and PaddlePaddle. Another interesting library is IDRLnet [13]. It uses pytorch, numpy, and Matplotlib. This library is illustrated on four different examples, namely the wave equation, Allan-Cahn equations, Volterra integrodifferential equations, and variational minimization problems.

What is the novelty of our library? We do not claim to be better than these alternative high quality and multi-functionality libraries. The main point of our library is its simplicity of use and straight compatibility with Google Colab. It is a natural "copy" of the functionality of the IGA-ADS library [5] into the PINN methodology. It contains a simple, straightforward interface for solving different time-dependent problems. Our library can be executed without accessing the HPC center just by using the Google Colab.

The structure of the paper is the following. In Sect. 2, we recall the general idea of PINN on the example of the heat transfer problem. Section 3 is devoted to our code structure, from Colab implementation, model parameters, basic Python classes, how we define initial and boundary conditions, loss functions, how we run the training, and how we process the output. Section 4 provides four examples from heat transfer, wave equation, thermal inversion and the process of pollution removal by artificially generated shock waves, and tumor growth simulations. We conclude the paper in Sect. 5.

2   Physics Informed Neural Network for Transient Problems on the Example of Heat Transfer Problem

Let us consider a strong form of the exemplary transient PDE, the heat transfer problem. Find $u \in C^2(0, 1)$ for $(x, y) \in \Omega = [0, 1]^2$, $t \in [0, T]$ such that

$$\underbrace{\frac{\partial u(x, y, t)}{\partial t}}_{\text{time evolution}} \underbrace{-\epsilon \frac{\partial^2 u(x, y, t)}{\partial x^2} - \epsilon \frac{\partial^2 u(x, y, t)}{\partial y^2}}_{\text{diffusion term}} = \underbrace{f(x, y, t)}_{\text{forcing}}, (x, y, t) \in \Omega \times [0, T], (1)$$

with initial condition $u(x, y, 0) = u_0(x, y)$, and zero-Neumann boundary condition $\frac{\partial u}{\partial n} = 0 (x, y) \in \partial\Omega$. In the PINN approach, the neural network is the solution, namely

$$u(x, y, t) = PINN(x, y, t) = A_n \sigma \left(A_{n-1} \sigma(...\sigma(A_1[x, y, t] + B_1)...) + B_{n-1}\right) + B_n,$$

where A_i are matrices representing DNN layers, B_i represent bias vectors, and σ is the sigmoid, which as we have shown in [2], is the best choice for PINN. We define the loss function as the residual of the PDE

$$LOSS_{PDE}(x, y, t) =$$
$$\left(\frac{\partial PINN(x, y, t)}{\partial t} - \epsilon \frac{\partial^2 PINN(x, y, t)}{\partial x^2} - \epsilon \frac{\partial^2 PINN(x, y, t)}{\partial y^2} - f(x, y, t)\right)^2 .(2)$$

We also define the initial condition loss $LOSS_{Init}(x, y, 0) = (PINN(x, y, 0) - u_0(x, y))^2$, as well as the loss of the residual of the boundary condition $LOSS_{BC}(x, y, t) = \left(\frac{\partial PINN(x,y,t)}{\partial n}(x, y, t) - 0 \right)^2$.

3   Structure of the Code

Our code is available at https://github.com/pmaczuga/pinn-notebooks

The code can be downloaded, openned in Google Colab, and executed in the fully automatic mode. There are the following model parameters to define

- LENGTH, TOTAL_TIME. The code works in the space-time domain, where the training is performed by selecting point along x, y and t axes. The LENGTH parameter defines the dimension of the domain along x and y axes. The domain dimension is [0,LENGTH]x[0,LENGTH]x[0,TOTAL_TIME]. The TOTAL_TIME parameter defines the length of the space-time domain along the t axis. It is the total time of the transient phenomena we want to simulate.
- N_POINTS. This parameter defines the number of points used for training. By default, the points are selected randomly along x, y, and t axes. It is easily possible to extend the code to support different numbers of points or different distributions of points along different axes of the coordinate system.
- N_POINTS_PLOT. This parameter defines the number of points used for probing the solution and plotting the output plots after the training.
- WEIGHT_RESIDUAL, WEIGHT_INITIAL, WEIGHT_BOUNDARY. These parameters define the weights for the training of residual, initial condition, and boundary condition loss functions.
- LAYERS, NEURONS_PER_LAYER. These parameters define the neural network by providing the number of layers and number of neurons per layer.
- EPOCHS, and LEARNING_RATE provide a number of epochs and the training rate for the training procedure.

Inside the Loss class, we provide interfaces for the definition of the loss functions. Namely, we define the residual_loss, initial_loss and boundary_loss. Since the initial and boundary loss is universal, and residual loss is problem specific, we provide fixed implementations for the initial and boundary losses, assuming that the initial state is prescribed in the initial_condition routine and that the boundary conditions are zero Neumann. The code can be easily extended to support different boundary conditions. We provide examples of loss functions in the next section.

We provide several routines for plotting the convergence of the loss function and for plotting the running average of the loss (see Fig. 1), for plotting the initial conditions in 2D and for plotting snapshots of the solution (see Fig. 2).

4   Examples of the Instantiation

Heat transfer. We first present the instance of our library for the heat transfer problem described in Sect. 2. The residual loss function $LOSS_{PDE}(x, y, t) =$

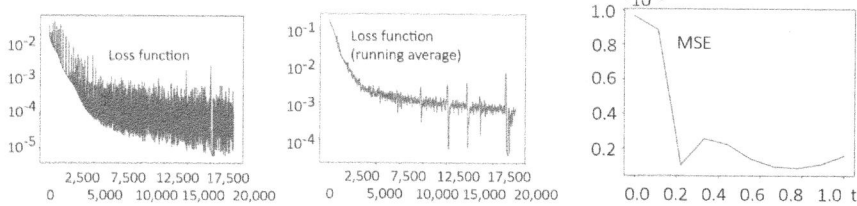

Fig. 1. Heat equation. Convergence of the residual loss function. Running average from the convergence of the residual loss function. Numerical error of the trained PINN solution to the heat transfer problem with manufactured solution.

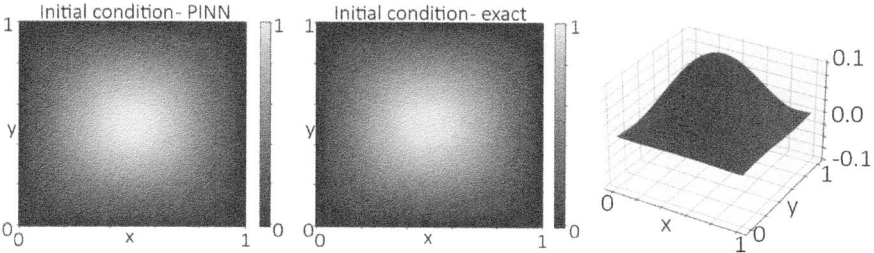

Fig. 2. Heat equation. Initial conditions in 2D. Snapshot from the simulation

$$\left(\frac{\partial PINN(x,y,t)}{\partial t} - \frac{\partial^2 PINN(x,y,t)}{\partial x^2} - \frac{\partial^2 PINN(x,y,t)}{\partial y^2} - f(x,y,t)\right)^2$$ translates into the following code

```
def residual_loss(self, pinn: PINN):
    x,y,t=get_interior_points(self.x_domain,
    self.y_domain,self.t_domain,self.n_points,pinn.device())
    u = f(pinn, x, y, t); z = self.floor(x, y)
    loss = dfdt(pinn, x, y, t, order=1) - \
        dfdx(pinn, x, y, t)**2-dfdy(pinn, x, y, t)**2
```

We employ the manufactured solution technique, where we assume the solution of the following form $u(x,y,t) = \exp^{-2\Pi^2 t}\sin\Pi x\sin\Pi y$, for $(x,y,t) \in [0,1]^2 \times [0,T]$. To obtain this particular solution, we set up the zero Dirichlet boundary conditions, which require the following code

```
def boundary_loss_dirichlet(self, pinn: PINN):
    down,up,left,right=get_boundary_points(self.x_domain,
    self.y_domain,self.t_domain,self.n_points,pinn.device())
    x_down,   y_down,   t_down    = down
    x_up,     y_up,     t_up      = up
    x_left,   y_left,   t_left    = left
    x_right,  y_right,  t_right   = right
    loss_down = f( pinn, x_down,  y_down,  t_down  )
    loss_up   = f( pinn, x_up,    y_up,    t_up    )
```

```
loss_left  = f( pinn, x_left , y_left ,  t_left  )
loss_right = f( pinn, x_right, y_right,  t_right )
return loss_down.pow(2).mean()+loss_up.pow(2).mean()+ \
  loss_left.pow(2).mean()+loss_right.pow(2).mean()
```

We also setup the initial state $u_0(x,y) = \sin(\Pi x)\sin(\Pi y)$ which translates into the following code

```
def initial_condition(x:torch.Tensor,y:torch.Tensor)->
  torch.Tensor:
  res = torch.sin(torch.pi*x) * torch.sin(torch.pi*y)
  return res
```

The default setup of the parameters for this simulation is the following:

```
LENGTH = 1.                TOTAL_TIME = 1.
N_POINTS = 15              N_POINTS_PLOT = 150
WEIGHT_RESIDUAL = 1.0      WEIGHT_INITIAL = 1.0
WEIGHT_BOUNDARY = 1.0      LAYERS = 4
NEURONS_PER_LAYER = 80     EPOCHS = 20_000
LEARNING_RATE = 0.002
```

The convergence of the loss function and the running average of the loss are presented in Fig. 1. The comparison of exact and trained initial conditions and the snapshot from the simulation is presented in Fig. 2 for time moment $t = 0.1$. The mean square error of the computed simulation is presented in Fig. 1. We can see the high accuracy of the trained PINN results.

Thermal inversion and the process of pollution removal by artificially generated shock waves. In this example, we aim to model the thermal inversion effect. The numerical results presented in this section are the PINN version of the thermal inversion simulation performed using isogeometric finite element method code [5] described in [9]. The scalar field u in our simulation represents the water vapor forming a cloud. The source represents the evaporation of the cloud evaporation of water particles near the ground. The thermal inversion effect is obtained by introducing the advection field as the gradient of the temperature. Following [10] we define $\frac{\partial T}{\partial y} = -2$ for lower half of the domain ($y < 0.5$), and $\frac{\partial T}{\partial y} = 2$ for upper half of the domain ($y > 0.5$).

We focus on advection-diffusion equations in the strong form. We seek the cloud vapor concentration field $[0,1]^2 \times [0,1] \ni (x,y,t) \to u(x,y,t) \in \mathcal{R}$

$$
\frac{\partial u(x,y,t)}{\partial t} + b(x,y,t) \cdot \nabla u(x,y,t) - \nabla \cdot (K(x,y)\ \nabla u(x,y,t))
$$

$$
\begin{aligned}
&= f(x,y,t), & (x,y,t) \in \Omega \times (0,T], \\
\nabla u \cdot n &= 0, & (x,y,t) \in \partial\Omega \times (0,T], \\
u(x,y,0) &= u_0(x,y), & (x,y,t) \in \Omega \times 0.
\end{aligned}
\tag{3}
$$

This PDE translates into

$$\frac{\partial u(x,y,t)}{\partial t} + \frac{\partial T(y)}{\partial y}\frac{\partial u(x,y,t)}{\partial y} - 0.1\frac{\partial u(x,y,t)}{\partial x^2} - 0.01\frac{\partial u(x,y,t)}{\partial y^2}$$

$$= f(x,y,t), (x,y,t) \in \Omega \times (0,T], \qquad (4)$$

$$\nabla u \cdot n = 0, \qquad (x,y,t) \in \partial\Omega \times (0,T],$$

$$u(x,y,0) = u_0(x,y), (x,y,t) \in \Omega \times 0.$$

We define the loss function as the residual of the PDE

$$LOSS_{PDE}(x,y,t) = \left(\frac{\partial PINN(x,y,t)}{\partial t} + \frac{\partial T(y)}{\partial y}\frac{\partial PINN(x,y,t)}{\partial y}\right.$$

$$\left. -0.1\frac{\partial PINN(x,y,t)}{\partial x^2} - 0.01\frac{\partial PINN(x,y,t)}{\partial y^2} - f(x,y,t)\right)^2. \qquad (5)$$

Followng [9] we modify the loss function to introduce the term modeling the artificially generated shock wave. The convergence of the loss function is summarized in Fig. 3. The snapshots from the simulations are presented in Fig. 4. In the thermal inversion, the cloud vapor that evaporated from the ground stays close to the ground, due to the distribution of the temperature gradients.

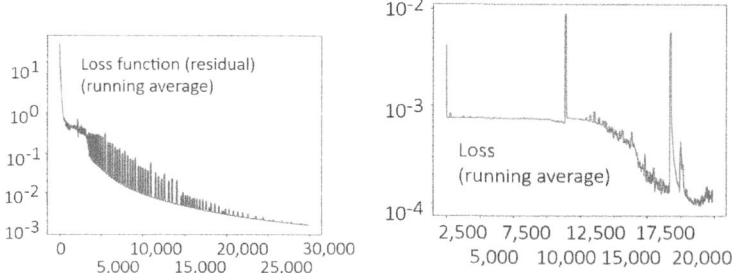

Fig. 3. Thermal inversion: Convergence of the loss function. Tumor growth: Convergence of the loss function.

To model the generation of the artificial shock waves, following the idea described in [9], we modify the advection by adding the term responsible for the generated shock wave.

Tumor growth. The next example concerns the brain tumor growth, as described in [11]. We seek the tumor cell density $[0,1]^2 \times [0,1] \ni (x,y,t) \rightarrow u(x,y,t) \in \mathcal{R}$, such that

$$\frac{\partial u(x,y,t)}{\partial t} = \nabla \cdot (D(x,y)\nabla u(x,y,t)) + \rho u(x,y,t)(1 - u(x,y,t))$$

$$(x,y,t) \in \Omega \times (0,T], \qquad (6)$$

$$\nabla u \cdot n = 0, (x,y,t) \in \partial\Omega \times (0,T], \quad u(x,y,0) = u_0(x,y), (x,y,t) \in \Omega \times 0,$$

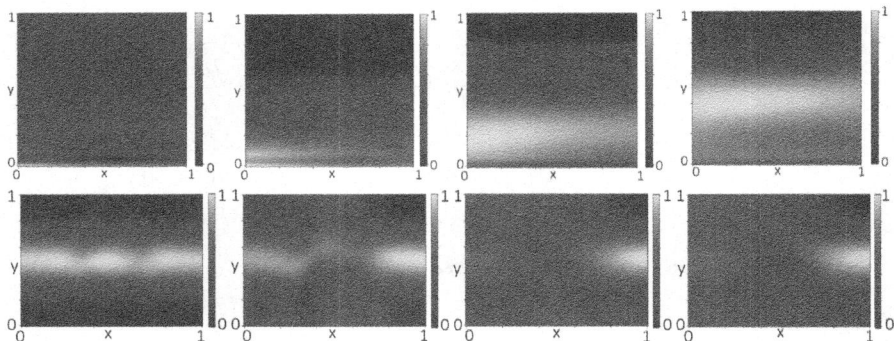

Fig. 4. Top panel: Thermal inversion. **Bottom panel**: Hail cannon simulation.

which translates into

$$\frac{\partial u(x,y,t)}{\partial t} - \frac{\partial D(x,y)}{\partial x}\frac{\partial u(x,y,t)}{\partial x} - D(x,y)\frac{\partial^2 u(x,y,t)}{\partial x^2}$$
$$- \frac{\partial D(x,y)}{\partial y}\frac{\partial u(x,y,t)}{\partial y} - D(x,y)\frac{\partial^2 u(x,y,t)}{\partial x^y} - \rho u(x,y,t)\left(1 - u(x,y,t)\right) = 0.$$

Here, $D(x,y)$ represents the tissue density coefficient, where $D(x,y) = 0.13$ for the white matter, $D(x,y) = .013$ for the gray matter, and $D(x,y) = 0$ for the cerebrospinal fluid (see [11] for more details). Additionally, $\rho = 0.025$ denotes the proliferation rate of the tumor cells. We simplify the model

$$\frac{\partial u(x,y,t)}{\partial t} - D(x,y)\frac{\partial^2 u(x,y,t)}{\partial x^2} - D(x,y)\frac{\partial^2 u(x,y,t)}{\partial x^y} \qquad (7)$$
$$- \rho u(x,y,t)\left(1 - u(x,y,t)\right) = 0.$$

We define the loss function as the residual of the PDE

$$LOSS_{PDE}(x,y,t) + \left(\frac{\partial u(x,y,t)}{\partial t} - \frac{\partial D(x,y)}{\partial x}\frac{\partial u(x,y,t)}{\partial x} - D(x,y)\frac{\partial^2 u(x,y,t)}{\partial x^2}\right.$$
$$\left. - \frac{\partial D(x,y)}{\partial y}\frac{\partial u(x,y,t)}{\partial y} - D(x,y)\frac{\partial^2 u(x,y,t)}{\partial x^y} - \rho u(x,y,t)\left(1 - u(x,y,t)\right)\right)^2 (8)$$

We summarize in Fig. 3 the convergence of the loss function. Additionally, Fig. 5 presents the snapshots from the simulation.

Stationary non-linear Navier-Stokes problem. Let us focus on the stationary cavity flow problem [16] described with the stationary non-linear Navier-Stokes equation for the incompressible fluid; see Fig. 7. The Dirichlet boundary condition drives the cavity flow for the velocity $u_x = 1$, $u_y = 0$ on the top boundary. On the remaining parts of the boundary, the velocity is equal to 0, and the ϵ thick transition zone in the left and right top corners ensures the possibility

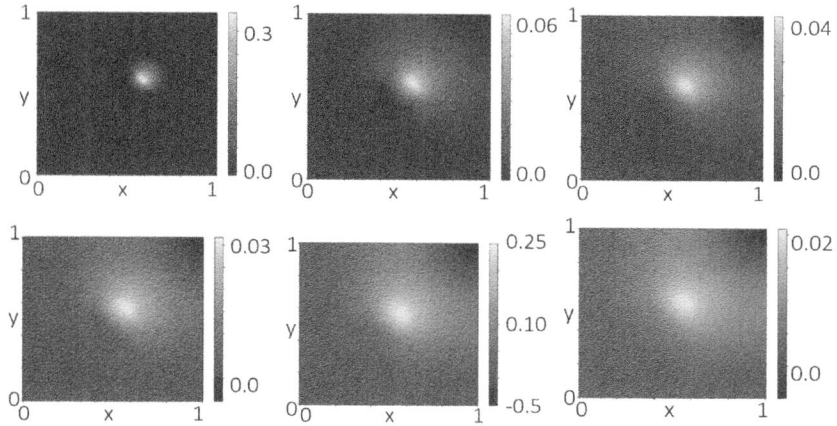

Fig. 5. Tumor growth. Snapshots from the simulation.

of a weak formulation. This problem exhibits pressure singularities at the two corners. Let $\Omega = (0,1)^2$ be the open boundary. The cavity flow problem reads: Find velocity u and pressure field p such that:

$$\begin{cases} -\dfrac{\Delta \mathbf{u}}{Re} + u \cdot \nabla u + \nabla p = 0 & \text{in } \Omega, \\ \nabla \cdot \mathbf{u} = 0 & \text{in } \Omega, \\ \mathbf{u} = h & \text{in } \Gamma \end{cases} \quad h = \begin{cases} 0 & x \in (0,1),\ y = 0 \\ 0 & x \in \{0,1\},\ y \in (0, 1-\epsilon) \\ 1 & x \in (0,1),\ y = 1 \\ \dfrac{\epsilon - y + 1}{\epsilon} & x \in \{0,1\},\ y \in (1-\epsilon, 1) \end{cases}$$
$$(9)$$

System (9) can be rewritten as

$$w_1(x_1, x_2) = \frac{\partial u_1(x_1, x_2)}{\partial x_1}, \quad w_2(x_1, x_2) = \frac{\partial u_1(x_1, x_2)}{\partial x_2}$$

$$z_1(x_1, x_2) = \frac{\partial u_2(x_1, x_2)}{\partial x_1}, \quad z_2(x_1, x_2) = \frac{\partial u_2(x_1, x_2)}{\partial x_2}$$

$$-\frac{\partial w_1(x_1, x_2)}{\partial x_1} - \frac{\partial w_2(x_1, x_2)}{\partial x_2} + \frac{\partial p(x_1, x_2)}{\partial x_1} +$$
$$u_1(x_1, x_2)w_1(x_1, x_2) + u_2(x_1, x_2)w_2(x_1, x_2) = 0,$$
$$-\frac{\partial z_1(x_1, x_2)}{\partial x_1} - \frac{\partial z_2(x_1, x_2)}{\partial x_2} + \frac{\partial p(x_1, x_2)}{\partial x_2} +$$
$$u_1(x_1, x_2)z_1(x_1, x_2) + u_2(x_1, x_2)z_2(x_1, x_2) = 0,$$
$$\frac{\partial u_1(x_1, x_2)}{\partial x_1} + \frac{\partial u_2(x_1, x_2)}{\partial x_2} = 0.$$

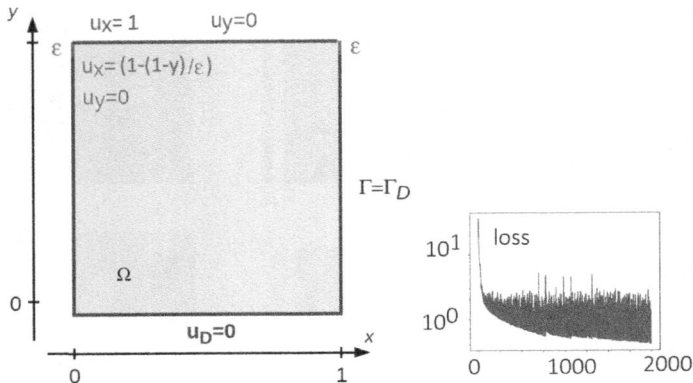

Fig. 6. Non-stationary cavity flow problem. Convergence of PINN training.

Despite the non-linear problem, we easily define the following residuals

$$RES_0 = \frac{\partial u_1}{\partial x_1} - w_1, \quad RES_1 = \frac{\partial u_1}{\partial x_2} - w_2, \quad RES_2 = \frac{\partial u_2}{\partial x_1} - z_1,$$

$$RES_3 = \frac{\partial u_2}{\partial x_2} - z_2, \quad RES_4 = -\frac{\partial w_1}{\partial x_1} - \frac{\partial w_2}{\partial x_2} + \frac{\partial p}{\partial x_1} + u_1 w_1 + u_2 w_2 - f_1,$$

$$RES_5 = -\frac{\partial z_1}{\partial x_1} - \frac{\partial z_2}{\partial x_2} + \frac{\partial p}{\partial x_2} + u_1 z_1 + u_2 z_2 - f_2,$$

$$RES_6 = \frac{\partial u_1(x_1, x_2)}{\partial x_1} + \frac{\partial u_2(x_1, x_2)}{\partial x_2}.$$

The Stokes code is available at `https://colab.research.google.com/drive/15eDVY7DyRfu5ugJRISETPjk-A-W6wA88`

The obtained solution is shown in Fig. 7. The Colab execution time on L4 graphic card, for 20,000 iterations with 4 layers of 200 neurones (a total of $200 \times 200 \times 3 = 24,000,000$ parameters), with 100×100 integration points and learning rate 0.005 is around 15 min. The benefit of PINN solution with respect to traditional finite element method solution is that it does not require any linearization or special stabilization methods. The non-linear PDEs are (split into a first order system though) are directly implemented in the loss function.

5   Conclusions

We have created a code `https://github.com/pmaczuga/pinn-notebooks` that can be downloaded and opened in the Google Colab. It can be automatically executed using Colab functionality. The code provides a simple interface for running two-dimensional time-dependent simulations on a rectangular grid. It provides an interface to define residual loss, initial condition loss, and boundary condition loss. It provides examples of Dirichlet and Neumann boundary conditions. The

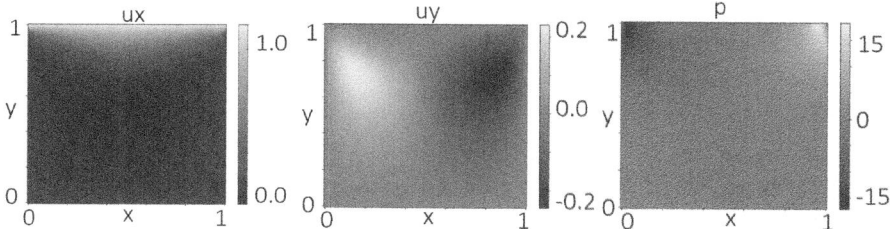

Fig. 7. The first and second components of the velocity vector field for the cavity flow problem. The pressure scalar field for the cavity flow problem.

code also provides routines for plotting the convergence, generating snapshots of the simulations, verifying the initial condition, and generating the animated gifs. We also provide four examples, the heat transfer, the thermal inversion from advection-diffusion equations, the brain tumor model, and the Stokes problem. The future work may involve development of the Variational PINN library with adaptive test space, following [17–22]. The PINN methods are naturally slower than finite element method codes, but they enable easy approximation of non-linear PDEs, without the necessity of linearization of the formulation. The non-linear problem can be directly incorporated into the residual.

Acknowledgements. This work was supported by the program "Excellence initiative - research university" for the AGH University of Science and Technology. The visit of Maciej Paszyński at Oden Institute was partially supported by J. T. Oden Research Faculty Fellowship.

A Code sniplets

Thermal inversion and shock wave generation The residual loss includes now the vertical temperature gradient dTy, the diffusion variables Kx and Ky, and the source term source.

```
def residual_loss(self, pinn: PINN):
    x, y, t = get_interior_points(self.x_domain, self.
        y_domain, self.t_domain,
    self.n_points, pinn.device())
    loss = dfdt(pinn, x, y, t).to(device)
    - self.dTy(y, t)*dfdy(pinn, x, y, t).to(device)
    - self.Kx*dfdx(pinn, x, y, t,order=2).to(device)
    - self.Ky*dfdy(pinn, x, y, t, order=2).to(device)
    - self.source(y,t).to(device)
    return loss.pow(2).mean
def source(self,y,t):
    d=0.7; res=torch.clamp((torch.cos(t*math.pi)-d)*1/(1-
        d), min=0)
    res2 = (150 - 1200 * y) * res
```

```
      res3 = torch.where(t <= 0.3, res2, 0)
      res4 = torch.where(y <= 0.125, res3, 0)
      return res4.to(device)
```

During the training, we use the following global parameters

```
LENGTH = 1.                WEIGHT_BOUNDARY = 10.0
TOTAL_TIME = 1.            LAYERS = 2
N_POINTS = 15              NEURONS_PER_LAYER = 600
N_POINTS_PLOT = 150        EPOCHS = 30_000
WEIGHT_RESIDUAL = 20.0     LEARNING_RATE = 0.002
WEIGHT_INITIAL = 1.0
```

The shock wave implementation in the advection term involves the following modifications to dTy routine:

```
def dTy(self,x:torch.Tensor,y:torch.Tensor,t:torch.Tensor)
    ->torch.Tensor:
  sin_term=torch.sin(torch.pi*t/2)*torch.sin(torch.pi*x)
      -0.8*torch.sin(torch.pi*t/2)
  es_threshold=torch.maximum(5*sin_term,torch.tensor(0.0,
      device=x.device))
  mask=(y<res_threshold);peak_value=torch.tensor(-3.0,
      device=device,dtype=x.dtype)
  def_val=torch.tensor(0.0,device=device,dtype=x.dtype)
  foff_rate=torch.tensor(1.0,device=device,dtype=x.dtype)
  inverted_mask_np=(~mask).cpu().numpy()
  distances_np=scipy.ndimage.distance_transform_edt(
      inverted_mask_np)
  distances=torch.from_numpy(distances_np).to(device=
      device,dtype=x.dtype)
  falloff_values=torch.clamp(peak_value+foff_rate*
      distances,max=def_val)
  final_grid=torch.where(mask.to(device),peak_value.to(
      device),falloff_values.to(device))
  return final_grid.to(device)*(-2)
```

Tumor growth simulations The loss function involves now the variable diffusion terms and the proliferation terms

```
def residual_loss(self, pinn: PINN):
    x, y, t = get_interior_points(
        self.x_domain, self.y_domain, self.t_domain,
        self.n_points, pinn.device())
    rho = 0.025
    def D_fun(x, y) -> torch.Tensor:
        res = torch.zeros(x.shape, dtype=x.dtype, device=
            pinn.device())
        dist = (x-0.5)**2 + (y-0.5)**2
        res[dist < 0.25] = 0.13; res[dist < 0.02] = 0.013
        return res
    D = D_fun(x, y); u = f(pinn, x, y, t)
```

```
loss = dfdt(pinn,x,y,t)-D*dfdx(pinn,x,y,t,order=2) \
        - D*dfdy(pinn,x,y,t,order=2)-rho*u*(1-u)
return loss.pow(2).mean()
```

The initial and boundary condition loss functions are unchanged. The initial state is given as follows:

```
def initial_condition(x: torch.Tensor, y: torch.Tensor) ->
    torch.Tensor:
 d = torch.sqrt((x-0.6)**2 + (y-0.6)**2)
 res = -d**2 - 4*d + 0.4; res = res * (res > 0)
 return res
```

We setup the following model parameters

```
LENGTH = 1.             WEIGHT_BOUNDARY = 1.0
TOTAL_TIME = 1.         LAYERS = 4
N_POINTS = 20           NEURONS_PER_LAYER = 80
N_POINTS_PLOT = 150     EPOCHS = 50_000
WEIGHT_RESIDUAL = 1.0   LEARNING_RATE = 0.005
WEIGHT_INITIAL = 1.0
```

Non-linear stationary Navier-Stokes The residual loss for non-linear stationary Navier-Stokes is the following

```
def calculate_loss(pinns: list[PINN], x: torch.Tensor, y:
    torch.Tensor) -> torch.Tensor:
 ux = pinns[0]; uy = pinns[1]; p = pinns[2]
 duxdx = pinns[3]; duxdy = pinns[4]; duydx = pinns[5]
 duydy = pinns[6]; u1 =  f(ux, x, y); u2 =  f(uy, x, y)
 du1dx = f(duxdx, x, y); du1dy = f(duxdy, x, y)
 du2dx = f(duydx, x, y); du2dy = f(duydy, x, y)
 uxdotgradux = u1 * du1dx + u2 * du1dy
 uydotgraduy = u1 * du2dx + u2 * du2dy
 d2uxdx = dfdx(duxdx,x,y); d2uxdy = dfdy(duxdy,x,y)
 dpdx = dfdx(p,x,y,order=1); d2uydx = dfdx(duydx, x, y)
 d2uydy = dfdy(duydy, x, y); dpdy = dfdy(p,x,y)
 loss1 = -d2uxdx - d2uxdy + dpdx + uxdotgradux
 loss2 = -d2uydx - d2uydy + dpdy + uydotgraduy
 loss3 = duxdx(x, y) + duydy(x, y)
 loss_duxdx = duxdx(x, y) - dfdx(ux, x, y)
 loss_duxdy = duxdy(x, y) - dfdy(ux, x, y)
 loss_duydx = duydx(x, y) - dfdx(uy, x, y)
 loss_duydy = duydy(x, y) - dfdy(uy, x, y)
 return loss1.pow(2).mean()+loss2.pow(2).mean()+ \
     loss3.pow(2).mean()+loss_duxdx.pow(2).mean()+ \
     loss_duxdy.pow(2).mean()+loss_duydx.pow(2).mean()+ \
     loss_duydy.pow(2).mean()
```

The boundary conditions and zero pressure value at the center point are enforced using the hard constraints.

```
def zero_dirich(x,y):return 20*x*(1-x)*y*(1-y)
```

```
def force_up_stream(x,y):return torch.exp(-1000*(y-1)**2)
def zero_middle(x,y):return -torch.exp(-1000*(y-0.5)**2)*
    torch.exp(-1000*(x-0.5)**2)+1.0
def ux_constraint(logs,x,y):return logs*zero_dirich(x,y)
    +force_up_stream(x,y)
def uy_constraint(logs,x,y):return logs*zero_dirich(x,y)
def p_constraint(logs,x,y):return logs*zero_middle(x,y)
```

The code requires an extension to support PINN with vector output

```
ux_pinn = PINN(LAYERS, NEURONS_PER_LAYER, hard_constraint=
    ux_constraint)
uy_pinn = PINN(LAYERS, NEURONS_PER_LAYER, middle_layers=
    ux_pinn.middle_layers, hard_constraint=uy_constraint)
p_pinn = PINN(LAYERS, NEURONS_PER_LAYER, middle_layers=
    ux_pinn.middle_layers, hard_constraint=p_constraint)
duxdx_pinn = PINN(LAYERS, NEURONS_PER_LAYER,
    middle_layers=ux_pinn.middle_layers)
duxdy_pinn = PINN(LAYERS, NEURONS_PER_LAYER,
    middle_layers=ux_pinn.middle_layers)
duydx_pinn = PINN(LAYERS, NEURONS_PER_LAYER,
    middle_layers=ux_pinn.middle_layers)
duydy_pinn = PINN(LAYERS, NEURONS_PER_LAYER,
    middle_layers=ux_pinn.middle_layers)
pinns = [ux_pinn, uy_pinn, p_pinn, duxdx_pinn, duxdy_pinn,
    duydx_pinn, duydy_pinn]
multiPINN = MultiPINN(LAYERS, NEURONS_PER_LAYER, pinns=
    pinns).to(DEVICE)
```

References

1. Raissi, M., Perdikaris, P., Karniadakis, G.E.: Physics-informed neural networks: a deep learning framework for solving forward and inverse problems involving nonlinear partial differential equations. J. Comput. Phys. **378**, 686–707 (2019)
2. Maczuga, P., Paszyński, M.: Influence of activation functions on the convergence of physics-informed neural networks for 1D wave equation. In: Computational Science – ICCS,: 23rd International Conference, Prague, July 3–5, 2023. Proceedings, Part I, pp. 74–88 (2023)
3. Kingma, D.P., Lei Ba, J.: ADAM: a method for stochastic optimization. arXiv:1412.6980 (2014)
4. Chen, Y., Xu, Y., Wang, L., Li, T.: Modeling water flow in unsaturated soils through physics-informed neural network with principled loss function. Comput. Geotech. **161**, 105546 (2023)
5. Łoś, M., Woźniak, M., Paszyński, M., Lenharth, A., Pingali, K.: IGA-ADS: Isogeometric analysis FEM using ADS solver, Computer & Physics. Communications **217**, 99–116 (2017)
6. Łoś, M., Paszyński, M., Kłusek, A.: Witold Dzwinel, Application of fast isogeometric L2 projection solver for tumor growth simulations. Comput. Methods Appl. Mech. Eng. **316**, 1257–1269 (2017)

7. Łoś, M., Kłusek, A., Amber Hassaan, M., Pingali, K., Dzwinel, W., Paszyński, M.: Parallel fast isogeometric L2 projection solver with GALOIS system for 3D tumor growth simulations. Comput. Methods Appl. Mech. Eng. **343**, 1–22 (2019)
8. Maczuga, P., Oliver-Serra, A., Paszyńska, A., Valseth, E., Paszyński, M.: Graph-grammar based algorithm for asteroid tsunami simulations. J. Comput. Sci. **64**, 101856 (2022)
9. Misan, K., et al.: Fast isogeometric analysis simulationsof a process of air pollution removal byartificially generated shock waves. In: Computational Science – ICCS,: 22rd International Conference, London, June 21–23, 2022. Proceedings, Part I, pp. 298–311 (2022)
10. U.S. Standard Atmosphere vs. Altitude, Engineering ToolBox (2003). https://www.engineeringtoolbox.com/standard-atmosphere-d_604.html
11. Zhu, A., Vo, J., Lowengrub, J.: Accelerating Parameter Inference in Diffusion-Reaction Models of Glioblastoma Using Physics-Informed Neural Networks, SIAM Undergraduate Research Online (2022)
12. Lu, L., Meng, X., Mao, Z., Em Karniadakis, G.: DeepXDE: a deep learning library for solving differential equations. SIAM Rev. **63**(1), 208–228 (2021)
13. Peng, W., Zhang, J., Zhou, W., Zhao, X., Yao, W., Chen, X.: IDRLnet: a Physics-Informed Neural Network Library arxiv2107.04320 (2021)
14. Goswami, S., Anitescu, C., Chakraborty, S., Rabczuk, T.: Transfer learning enhanced physics informed neural network for phase-field modeling of fracture. Theoret. Appl. Fract. Mech. **106**, 102447 (2020)
15. Haghighat, E., Raissi, M., Moure, A., Gomez, H., Juanes, R.: A physics-informed deep learning framework for inversion and surrogate modeling in solid mechanics. Comput. Methods Appl. Mech. Eng. **379**, 1879–2138 (2021)
16. Matuszyk, P., Paszyński, M.: Fully automatic hp adaptive finite element method for the stokes problem in two dimensions. Comput. Methods Appl. Mech. Eng. **197**(51–52), 4549–4558 (2008)
17. Demkowicz, L.: Computing with hp-adaptive finite elements, vol. 1, Wiley (2006)
18. Demkowicz, L., Kurtz, J., Pardo, D., Paszynski, M., Rachowicz, W., Zdunek, A.: Computing with hp-Adaptive Finite Elements: Volume II Frontiers: Three Dimensional Elliptic and Maxwell Problems with Applications (1st ed.), Chapman and Hall/CRC (2007)
19. Paszyńska, A., Paszyński, M., Grabska, E.: Graph transformations for modeling hp-adaptive finite element method with triangular elements. Lect. Notes Comput. Sci. **5103**, 604–613 (2008)
20. Goik, D., Jopek, K., Paszyński, M., Lenharth, A., Nguyen, D., Pingali, K.: Graph grammar based multi-thread multi-frontal direct solver with Galois scheduler. Procedia Comput. Sci. **29**, 960–969 (2014)
21. Paszyński, M., Paszyńska, A.: Graph transformations for modeling parallel *hp*-adaptive finite element method. In: Parallel Processing and Applied Mathematics: 7th International Conference, pp. 1313–1322 (2007)
22. Paszyńska, A., Paszyński, M., Jopek, K., Woźniak, M., Goik, D., Gurgul, P., AbouEisha, H., Moshkov, M., Calo, V.M., Lenharth, A., Nguyen, D., Pingali, K.: Quasi-optimal elimination trees for 2D grids with singularities. Sci. Program. **1**, 303024 (2015)

Dynamic Neural Network with Matrix-Extended Residual Connections

Szymon Świderski[1] and Agnieszka Jastrzębska[1,2(✉)]

[1] Warsaw University of Technology, Warsaw, Poland
A.Jastrzebska@mini.pw.edu.pl
[2] The John Paul II Catholic University of Lublin, Lublin, Poland

Abstract. The issue of adjusting neural network structure is one of the core problems in artificial intelligence. The issue of adjusting neural network structure is one of the core problems in artificial intelligence. A highly desirable scenario is a dynamic architecture that evolves structurally during the training process. In this paper, we propose a new and powerful tool that facilitates dynamic changes in network structure. We introduce a novel form of residual connections based on matrix extensions, enabling adaptable weight matrices and enhancing structural flexibility. The approach enhance the potential for structural modifications. We conducted a series of comprehensive experiments confirming that the new residual connections scheme behaves very well. The new type of connection improves performance by enabling better error flow during the error backpropagation phase, resulting in more efficient training. Our method demonstrates superior performance and enhanced trackability during the training process. The paper is supplemented by Python source code to ensure reproducibility. This method marks a significant starting point, showing immense potential for more advanced dynamic neural network models and transfer learning with dynamic models.

Keywords: dynamic neural network · changing architecture · training · shrinking · growing · residual connections with matrix extensions · self-changing neural networks

1 Introduction

The architecture of a neural network plays a critical role in determining its efficiency and overall performance. Designing an effective architecture requires making decisions about the number of layers, their connections, and other hyperparameters. While many networks rely on fixed structures, our focus is on models that can modify their architecture during training—a challenging task that

The research was funded by the National Science Centre, Poland, grant number 2024/53/B/ST6/00021.

requires innovative solutions. To support this dynamic adaptation, we introduce a new way of connecting layers, called *matrix-extended residual connections*. This method significantly improves the process of dynamically modifying network structures during training.

Inspired by the mechanism of *skip connections* present in ResNet [6], this method has been designed to better accommodate networks with evolving structures. While the primary goal is to improve the performance of dynamically changing architectures, these connections are also valuable for static architectures, providing broader applicability. They enhance the traceability and interpretability of the learning process, offering deeper insights into model behavior.

The proposed modification to the classical skip connections is particularly effective for dynamic neural networks, where the structure evolves during training. The newly designed connections help preserve learned information while seamlessly integrating new layers, essential for maintaining stability and performance in such networks.

In our previous work, titled "Dynamic Growing and Shrinking of Neural Networks with Monte Carlo Tree Search" [14], we presented a basic method for modifying network architecture during training. That approach used standard skip connections, which performed as initially described in the ResNet paper [6]. In this paper, we introduce a new way of connecting layers that significantly enhances the efficiency of dynamic structural changes. Beyond dynamic networks, these connections show promise for applications such as transfer learning, where preserving and extending learned information is critical. The proposed design for residual connections handling enables seamless architecture expansion without any loss of previously learned knowledge, offering a powerful tool for both dynamic and static neural networks.

To support reproducibility and practical use, we have released our implementation as an open-source package named `growingnn`. The package is available on PyPi at https://pypi.org/project/growingnn/, where all relevant links, including the GitHub repository with the source code and documentation, can be found. This tool aims to simplify the development and experimentation of dynamic neural networks for researchers and practitioners.

2   Literature Survey

The design of neural network architectures has always been challenging for developers. This led to the development of many new approaches over the years. Two key papers were particularly influential for this work, namely "Deep Residual Learning for Image Recognition" [6] and "Dynamic Growing and Shrinking of Neural Networks with Monte Carlo Tree Search" [14]. The first paper introduced the concept of skip connections, also known as residual connections, in neural networks.

The second paper, published by our team, presented an algorithm that allows dynamic modification of the network structure during training in a highly efficient manner. By integrating the best aspects of these two approaches, we developed a powerful new method that introduces exciting possibilities in the field of growing neural networks.

One of the most well-known methods for dynamically changing network structures is GradMax [3]. This method can adjust the architecture during training without losing previously learned data. While the general concept is similar to our approach, the procedures for modifying the network architecture differ significantly. In GradMax, changes are based on SVD (Singular Value Decomposition). This approach ensures that new neurons do not interfere with existing knowledge. In contrast, our method uses specialized connections, referred to as residual connections with matrix extensions.

A recent paper, "Dynamic Neural Network Structure: A Review of Its Theories and Applications" [5], provides a detailed overview of dynamic neural network structures. This paper describes a taxonomy and associated naming convention for dynamic neural networks. The group most relevant to our approach is the *adaptive layer* group, within which three main types of methods exist.

The first type of an adaptive layer method is called *stacking layer*, where the architecture is modified by adding layers sequentially [12]. The second type is called *residual approximation*. It originates from ResNet [6]. We can classify our new contribution to this subgroup. They adjust residual block connections to alter the network structure dynamically. The third type, *shortcut connections*, originates from the ResNeXt [17] architecture. These building blocks are more complex, with an emphasis on connections that bypass multiple layers. The paper "Going Deeper with Convolutions" [15] introduced a method that enables support for very deep architectures with residual-like connections. However, this approach is limited by concatenation layers, which combine inputs from various layers into a unified structure. Our approach addresses these limitations by introducing new types of connections. Additionally, shortcut-based structures often rely on predefined building blocks that restrict their ability to expand flexibly.

We believe that the methods belonging to the adaptive layer group [5] are all effectively unified in our algorithm. This highlights the significance of our work, as it provides a straightforward and efficient way to integrate these ideas. In our algorithm, we generate a set of actions in each iteration. These actions allow us to add sequential or residual layers, encompassing the capabilities of all the methods described in the adaptive layer group. The method presented in this paper is not limited to building blocks or any type of concatenate layer. The new way of residual connections allows to easily create new connections without any data loss on the connection.

In the field of dynamic neural networks, two key approaches focus on optimizing network structure. The first is reinforcement learning, which has demonstrated auspicious results [18]. The core idea behind this approach involves evaluating and grading structural changes to determine the most optimal adjustments. The second approach is evolutionary algorithms. Neuroevolution [13] is an approach to developing neural networks by leveraging evolutionary algorithms, optimizing not just the weights but also the architectures, hyperparameters, activation functions, and even learning rules of neural networks.

It is also worth mentioning papers that present non-standard, interesting ideas, such as the Cascade-Correlation Architecture [4], which adds new neurons for each unrecognized pattern; the work by Kilcher et al. [8], which explores escaping flat regions through structural modifications; and the Convex Neural Networks [1], which demonstrates how neural networks can adapt to various linear structures by adding neurons to a single hidden layer at each step.

3   The Method

The described algorithm is based on Stochastic Gradient Descent (SGD) and Adam optimizer [10]. It operates with a model represented as a directed acyclic graph of layers. Each layer is an independent node that manages its own incoming and outgoing connections to other layers. In this algorithm, the training procedure is divided into generations and then further into epochs. One epoch consists of one forward and backward propagation.

A generation consists of a training phase with multiple epochs, followed by a structure modification phase. After training, all possible changes to the current structure are generated. We refer to these changes as actions in this paper. The simulation provides information on which action is the best, and the best action is then applied to the structure. The overall flow of our method is in Fig. 1.

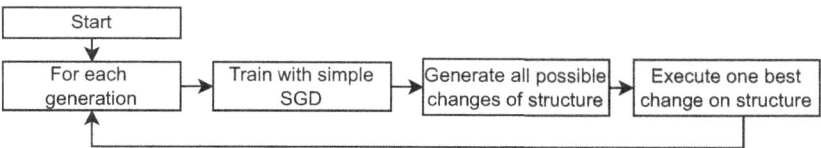

Fig. 1. Block diagram showing the flow of the method addressed in this study.

We modify the architecture using three types of actions. The first type involves adding a sequential layer between two directly connected layers. The second type adds a layer with residual connections; a layer is added between two not directly connected layers through skip connections. The third type consists of actions that remove an existing layer. In each generation, the algorithm generates all possible actions for the current state and evaluates them using a Monte Carlo simulation to determine and execute the best action.

The graph starts from the smallest possible setup: an input and output layer linked by a single connection and evolves over generations by adding or removing layers. Architecture changes in a given generation only if there is no improvement in accuracy, which means that if a given structure is capable of learning a given dataset, training continues without a change in the structure.

3.1 Data Flow in Dynamic Neural Structure

Residual connections in deep learning bring a significant change in training capabilities, but these solutions have their own limitations. If we analyze the data flow in residual connections [6], we can see that building blocks are limited in terms of learning capabilities. The error that is backpropagated is partially lost due to the building block structure. Data flowing through the building block at some point needs to go through a summation phase to combine outputs from two different layers into one.

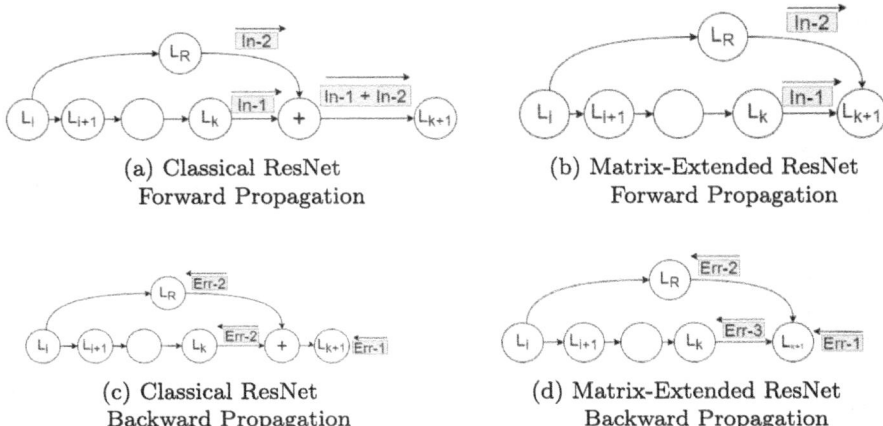

(a) Classical ResNet
Forward Propagation

(b) Matrix-Extended ResNet
Forward Propagation

(c) Classical ResNet
Backward Propagation

(d) Matrix-Extended ResNet
Backward Propagation

Fig. 2. Comparison of data flow in forward and backward propagation in Classical ResNet and Matrix-Extended ResNet.

The building block structure in ResNet consists of one residual (skip) connection that skips a given number of layers. We will use examples illustrated in Fig. 2 to present the limitations of classical ResNet connections. In those examples, layers $l_i, l_{i+1}, \ldots, l_{k+1}$ are connected sequentially, with each l_i connected to l_{i+1}. The skip connection is connected from l_i to an additional layer l_R and from l_R to l_{k+1}.

The signal flow is depicted by blue rectangles labeled with In and arrows, as shown in Fig. 2. During the forward propagation phase, the output from layer L_i is forwarded to two subsequent layers: one via a skip connection and the other sequentially, as illustrated in Fig. 2a. After this step, data flows through all layers without loss, except at layer l_{k+1}. In this layer, inputs from l_k and l_R are combined through summation, expressed as $In1 + In2$. Although this input combination is valid for layer l_{k+1}, we hypothesize that a significant portion of the interpretable data for l_{k+1} is lost. In our proposed approach, referred to as the *residual matrix extended connection* (Fig. 2b), we omit the summation step. Instead, the matrices are integrated to generate a new input signal for layer l_{k+1}, as visualized in Eq. 3.

In the back-propagation phase for the classical ResNet, layer l_{k+1} sends the same error to both l_R and l_{k+1}. When both layers receive the same error, their potential to detect distinct features is limited. While this approach effectively addresses exploding or vanishing gradients, allowing the network to have a much deeper structure, we aim to retain these advantages while enhancing efficiency. Our method does not backpropagate the same error to all connected layers. After extending the matrix for the new connection, we know which part belongs to each layer and send only the error assigned to that part. The layer that back-propagates the error calculates it individually for each connected layer, based on the corresponding part of the main weight matrix, as shown in Eq. 3.

3.2 Matrix-Extended Residual Connections

New residual connections with matrix extension are based on reshaping matrices. This helps the neural network adjust to the new connections without losing information.

In the first epoch, after adding a new connection, the neural network will return exactly the same output, but the number of all trainable weights will increase. This means that the layer will not forget any data it has already learned, but it will have a larger number of weights that are initially set to zero.

– **Before adding new connection:**

$$\mathbf{z}^{[l]} = \mathbf{W}^{[l]}\mathbf{a}^{[0]} + \mathbf{b}^{[l]} \tag{1}$$

– **After adding n connections with classical ResNet approach:**

$$\mathbf{z}^{[l]} = \mathbf{W}^{[l]} \sum_{i=0}^{n} \mathbf{a}^{[i]} + \mathbf{b}^{[l]} \tag{2}$$

– **After adding n connections with Matrix-extended ResNet:**

$$\mathbf{z}^{[l]} = \begin{bmatrix} W^{[l]} \\ \vdots \\ 0 \end{bmatrix} \begin{bmatrix} a^{[0]} \cdots a^{[n]} \end{bmatrix} + \mathbf{b}^{[l]} \tag{3}$$

In the presented formulas, $\mathbf{a}$ represents the data forwarded from the preceding layer, $\mathbf{W}$ is the weight matrix, and $\mathbf{b}$ is the bias. $\mathbf{z}^{[l]}$ denotes the output of a layer l, which is subsequently processed by an activation function and passed to the next layers. To handle multiple input signals without data loss, we extend the weight matrix by adding rows filled with zero values and combining all incoming signals into a single matrix in a column-wise manner. The output remains the same as before during the first epoch after introducing this change (i.e., after adding the new connections). However, in subsequent backpropagation steps, these newly added zero weights become trainable, enabling additional flexibility for the layer. This approach allows us to add or remove a specific number of layers without losing data or functionality.

With our approach, the backpropagated error through multiple residual connections does not need to be duplicated. Instead, only the error related to the specific part of the weight matrix corresponding to a given layer's input is propagated. This ensures that each layer receives a backpropagated error that is directly calculated for it.

Our new approach can be interpreted as increasing the number of connections but not the number of neurons, which brings a new field of possibilities to the algorithms that change the structure while training. Adding those connections can not only bring changes to the structure but also increase the number of trainable weights inside the layer without causing any loss in network memory.

3.3   Trackability of Neural Network Learning Process

We extend traditional residual connections with a novel approach called matrix-extended residual connections. This method offers improved traceability by distributing the back-propagated error more effectively. Our approach ensures that each layer receives a unique error signal that is tailored specifically to it. This allows for a more precise analysis of the error propagation and enables a deeper understanding of the learning process. This brings a significant change in our algorithm since the structure can grow to extensive and complicated graphs. Under the assumption that each layer analyzes a distinct set of features, our method facilitates a clearer interpretation of learning dynamics.

Figure 3 illustrates the evolution of the network structure on the CIFAR-10 dataset. Each dot represents a single layer. The color of the dot indicates the error propagated through that layer: red corresponds to the highest propagated error, while blue indicates the smallest error. The error for each layer is calculated and normalized according to the following formula: $\min\left(\max\left(\frac{\sum|E|}{E_{\max}}, 0\right), 1\right)$, where E is the error received in a given epoch by a given layer. In the presented images, we only show the state from the last epoch of a given generation, just before the change in structure, and the maximum error $E_{\max}$ is defined as $E_{\max} = \prod(\text{shape of } E)$.

After analyzing how the error evolves during training and structural changes, we can apply methods such as MDA [7] to visualize and understand the features learned by each layer. This approach provides a deeper insight into the overall learning process of the neural network.

4   Empirical Analysis

4.1   Dataset and Empirical Setup

In the conducted experiments, we used three widely recognized datasets: MNIST [2], Fashion-MNIST (FMNIST) [16], and CIFAR-10 [11]. MNIST contains handwritten digits. FMNIST consists of images of various clothing items, providing a more complex challenge. CIFAR-10 includes images of objects from 10 different classes, such as animals and vehicles, offering a more diverse and challenging dataset.

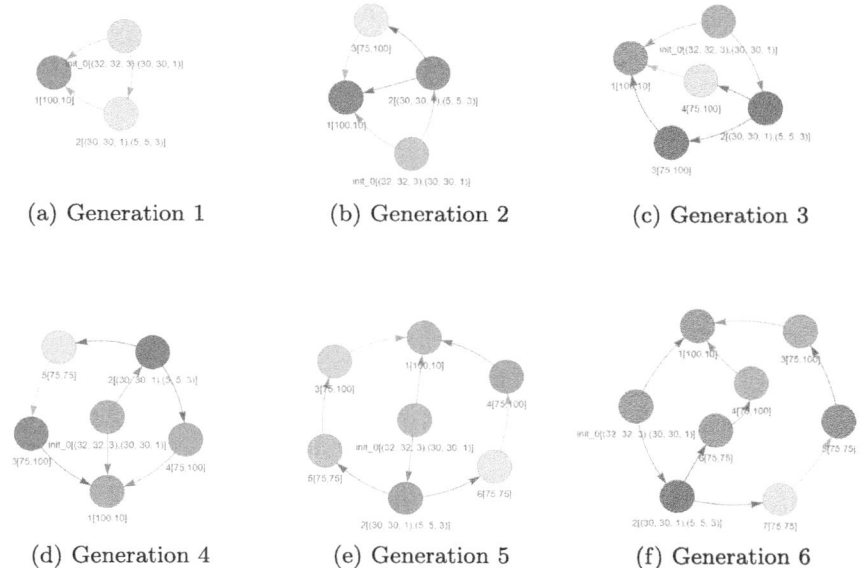

(a) Generation 1   (b) Generation 2   (c) Generation 3

(d) Generation 4   (e) Generation 5   (f) Generation 6

Fig. 3. History of structural changes during CIFAR-10 training. Error magnitudes are shown in red (high) and blue (low). Classical ResNet connections were used. (Color figure online)

We conducted a series of experiments to validate the three hypotheses outlined in Sect. 4.2. To encourage structural growth, each experiment started with a minimal configuration: a single 3×3 convolutional input layer and a dense output layer with 10 neurons. For MNIST and FMNIST, we used three generations with 10 epochs per generation and a hidden size of 50; for CIFAR-10, six generations with 50 epochs per generation and a hidden size of 100. All convolutional layers used 3×3 kernels, ReLU activations in hidden layers, and a sigmoid output. Training used a batch size of 128 and multiclass cross-entropy loss. A simulation scheduler guided the search, with each simulation running for 60 min to find the best action. Proposed actions were graded via 20-epoch training runs, using 20 random samples per class. The simulation algorithm was a modified Monte Carlo Tree Search (MCTS), optimized for efficient action space exploration.

In our experiments, we partitioned each dataset into training and testing subsets, with the testing set comprising 20% of the data. We ensured an even distribution of images per class using stratified sampling via the *train_test_split* function from the sklearn library. Consequently, for MNIST, the training set contained 48,000 images, and the testing set had 12,000 images. Fashion MNIST was divided similarly, with 48,000 training and 12,000 testing images. For CIFAR-10, which consists of 60,000 images in total, we allocated 40,000 images for training and 10,000 for testing.

4.2 Classification Quality of the New Approach

The conducted experiments aimed to empirically validate three hypotheses concerning the proposed residual connection mechanism. We compare it with the classical residual connections from the ResNet model.

RH1 The first hypothesis states that dynamically modifiable residual connections with matrix extensions tend to guide the weight distribution in the network toward a normal distribution.

RH2 The second hypothesis states that residual connections with matrix extensions improve learning capabilities by allowing the error signal to propagate more effectively through the network.

RH3 The third hypothesis states that dynamically modifiable residual connections with matrix extensions perform better than traditional residual connections and are particularly effective in networks that dynamically change their architecture during training.

The first hypothesis (**RH1**) proposes that dynamically modifiable residual connections with matrix extensions guide the weight distribution toward a normal distribution. In this experiment, we initialized neural networks with three types of general weight distributions: uniform, Gaussian, and gamma. After this, we run the experiments with our algorithm, which modifies the network structure.

Table 1. Comparison of mean values for three repetitions across three datasets and distribution types between classical ResNet and matrix-extended ResNet, based on skewness and kurtosis of the weight distribution in the final epoch.

Dataset	Distribution Type	Classical ResNet		Matrix-extended ResNet	
		Skewness	Kurtosis	Skewness	Kurtosis
CIFAR-10	Gamma	0.11	−1.48	3.67	12.84
	Normal	0.34	−0.86	0.85	0.31
	Uniform	0.26	−1.32	1.45	2.25
FMNIST	Gamma	0.20	−1.33	3.60	12.74
	Normal	0.58	−0.90	0.51	−0.70
	Uniform	−0.08	−1.31	1.02	1.13
MNIST	Gamma	0.38	−1.12	0.45	−0.47
	Normal	0.34	−1.08	2.06	4.39
	Uniform	−0.06	−1.48	1.70	4.89
Mean		0.23	−1.21	1.70	4.15

The graphs in Fig. 4 illustrate how weight distribution changes over the course of training. Each line in a graph represents the weight distribution for an epoch. The earliest epoch is displayed at the bottom, and the final distribution is shown

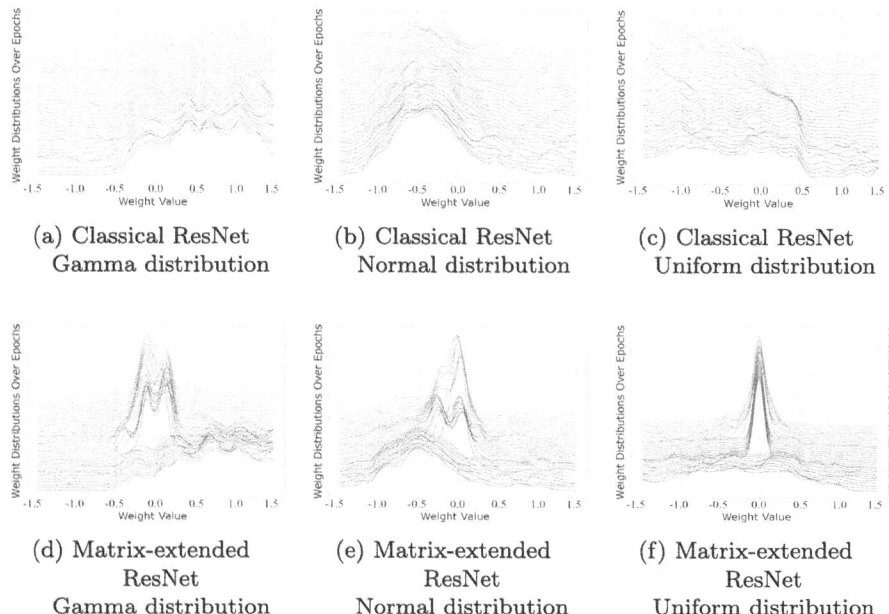

(a) Classical ResNet
Gamma distribution

(b) Classical ResNet
Normal distribution

(c) Classical ResNet
Uniform distribution

(d) Matrix-extended
ResNet
Gamma distribution

(e) Matrix-extended
ResNet
Normal distribution

(f) Matrix-extended
ResNet
Uniform distribution

Fig. 4. Comparison of weight distribution changes throughout the neural network structure between the classical ResNet and the matrix-extended ResNet connections on MNIST dataset.

at the top. Our analysis focuses on the input and output layers, as they consistently reflect changes in the overall weight distribution of the network.

As illustrated in Fig. 4, the proposed method consistently exhibits spikes near zero in the weight distributions, with the number of weights close to zero increasing over the course of training. In comparison, the classical ResNet approach shows minimal variation in its weight distribution across epochs. Table 1 presents the average distribution metrics across all seeds. While both methods display some asymmetry, the proposed approach yields a mean kurtosis of 4.15, which is closer to the value of a normal distribution [9]. In contrast, the classical method has an average kurtosis of -1.21, indicating a significantly flatter distribution. These findings suggest that our method produces distributions that more closely resemble normality, a result influenced by the growing number of zero weights introduced after each structural modification.

The second hypothesis (**RH2**) states that residual connections with matrix extensions improve learning capabilities by propagating the error signal more effectively through the network. During each epoch, we calculated a normalized error for every layer. The formula for calculating this normalized error is detailed in Sect. 3.3. This error does not directly reflect the overall performance of the network but rather indicates the magnitude of the error passing through a specific layer.

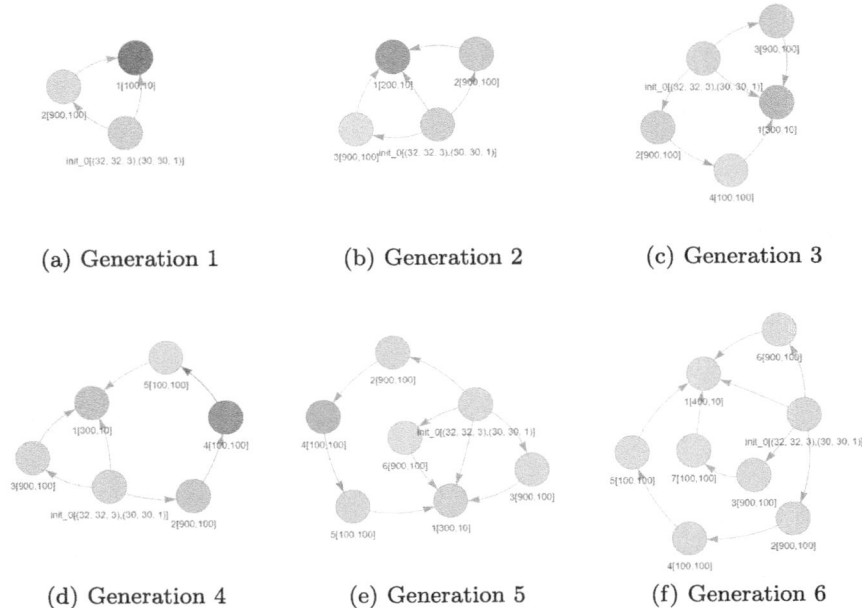

(a) Generation 1   (b) Generation 2   (c) Generation 3

(d) Generation 4   (e) Generation 5   (f) Generation 6

Fig. 5. History of structural changes during CIFAR-10 training. Error magnitudes are shown in red (high) and blue (low). Matrix-extended ResNet connections were used. (Color figure online)

Figure 5 illustrates the learning process in the proposed solution, where the error stabilizes more effectively, as indicated by the predominance of blue nodes representing low error. In contrast, Fig. 3 shows the traditional approach, showing a higher overall error between layers. In the standard ResNet architecture in Fig. 3, each layer transmits identical error signals to all connected layers via residual connections. In contrast, our approach utilizes a matrix-extended connection in Fig. 5, allowing for layer-specific error propagation. In the conventional approach, when a high error occurs in one layer, it propagates through the residual connections to all subsequent layers, causing a cascade of high errors across many layers. This results in inefficient error transmission, making it harder for the network to learn effectively. This is supported by Table 2, which shows that the mean and variance of errors across all layers are lower for the matrix-extended ResNet connections.

The third hypothesis (**RH3**) states that dynamically modifiable residual connections with matrix extensions outperform traditional residual connections and are particularly effective in networks that change their architecture dynamically during training. The experiments for the first and second hypotheses demonstrated the benefits of our method. To further support this, we compared the accuracy of a model trained with matrix-extended connections to that of a classical ResNet with skip connections.

Table 2. Mean and variance for normalized error distribution across the whole neural network structure.

Dataset	Seed	Classical ResNet		Matrix-extended ResNet	
		Mean	Variance	Mean	Variance
CIFAR-10	0	0.485	0.058	0.007	0.000
	1	0.199	0.010	0.000	0.000
	2	0.011	0.000	0.074	0.014
FMNIST	0	0.081	0.003	0.029	0.001
	1	0.056	0.003	0.052	0.004
	2	0.079	0.016	0.022	0.001
MNIST	0	0.027	0.001	0.009	0.000
	1	0.013	0.000	0.011	0.000
	2	0.031	0.001	0.007	0.000
Mean		0.109	0.010	0.024	0.002

Table 3. Comparison of training accuracy between the classical ResNet model and the model with matrix-extended connections.

Dataset	Seed	Classical ResNet	Matrix-extended ResNet
CIFAR-10	0	64.16%	99.96%
	1	82.20%	100.0%
	2	98.15%	98.13%
FMNIST	0	89.30%	88.19%
	1	91.68%	86.66%
	2	78.14%	90.75%
MNIST	0	98.14%	97.83%
	1	97.84%	98.31%
	2	97.36%	98.06%
Mean		88.55%	95.32%

The most significant differences were observed with the CIFAR-10 dataset. In this case, the matrix-extended ResNet connections demonstrated superior learning capabilities, delivering better results. As shown in Table 3, our method generally led to better results. The training history, illustrated in Figs. 6a and 6b, shows that each structural change in the classical approach caused considerable instability. In contrast, our method exhibited visible instability only in seed 0 of the third generation. In the matrix-extended connections, we observe several smooth spikes following each generation, which occur every 50 epochs. There is a significant improvement compared to the old method, indicating that the new approach has substantial potential for modifying the learning structure without losing any previously acquired data. The results shown in Table 3 indicate

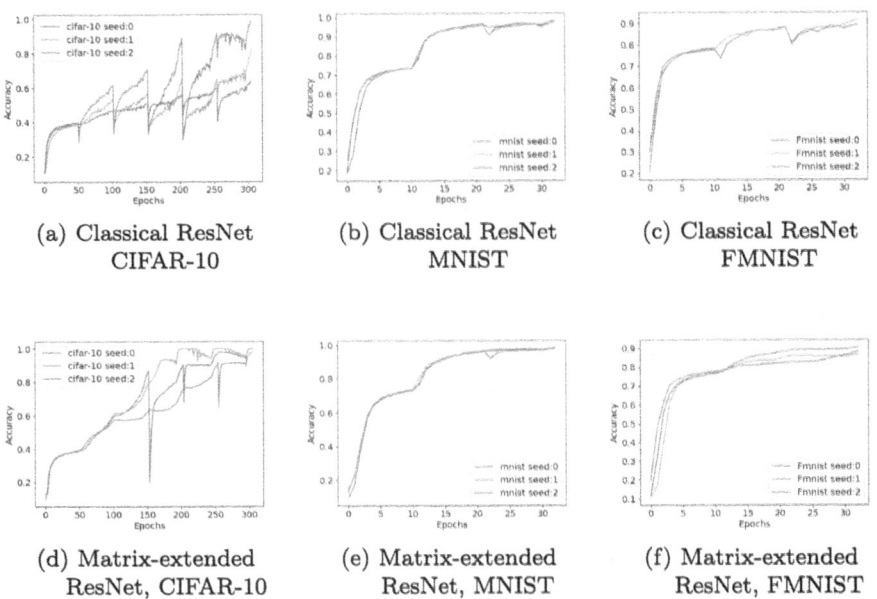

(a) Classical ResNet
CIFAR-10

(b) Classical ResNet
MNIST

(c) Classical ResNet
FMNIST

(d) Matrix-extended
ResNet, CIFAR-10

(e) Matrix-extended
ResNet, MNIST

(f) Matrix-extended
ResNet, FMNIST

Fig. 6. Comparison of training history with accuracy per epoch between the classical ResNet and the matrix-extended ResNet connections.

that the matrix-extended ResNet connections outperform the classical ResNet connections. The mean accuracy for the matrix-extended ResNet connections is 95.32%, compared to 88.55% for the classical ResNet connections. It is important to note that all training was conducted using our code written from scratch and that we employed basic tools for training, such as Stochastic Gradient Descent. These results are based on the training dataset, as our focus is on the training method rather than the model's performance on the test dataset. Our goal is to analyze the training procedure itself, not the test performance. To validate the method, we had to address specific features that are observable when a model is trained.

5   Conclusion

This paper presents a novel solution to common challenges in dynamic neural networks. We introduce matrix-extended ResNet connections, allowing structural changes without losing previously learned knowledge. These connections improve the efficiency of information flow by eliminating the summation step typically found in ResNet skip connections. This design enables easier and more flexible modification of the network structure during training, supporting dynamic adaptation without disrupting existing representations.

Moreover, our approach goes beyond simply addressing the summation step. Many other methods discussed in the literature also involve some form of merging, which has similar limitations. Our method simplifies this process by directly modifying the matrix, offering several advantages. No additional step is required to connect layers; the layer connections are straightforward and more closely resemble natural neural behaviors. The summation steps do not align with the foundational concept of artificial neurons, which was to emulate the behavior of natural neurons. In our approach, connecting a new layer allows neurons to establish additional connections that include the newly added layer. This opens new possibilities for algorithms that dynamically adjust their structure during training.

The conducted experiments validate the proposed thesis. Dynamically modifiable structures with matrix-extended residual connections enhance learning capabilities by providing better-adjusted error flow during backpropagation. Conducted experiments show that our method outperformed classical residual connections, achieving higher accuracy. When comparing the training history, it is clear that in the traditional approach, each change resulted in a visible drop in performance for the CIFAR-10 dataset. On the other hand, the training history for matrix-extended connections showed no negative impact in most changes, as there was no summation point or any additional step that could negatively affect the results. We also analyzed the error propagation through the network for both methods. In the classical approach, the summation step forces the error to be replicated, leading to significantly higher errors in most layers of the structure. In contrast, our solution maintains a low error throughout most of the structure, thereby making the learning process more efficient.

Additionally, our method offers a novel way to track the training process, enabling a more precise error propagation analysis and a deeper understanding of the learning dynamics. This is especially beneficial for large, complex networks.

All training code was written from scratch without additional frameworks, focusing on simple datasets. This demonstrates that effective and fast training can be achieved using only a CPU. The open-source implementation is available as a Python package named *growingnn* on PyPi: https://pypi.org/project/growingnn.

References

1. Bach, F.R.: Breaking the curse of dimensionality with convex neural networks. J. Mach. Learn. Res. **18**, 1–53 (2017). https://jmlr.org/papers/volume18/14-546/14-546.pdf
2. Deng, L.: The MNIST database of handwritten digit images for machine learning research. IEEE Signal Process. Mag. **29**(6), 141–142 (2012)
3. Evci, U., Vladymyrov, M., Unterthiner, T., van Merriënboer, B., Pedregosa, F.: GradMax: growing neural networks using gradient information. In: Proceedings of ICLR 2022 (2022). https://iclr.cc/virtual/2022/poster/7131
4. Fahlman, S.E., Lebiere, C.: The Cascade-Correlation Learning Architecture, pp. 524–532. Morgan Kaufmann Publishers Inc., San Francisco (1990)
5. Guo, J., Chen, C.L.P., Liu, Z., Yang, X.: Dynamic neural network structure: a review for its theories and applications. IEEE Trans. Neural Netw. Learn. Syst. 1–21 (2024). https://doi.org/10.1109/TNNLS.2024.3377194

6. He, K., Zhang, X., Ren, S., Sun, J.: Deep residual learning for image recognition. In: 2016 IEEE Conference on Computer Vision and Pattern Recognition (CVPR), pp. 770–778 (2016). https://doi.org/10.1109/CVPR.2016.90

7. Islam, M.T., et al.: Revealing hidden patterns in deep neural network feature space continuum via manifold learning. Nat. Commun. **14** (2023). https://doi.org/10.1038/s41467-023-43958-w

8. Kilcher, Y., Bécigneul, G., Hofmann, T.: Escaping flat areas via function-preserving structural network modifications. In: Proceedings of ICLR 2019 (2019). https://openreview.net/forum?id=H1eadi0cFQ

9. Kim, T.H., White, H.: On more robust estimation of skewness and kurtosis. Financ. Res. Lett. **1**(1), 56–73 (2004). https://doi.org/10.1016/S1544-6123(03)00003-5, https://www.sciencedirect.com/science/article/pii/S1544612303000035

10. Kingma, D., Ba, J.: Adam: a method for stochastic optimization. In: International Conference on Learning Representations (2014)

11. Krizhevsky, A., Hinton, G., et al.: Learning multiple layers of features from tiny images (2009)

12. Simonyan, K., Zisserman, A.: Very deep convolutional networks for large-scale image recognition. In: 3rd International Conference on Learning Representations (ICLR 2015), pp. 1–14. Computational and Biological Learning Society (2015)

13. Stanley, K., Clune, J., Lehman, J., Miikkulainen, R.: Designing neural networks through neuroevolution. Nat. Mach. Intell. **1** (2019). https://doi.org/10.1038/s42256-018-0006-z

14. Świderski, S., Jastrzębska, A.: Dynamic growing and shrinking of neural networks with Monte Carlo tree search. In: Franco, L., de Mulatier, C., Paszynski, M., Krzhizhanovskaya, V.V., Dongarra, J.J., Sloot, P. (eds.) Computational Science - ICCS 2024, pp. 362–377. Springer, Cham (2024)

15. Szegedy, C., et al.: Going deeper with convolutions. In: 2015 IEEE Conference on Computer Vision and Pattern Recognition (CVPR), pp. 1–9. IEEE Computer Society, Los Alamitos (2015). https://doi.org/10.1109/CVPR.2015.7298594, https://doi.ieeecomputersociety.org/10.1109/CVPR.2015.7298594

16. Xiao, H., Rasul, K., Vollgraf, R.: Fashion-mnist: a novel image dataset for benchmarking machine learning algorithms. CoRR abs/1708.07747 (2017). http://arxiv.org/abs/1708.07747

17. Xie, S., Girshick, R., Dollár, P., Tu, Z., He, K.: Aggregated residual transformations for deep neural networks. In: 2017 IEEE Conference on Computer Vision and Pattern Recognition (CVPR), pp. 5987–5995 (2017). https://doi.org/10.1109/CVPR.2017.634

18. Zoph, B., Vasudevan, V., Shlens, J., Le, Q.V.: Learning transferable architectures for scalable image recognition. In: 2018 IEEE/CVF Conference on Computer Vision and Pattern Recognition, pp. 8697–8710 (2018). https://doi.org/10.1109/CVPR.2018.00907

Proof of Training: Obtaining Verifiable ML Models by Delegating Training to a Blockchain Network

Łukasz Krzywiecki[1] and Gabriel Wechta[2(✉)]

[1] Department of Fundamentals of Computer Science,
Wroclaw University of Science and Technology, Wroclaw, Poland
lukasz.krzywiecki@pwr.edu.pl
[2] Department of Cryptology, NASK National Research Institute Location,
Warsaw, Poland
gabriel.wechta@nask.pl

Abstract. The recent rise of Bitcoin has sparked an unprecedented trend of enthusiasts acquiring expensive hardware for mining. This self-perpetuating race, driven by Bitcoin's PoW consensus mechanism, has led to the creation of massive computational centers dedicated solely to solving impractical hash inversions. However, these computational resources could be redirected toward more meaningful tasks if nodes were properly incentivized.

In this paper, we introduce Proof of Training (PoT), a novel consensus mechanism that offers two key advantages over previous approaches. First, it replaces wasteful computations with ML training. Second, by aligning the inherent distrust between nodes in blockchain networks with distributed model training, PoT not only achieves consensus but also produces a trained model along with proof that it was trained on the client-provided data.

PoT enables clients to hire the blockchain network to provably train arbitrary models using their provided datasets and architectures. Meanwhile, nodes that participate in training are rewarded with PoT cryptocurrency based on their computational contributions.

Keywords: Trustworthy ML · Delegated ML · Blockchain networks · Consensus mechanism

1 Introduction

In recent years, machine learning (ML) and cryptocurrencies have entered the mainstream spotlight. Both have shown immense potential to transform industries, yet they also face growing criticism for their environmental impact [19, 27]. While the computational resources required for ML model training are often justified by the functionality and value they deliver, the same cannot be said for blockchain networks. Consensus mechanisms such as Bitcoin's Proof of Work

© The Author(s), under exclusive license to Springer Nature Switzerland AG 2025
M. H. Lees et al. (Eds.): ICCS 2025, LNCS 15904, pp. 207–221, 2025.
https://doi.org/10.1007/978-3-031-97629-2_15

(PoW) [29] rely on solving cryptographic puzzles, where significant energy consumption serves no purpose beyond guaranteeing the immutability of the transaction ledger, making the process highly wasteful.

However, recent research suggests that the long-term sustainability of cryptocurrencies must rely on consensus frameworks that preserve computational effort rather than eliminating it altogether [12,35,36]. Therefore, we argue that the solution to blockchain inefficiency is not to remove complex computation but to design a consensus mechanism that harnesses global computing power for a tangible, desirable and practical purpose while also remaining profitable for participants [3,16,21].

One of the key challenges in ML model creation is the significant computational power required for training models. Lately, even moderately complex models often demand more resources than most organizations can provide. As a result, many organizations must shift to Cloud Computing Services (CCS) such as Google Cloud Platform, Amazon AWS, or Microsoft Azure which provide significant computing resources often dedicated to ML training under some fees. However, in general, CCSs do not offer any provable guarantees that user-provided models have been trained honestly, that is without any malicious modifications to architecture or dataset, that includes data mislabeling, injections, dropping, etc.

1.1   Our Contribution

In this work, we introduce a novel consensus mechanism, Proof of Training (PoT), aimed at making cloud-like ML training verifiable and reducing wasteful computation. Unlike existing usability-focused consensus mechanisms, PoT emphasizes real-world functionality and leverages other established and trusted consensus mechanisms. We accomplish those goals through distributed protocol during which nodes independently create proofs that they performed training, while simultaneously verifying that their results align with those of other participating nodes. Each confirmation is recorded on the public blockchain, enabling client and third-party verification of the training process. Once a specified number of confirmations is reached, nodes proceed to the next round, which could be a different stage of the previous model or a completely fresh model. Consequently, PoT can be viewed as a secure multi-party computation alternative for *ML as a Service* and CCS platform providers.

Notably, PoT can facilitate model training from the beginning or verify the correctness of externally trained model, additionaly PoT in the context of blockchain networks ia a comprehensive consensus mechanism, allowing it to be substitute for any existing consensus mechanism in various blockchain networks.

In Sect. 2 we present ML assumptions and formal notation that allows us to define adn formalize malicious ML training. Section 3 is devoted to the PoT. We begin by highlighting the key aspect of PoT's design, that forces each participant to perform training by itself. Later we move to the design and protocol description. We also discuss each step and explain its impact on the the secuirty of the training.

1.2   Related Work

Attacks on ML Training. Data poisoning within the context of SVMs has been extensively examined in several studies [1, 5, 10]. Additionally, the authors of [28] investigated contaminated training data, shedding light on how such data can benefit attackers. Subsequent research delved into various other applications of data poisoning, including malware detection [14], sentiment analysis [13], collaborative filtering [6], and attacks on spam filters [11]. In [8], the authors proposed a general optimization framework for offline poisoning attacks, while another study [33] experimented with undermining license plate recognition models. Furthermore, [9] provided a comprehensive review of recent attacks and defenses spanning different data modalities. Lastly, the authors of [18] categorized a wide array of vulnerabilities associated with datasets in their study.

Consensus Mechanisms. Nakamoto in his seminal work [29] proposed Proof of Work, the first well-adopted consensus mechanism for permissionless blockchain networks. In response to the recognized limitations of PoW, the authors of [22] introduced a consensus mechanism mitigating wasteful computation called Proof of Stake (PoS). Building upon advancements of PoS, the authors of [4] presented a mechanism called Proof of Activity (PoA) incentivizing nodes to remain online. This is in contrast to PoS where nodes tend to go idle over time. For a comprehensive overview of blockchain technology, particularly focusing on consensus mechanisms, refer to [3, 35]. Latest attempts in shifting the attention of blockchain network creators towards useful computation resulted in new cryptocurrencies such as Primecoin [21] and in general frameworks [3, 16].

Similar to this Work. To our knowledge, only a handful of approaches use AI/ML training as a consensus mechanism foundation: Proof of Learning (PoL) [20, 26] and Li's Proof of Training (PoT) [24]. PoL makes nodes compete to create the most effective model for a task and independently determine accuracy and parameters. However, PoL neglects key aspects of permissionless blockchain networks, including transaction time realization (heavily dependent on model architecture) and blockchain data storage (requirements escalate rapidly). Recent work [15] weakens PoL's integrity by presenting a spoofing attack. Li's PoT similarly incorporates competition to improve model quality by rewarding better-performing nodes. However, it gives nodes client-specified time limits for training, causing unpredictable block publication times. Furthermore, it assumes a complicated network architecture, unrealistic in decentralized blockchain networks.

Other works explore blockchain and trustworthy ML training intersection. Navarro et al. [30] proposed methods for trustworthy neural network training via blockchain. Lihu et al. [25] introduced a proof of useful work for AI on blockchain. Our PoT differs through stronger focus on adversarial training integrity and its verification mechanism designed to ensure training integrity. The concept of verifying ML training via partial re-execution has been explored previously, though often outside blockchain contexts. For instance, VerDE [2] employs a similar strategy where trainers share intermediate checkpoints and verifiers re-execute training segments with the largest weight changes.

2 Malicious ML Training

Our discussion adopts a higher-level perspective, abstracting the ML training process from its specific implementation, thus allowing us to generalize our discourse across various types of models, with the only caveat being that a model use gradient-based learning methodologies. Other types of models may also be used with the PoT after defining an appropriate way of calculating a *training secret* (see Sect. 3). However, in this work, we limit our discussion to models utilizing gradient-based learning.

We use the notation $x \xleftarrow{\$} X$ to indicate that an element x is sampled uniformly at random from the set X. We use k, n, t to denote numbers from $\mathbb{N}^+$. We denote a sequence $(x_1, x_2, \ldots, x_t)$ as $(x_i)_1^t$ or simply as $x^{(t)}$. Let $I = \{i_1, \ldots, i_k\} \subset \{1, \ldots, n\}$ denote indices. We use A_I to denote subset of A, which elements are in I, that is $A_I = \{a_{i_1}, \ldots, a_{i_k}\} \subset A = \{a_1, \ldots, a_n\}$.

Clean Samples. Let $\mathcal{X}$, $\mathcal{Y}$ denote the set of input and output data, respectively. A process of training a parameterized function $f_S \colon \mathcal{X} \to \mathcal{Y}$ (e.g., a neural network) involves updating its parameters S (weights and biases) while minimizing the loss function L. L measures the difference between the output of the model and the desired output for a given input. Given a set of n samples $\mathcal{D} = \{(x, y)\} \subset \mathcal{X} \times \mathcal{Y}$, where $\mathcal{X}$ represents the input data and $\mathcal{Y}$ the corresponding set of labels (e.g., for the classification problem, $\mathcal{X}$ represents input vectors and $\mathcal{Y}$ the corresponding set of categories). Let $\mathcal{M} \subset \mathcal{D}$ denote the training dataset, and $\mathcal{D} \setminus \mathcal{M}$ denote a test set. The objective is to set the optimal parameters S such that for all fresh samples $(x, y) \in \mathcal{D} \setminus \mathcal{M}$ predicted values $y \leftarrow f_S(x)$ have the lowest possible error in terms of L.

Malicious Samples. Let $\hat{\mathcal{X}}$ denote the set of input data corresponding to *incorrect* output data $\hat{\mathcal{Y}}$. Let $\hat{\mathcal{D}} = \{(\hat{x}, \hat{y})\} \subset \hat{\mathcal{X}} \times \hat{\mathcal{Y}}$ denote the set of samples, where $(\hat{x}, \hat{y})$ is a pair of *incorrect* data. Let $\hat{\mathcal{M}} \subset \hat{\mathcal{D}}$ denote the malicious samples used during the malicious training process. Thus $\hat{\mathcal{D}} \setminus \hat{\mathcal{M}}$ will be used to evaluate the *malicious accuracy*.

Regular Training. Gradient descent-based training [31] is an iterative process of adjusting weights and biases of a model in the direction of the steepest descent of the loss function gradient $g \leftarrow \frac{1}{b} \nabla_{S_i} \sum_{(x,y) \in \mathcal{B}_i} L(f_{S_i}(x), y)$. Here, we consider a *regular training* (RT) with parameters $(\mathcal{D}, \mathcal{M}, f, S_0, L, (\xi_i)_1^t)$, where: $\mathcal{M} \subset \mathcal{D}$ denotes the training samples, f denotes a chosen architecture of the model, S_0 denotes the initial parameters of the model, L is a loss function, and $(\xi_i)_1^t$ denotes the randomness used. The process iteratively updates $S_{i+1} \leftarrow S_i - \eta \cdot g$, with learning rate η, sampling (x, y) from batches $\mathcal{B}_i \subset \mathcal{M}$ of size b (i.e. $|\mathcal{B}_i| = b$).

We denote $S_t \leftarrow \mathsf{RT}(f_{S_0}, (\mathcal{B}_i)_1^t, (\xi_i)_1^t)$, or $S_t \leftarrow f_{S_0}(\mathcal{B}^{(t)})$ to indicate that the model parameters S_t are obtained by applying the training sequence of t batches $(\mathcal{B}_i)_1^t = \mathcal{B}^{(t)}$, using randomness $(\xi_i)_1^t = \xi^{(t)}$ for all indices $i \in I = \{1, \ldots, t\}$, to the model architecture f with the initial parameters S_0. Moreover let $S_i \leftarrow f(S_{i-1}, \mathcal{B}_i, \xi_i)$ denote a single loop in RT, i.e. processing a single ith batch, where ξ_i denotes all randomness used in ith loop.

Malicious Training. Let $\hat{I} = \{i_1, \ldots i_k\} \subset \{1, \ldots, t\} = I$. (k,t)-*malicious* training involves the same process as RT, except that the k batches $\hat{\mathcal{B}}_i$ are taken from malicious samples $\hat{\mathcal{M}}$ for some indices $\hat{I}$. Thus we denote $\hat{\mathcal{B}}_{\hat{I}}^{(k,t)} = \mathcal{B}^{(t)} \backslash \mathcal{B}_{\hat{I}} \cup \hat{\mathcal{B}}_{\hat{I}}$ to indicate that some k out of t clean batches $\mathcal{B}_i \in \mathcal{B}^{(t)}$ were replaced with $\hat{\mathcal{B}}_j$ for $j \in \hat{I}$. We denote $\hat{S}_t \leftarrow \mathsf{RT}(f_{S_0^*}, \hat{\mathcal{B}}_{\hat{I}}^{(k,t)}, (\xi_i)_1^t)$, or $\hat{S}_t \leftarrow \hat{f}_{S_0^*}(\hat{\mathcal{B}}^{(k,t)}, \xi^{(t)})$ to indicate that $\hat{S}_t$ is obtained from initial state S_0^* using the training sequence $\hat{\mathcal{B}}^{(k,t)}$ with k malicious batches and the randomness sequence $\xi^{(t)}$.

Advantage of Malicious Training. Let $S_t \leftarrow f_{S_0}(\mathcal{B}^{(t)}, \xi^{(t)})$ and $\hat{S}_t \leftarrow \hat{f}_{S_0^*}(\hat{\mathcal{B}}^{(k,t)}, \xi^{(t)})$ denote clean and malicious training, respectively. Accordingly, let $(x, y) \xleftarrow{\$} \mathcal{D} \backslash \mathcal{M}$ and $(\hat{x}, \hat{y}) \xleftarrow{\$} \hat{\mathcal{D}} \backslash \hat{\mathcal{M}}$ denote sampling clean and malicious data, respectively.

Definition 1 ((α, β)-advantage of Malicious Training). *We say Malicious Training has (α, β)-advantage for:*

$$\alpha = \left| \Pr[y \leftarrow S_t(x)] - \Pr[y \leftarrow \hat{S}(x)] \right|,$$
$$\beta = \Pr[\hat{y} \leftarrow \hat{S}_t(\hat{x})].$$

The (α, β)-advantage is deemed successful when α is small and β is sufficiently large. We say that k malicious batches $\hat{\mathcal{B}}_{\hat{I}}$ are *injected* into the training process (instead of clean samples $\mathcal{B}_{\hat{I}}$) to achieve β accuracy on malicious data. From the adversary's perspective, achieving (α, β)-advantage implies that the selection of $\hat{\mathcal{B}}_{\hat{I}}$ and k yields satisfactory accuracy β on malicious data for the resulting model $\hat{S}_t$, simultaneously the resulting model remains α-close to the regular accuracy of S_t trained on clean data.

2.1 Distributed Deterministic Training and Verification

A potential countermeasure against malicious injections involves ensuring that the RT process is deterministically verifiable. Assume we have regular RT training sequence $S_t \leftarrow \mathsf{RT}(f_{S_0}, (\mathcal{B}_i)_1^t, (\xi_i)_1^t)$, which is a series of steps

$$S_1 \leftarrow f(S_0, \mathcal{B}_1, \xi_1), \ldots, S_t \leftarrow f(S_{t-1}, \mathcal{B}_t, \xi_t).$$

We can record all the parameters $S_0, \ldots, S_t$, batches $\mathcal{B}^{(t)}$ and randomness $\xi^{(t)}$. This preservation enables us to recreate the entire RT process later and verify whether the resulting model matches the one obtained during the initial training.

Distributed verifiable RT. Recreating the recorded RT process is indeed as complex as the initial training, requiring the repetition of all its steps over the recorded intermediate values S_0, $S^{(t)}$, $\mathcal{B}^{(t)}$, $\xi^{(t)}$.

2.2 Distributed Deterministic Training Requirements

Here we define assumptions about the training process that are essential for PoT:

- *Sequential*: Training of the model from the beginning to the end can be split into stages where each stage depends on the outcome of the preceding stage.
- *Deterministic*: The operation $S_i \leftarrow f(S_{i-1}, \mathcal{B}_i, \xi_i)$, for the same inputs, will always produce the same result. This in practice enforce usage of the same parameters during training e.g., the same seed, framework version etc.
- *Reproducible*: When the pseudorandom number generator used to obtain ξ is initialized with the same seed, it produces the same $\xi^{(t)}$ values. Due to this and the deterministic assumption, the training process can be replicated on different devices ensuring that the output $S_t \leftarrow \mathsf{RT}(f_{S_0}, (\mathcal{B}_i)_1^t, (\xi_i)_1^t)$ will remain the same.
- *Computationally expensive*: Computing $f(S_{i-1}, \mathcal{B}_i, \xi_i)$ is not trivial. This informal statement aims to limit possible f functions to practical training problems.
- *Well-defined*: f is commonly accepted and widely represented in ML libraries.

From now we are discussing only deterministic training, with randomness ξ deterministically reproduced on all involved nodes, we skip ξ in our notation. Thus, $f(S_{i-1}, \mathcal{B}_i)$ denotes $f(S_{i-1}, \mathcal{B}_i, \xi_i)$.

3   Proof of Training

Broadly speaking, the Proof of Training consensus mechanism allows unambiguous training verification, by constructing a solution to a consensus puzzle alongside recreating the recorded RT.

Training Secret. The core building block of PoT is the capability to prove that a node indeed performed training from S_i to state S_{i+1}. This has to be achieved using a value that can be learned *only* during the training process. In the text, we call that value the *training secret* of stage S_i, and in equations, we denote it using ρ_i. Note that all functions depending only on S_{i+1} will not work (for example $\mathcal{H}(S_{i+1})$, where $\mathcal{H}$ is a cryptographic hash function) and the training process has to be somehow entangled, since S_{i+1} has to be publicly available for the protocol to work. Now we explain how such value can be derived.

When a batch $\mathcal{B}_i$ of size b is fed to the model, the training process iterates over inputs and makes predictions. Let $\tilde{y}_j$ be a prediction for input x_j. $\tilde{y}_j$ is compared to the expected output label y_j and an error $l_j = L(y_j, \tilde{y}_j)$ is calculated. At the end of the batch, all samples' errors are accumulated

$$g_i = \frac{1}{b} \sum_{1 \leq j \leq b} L(y_j, \tilde{y}_j).$$

Based on this accumulated error and learning rate η_i, the update algorithm is used to improve the model's weights w using

$$w \leftarrow w - \eta_i g_i.$$

Intermediate errors l_j are forgotten after accumulation. When $b > 1$, retrieving these errors becomes infeasible, even with access to g_i, $\mathcal{B}$, and both the old and new values of w are available. l_js are essential for constructing the training secret, they must be treated as secrets. Consequently, PoT requires a batch size of at least two.

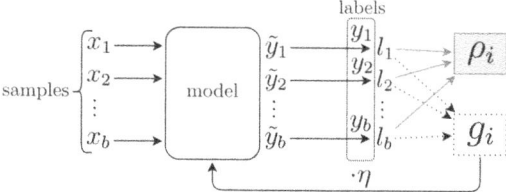

Fig. 1. Graphical representation of the process of incorporating the *training secret* creation into the basic training iteration.

Training secret is calculated in parallel to the training loop (see Fig. 1) by hashing the concatenation of current l_j with the hash of the predecessor. This can be expressed by the following equation

$$\rho_i = \mathcal{H}(l_b \parallel \mathcal{H}(l_{b-1} \parallel \ldots \mathcal{H}(l_2 \parallel \mathcal{H}(l_1)) \ldots)). \tag{1}$$

If $\rho_{S_{i+1}}$ is derived during calculating $f(S_i, \mathcal{B}_{i+1})$, we denote that as

$$S_{i+1} \xleftarrow[\rho_{S_{i+1}}]{} f(S_i, \mathcal{B}_{i+1}).$$

Training as Sequential Steps. Training a single ML model can take anywhere from a few seconds to several days. Since it is crucial to ensure that leader election time remains predictable and depends on the resources a node possesses [35] clearly, a single training process cannot serve as the backbone of a consensus mechanism without modifications.

To solve this problem a PoT user, henceforth called an *employer*, before submitting a training request to the network, is responsible for dividing the training process into sequential steps, with the single-step size being constructed based on functions τ and Δ.

The function τ estimates the total computational effort required to train the entire model on dataset D. One, popular way of doing this relies on analyzing the number of multiply-and-add operations based on the model architecture. Importantly, that metric does not depend on the underlying hardware, resulting in a fair metric for all nodes, irrelevant of the hardware they possess. We utilize the explicit formula provided in [32] as τ.

The function Δ, on the other hand, partitions the dataset into a set $\mathbb{D} = \{\mathcal{B}_1, \mathcal{B}_2, \ldots, \mathcal{B}_t\}$, where each $\mathcal{B}_i$ represents a separately downloadable batch of training data. Δ takes two inputs: $\tau(A, D)$, and the number of blocks published in the last 24 h, similar to how Bitcoin's self-adjusting difficulty function operates (see equation (1) of [34]). Both inputs are used to ensure that the expected time for a single training step, $S_{i+1} \leftarrow f(S_i, \mathcal{B}_{i+1})$, remains roughly constant, aligning with the protocol-defined time interval T. It is important to note that such a design does not restrict the model owner's decision-making regarding the number of epochs and the batch size (except the aforementioned $b > 1$). This approach also solves the issue of scaling over time with increasingly large datasets and more complex models. In the case of a large computational load, the dataset would be divided into more batches, effectively resulting in more stages.

Employer's Preparation. The employer is responsible for establishing a publicly accessible pull-push repository containing the model architecture A, partitioned dataset $\mathbb{D}$, a seed for deriving $\xi^{(t)}$, and C the number of confirmation votes that employer wishes to receive, all secured with a digital signature. Nodes, after downloading these resources, must verify their authenticity. Additionally, the employer must deposit the fee for node participation into a designated address that is non-retrievable by the employer, but which balance is publically accessible. This fee will later be partially distributed as rewards to nodes that successfully participate in the training process. Note that, although there is no strict authentication mechanism, as this is a permissionless blockchain network, what truly legitimizes the employer is the commitment of coins to the non-retrievable address. The employer then announces their intent to engage the network by broadcasting a message called a *training request*, which includes all necessary data.

3.1 PoT Design

Finally, we present the Proof of Training consensus mechanism. PoT is divided into two phases, namely:

- *Verifiable training*: The first phase is responsible for training the model and creating a *training declaration* message.
- *Block building*: The second phase is responsible for electing a leader based on *block header* and creating the final *wrapped block* that will be appended to the blockchain.

Both phases at the end utilize the PoS mechanism [22], while the second also utilizes the PoA mechanism [4]. Functions D_{PoS}^{td} and D_{PoS}^{bh} are difficulty functions for PoS mechanisms presented below in Algorithm 1 and Algorithm 2, respectively. The workflow of the PoT consensus mechanism is depicted in Fig. 2.

Verifiable Training Phase (Algorithm 1). Steps 1, 2—nodes independently select stages that meet their preferences. This part depends solely on the strategy of PoT's client implementation. Notably, a node may prefer to participate in the whole training process for some M even though it is not the most profitable strategy at the time to, for example, avoid download bottlenecks. Step 3—by performing $f(S_i, \mathcal{B}_{i+1})$ node acquires *training secret* $\rho_{S_{i+1}}$. Step 4—*training declaration* contains:

- ID_{td}: Unique identifier of the *training declaration*.

Algorithm 1. Verifiable training
___
1: Every node independently selects stage S_i of model M to train, based on overheard training requests and its own preferences.
2: Every node downloads the selected stage S_i and batch $\mathcal{B}_{i+1}$.
3: Every node performs $S_{i+1} \xleftarrow[\rho_{S_{i+1}}]{} f(S_i, \mathcal{B}_{i+1})$ and saves $\rho_{S_{i+1}}$.
4: Node creates *training declaration* message, where **rand** is a random 256-bit string.

$$training\ declaration := \{\text{ID}_{td}, \text{ID}_M, \text{ID}_{S_{i+1}}, \mathcal{H}(S_{i+1}), \texttt{timestamp},$$
$$\mathcal{H}(\rho_{S_{i+1}} \,\|\, \texttt{rand}), \texttt{rand}, \texttt{coinstake}, \texttt{pub\_key}\}$$

5: (PoS) Node calculates $h^{td} = \mathcal{H}(training\ declaration)$. If $h^{td} \leq D_{PoS}^{td}(\texttt{coinstake})$ is true, it calculates $\sigma_{td} = \text{Sign}_{\texttt{priv\_key}}(training\ declaration)$, appends it to the message and publishes it, otherwise it increments $\texttt{timestamp}$ and starts again.
___

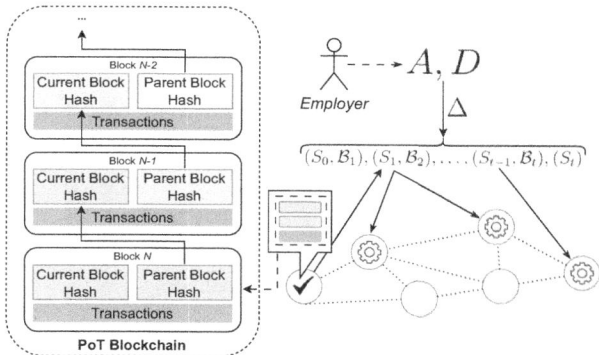

Fig. 2. Typical workflow of PoT consensus mechanism. An employer (right-hand size) publishes architecture A and dataset D. Using Δ, D is divided into $\{\mathcal{B}_1, \mathcal{B}_2, \dots, \mathcal{B}_t\}$. Participating node selects stage i, downloads $(S_i, \mathcal{B}_{i+1})$, performs $S_{i+1} \leftarrow f(S_i, \mathcal{B}_{i+1})$ and creates *training declaration*. A gear in the node indicates that a node is currently performing training. After a sufficient number of *training declarations* is published, one of the nodes creates *block header*, which leads to the creation of *wrapped block* and (left-hand size) effectively a new block. The elected leader (denoted by a checkmark) broadcasts a new block and uploads a new resulting state to the employer.

- ID_M, $\text{ID}_{S_{i+1}}$: Used for model and training stage identification. To reduce message size, a condensed identifier $\text{ID}_{\{M, S_{i+1}\}}$ can be employed (as we do in our implementation, refer to Sect. 3.1).
- $\mathcal{H}(S_{i+1})$: Allows nodes to verify if S_{i+1} held by them is equal to parameters held by the *training declaration* creator.
- $\texttt{timestamp}$: Used for PoS mechanism.
- $\mathcal{H}(\rho_{S_{i+1}} \,\|\, \texttt{rand})$: Known as *secret commitment*. It serves as a zero-knowledge proof of knowing $\rho_{S_{i+1}}$. Note that $\rho_{S_{i+1}}$ is not included in the message. Also, $\mathcal{H}(\rho_{S_{i+1}} \,\|\, \texttt{rand})$ reveals no information about $\rho_{S_{i+1}}$. At this point in the protocol, the only way to verify *secret commitment* is to derive $\rho_{S_{i+1}}$ individually

---

Algorithm 2. Block building

1: Node after receiving C *training declarations* for the S_{i+1} of M verifies that all meet $D_{PoS}^{td}(\text{coinstake})$, have correct signatures, and checks their *secret commitments* against his own *training secret*. If sound, the node creates *block header*, where $[\text{ID}_{td_1}, \text{ID}_{td_2}, \dots, \text{ID}_{td_C}]$ is a list of *training declarations* identifiers.

$$block\ header := \{\text{ID}_{bh}, \text{ID}_M, \text{ID}_{S_{i+1}}, \texttt{parent\_block\_hash}, \texttt{block\_index},$$
$$\texttt{timestamp}, \rho_{S_{i+1}}, \texttt{coinstake}, \texttt{pub\_key}, [\text{ID}_{td_1}, \text{ID}_{td_2}, \dots, \text{ID}_{td_C}]\}$$

2: *(PoS)* Node calculates $h^{bh} = \mathcal{H}(block\ header)$. If $h^{bh} \leq D_{PoS}^{bh}(\text{coinstake})$ is true it calculates $\sigma_{bh} = \text{Sign}(block\ header)$, appends it to the message and publishes it, otherwise it increments $\texttt{timestamp}$ and starts again.

3: Node uploads S_{i+1} to the employer.

4: * *(PoA)* Every node verifies C *training declaration* against $\rho_{S_{i+1}}$ included in *block header*. If valid, node acquires K stakeholders by passing $\mathcal{H}(block\ header)$ to *follow-the-coin*.

5: * Every online node checks whether it is one of K stakeholders. First $K-1$ stakeholders sign the hash of the *block header* and broadcast their signatures. Kth stakeholder extends the *block header* by creating a *wrapped block*, where $[\text{Tx}_1, \text{Tx}_2 \dots \text{Tx}_t]$ is a list of network transactions and $[\sigma_1, \sigma_2 \dots \sigma_{K-1}]$ is a list of $K-1$ stakeholders signatures. Kth stakeholder signs and broadcasts.

$$wrapped\ block := \{\text{ID}_{bh}, \text{Tx}_{\text{coinbase}}, [\text{Tx}_1, \text{Tx}_2, \dots, \text{Tx}_t], [\sigma_1, \sigma_2, \dots, \sigma_{K-1}]\}$$

6: * Every node after receiving *wrapped block*, checks stakeholder signatures and $\mathcal{H}(S_{i+1})$ against *training declarations* and if sound updates his local replica of the blockchain. Process for S_i is considered finished.

---

by appending the included $\texttt{rand}$ to $\rho_{S_{i+1}}$ and hashing them. $\rho_{S_{i+1}}$ will be revealed in Step 2 of Algorithm 2, after which everyone can verify the *secret commitment*.

- $\texttt{coinstake}$: The abbreviation for the coinstake transaction. The idea is borrowed from [22]. This value represents the cumulative age of all coins involved in the coinstake transaction. Coin age determines the difficulty of D_{PoS}^{td}. When a block is published, the age of all coins in the coinstake transaction resets to zero and are back in the possession of the issuer.

- $\texttt{pub\_key}$: Public key corresponding to private key $\texttt{priv\_key}$ used in the signature. Allows verification of σ_{td}.

Step 5—this step provides resilience against Sybil attacks [35] in the same manner as PoS achieves that. Also, PoS makes shared secret attacks, in which nodes silently cooperate and only one of them performs computation and shares results with others, infeasible. After publishing *training declaration*, a node can decide whether he wants to go back to Step 1 or continue to the next phase. If a node, during the PoS waiting mechanism, receives a block based on the stage that he is currently working on, he was late; he should dump the current state and start over.

Block Building Phase (Algorithm 2). Step 1—C is the number of *training declarations* chosen by an employer and fixed for M. Note that the node has to know $\rho_{S_{i+1}}$ to perform this step. Fields $\texttt{ID}_{bh}$, $\texttt{ID}_M$, $\texttt{ID}_{S_{i+1}}$, timestamp, coinstake, pub_key have the same purpose as in a *training declaration*.

- parent_block_hash, block_index: The *block header* serves as a vessel for the *wrapped block*, containing mandatory blockchain block fields.
- $\rho_{S_{i+1}}$: The node publishes the *training secret* so that others, who did not compute it, can verify the correctness of *training declarations*. Note that a lazy node cannot intercept $\rho_{S_{i+1}}$ and immediately broadcast their own *training declaration* or *block header* since they have to wait in Step 5 of Algorithm 1 or Step 2 of Algorithm 2.
- $[\texttt{ID}_{td_1}, \texttt{ID}_{td_2}, \ldots, \texttt{ID}_{td_C}]$: Each *training declaration* serves as an independent vote, confirming that the training of the model indeed yields that result. Note that in order to save space, only indices of *training declarations* are sent. If a node does not have access to them, it should request them from some nearby node, similar to the original idea for sharing transactions' data in [29].

Step 2—serves the same purpose as Step 5 in Algorithm 1. Step 3—node uploads S_{i+1} (i.e., weights, biases) to the employer. The employer's repository should provide a means for resolving the rare situation in which more than one node will upload S_{i+1} simultaneously. Note that the employer does not have to store every S_{i+1} and may choose to only keep the latest one. Steps denoted with an asterisk (*) are meant to be triggered by the event of receiving a *block header* for any M, and to be performed outside of normal Algorithm 1 or Algorithm 2 workflows. Steps 4, 5—typical last PoA steps, tailored for PoT. Step 6—the *wrapped block* created in Step 5 is appended to the blockchain, thus the global view of the blockchain is updated.

Discussion and Security. The main difficulty of providing the service of multi-party verifiable ML model training was assuring that some number of independent nodes would truly train the model without cheating, for example by stating random results, replaying others' messages, or by Sybil attacks.

We argue that introducing *training declaration* message as an additional communication round to a typical consensus mechanism serves as a solution. This communication overhead is justified by the fact that every *training declaration* can be viewed as an independent vote stating "I, pub_key, have trained the model M using batch $\mathcal{B}_{i+1}$ from stage S_i to stage S_{i+1}, I know the *training secret* and I present $\mathcal{H}(\rho_{S_{i+1}} \| \texttt{rand})$ as the proof of that." The more votes S_i has, the more confident the employer can be that the result is correct, hence, the number of votes C is also a parameter of the protocol and the decision about this value belongs to the employer. Correctness of *training declaration* can be verified by every node that has already performed $S_{i+1} \xleftarrow[\rho_{S_{i+1}}]{} f(S_i, \mathcal{B}_{i+1})$, and later, after the block is appended to the blockchain, by anyone, because $\rho_{S_{i+1}}$ is revealed in the *block header*.

Replay and Sybil attacks are prevented by the fact that every *training declaration* is sealed by the PoS mechanism. Additionally, since PoS checks are performed periodically, the process can run in the background, hence a node does not have to stay idle and can continue to train the next stage or move to another model. Note that forcing the network to append a block with an incorrect *training secret*, so as to perform data injection or just lazily participate in the protocol, is equivalent to performing a 51% attack.

After C *training declarations* are broadcasted to the network, any of the nodes that have already trained the model can start the *block building* phase. Note that since *block building* ends in PoA, there is no race between nodes to build the *block header* [4].

Another thing worth mentioning is that after C *training declarations* for stage S_{i+1} are broadcasted, it does not force the nodes to abandon a prepared *training declaration* which is waiting to meet $D^{td}_{PoS}(\texttt{coinstake})$ or quit training altogether. During PoS in the *block building* phase, nodes can actually increase their chances by taking C-length permutations of the total number of *training declarations* that they have heard. Therefore, the network will continue to perform the *block building* phase until Step 6 is performed by one of the nodes. Such design exponentially increases the probability of meeting $D^{bh}_{PoS}(\texttt{coinstake})$ in Step 2 and thus accelerates finalizing it. At this point in the protocol, the list of C *training declarations* can be thought of as PoW's *nonce* but from a very limited space [29]. This unfortunately opens up the possibility of producing more than one set of stakeholders for given S_{i+1} and M which may result in blockchain forks. Whether the advantages of this functionality outweigh its disadvantages will be determined in future work.

It is important to emphasize that a selfish node after receiving C *training declarations* in Step 2 may opt to delay and attempt to secure a more favorable set of stakeholders. This attack is mitigated by the aforementioned "permutation speed-up" that accelerates after every new *training declaration* gets published, making a selfish strategy extremely risky. Furthermore, if a node aims just to be selected as a Kth stakeholder, this does not provide much profit because PoT incentive design (due to space constraints, incentive design details cannot be included) keeps transaction fees at a low level.

Additionally, PoW's minting and verification times asymmetry (difficult to find, fast to verify) is also present in PoT. The *block building* phase is a time-consuming and depends on parameterized functions Δ, D^{td}_{PoS} and D^{bh}_{PoS} and through them is scaled to meet the expected time T. On the other hand, the block verification method consists of checking C *training declarations* against $\rho_{S_{i+1}}$ and checking K signatures against $\mathcal{H}(block\ header)$, thus block verification is a simple process that can be done in constant time.

Absence of the Employer. Model training requests are naturally queued, with priority determined by the reward offered by the employer. However, there may be periods when no active training requests are available. To address this, the network includes a dummy model architecture and dataset, whose sole purpose is to provide a minimal framework for nodes to continue participating in the protocol and continue handling incoming transactions. In this case, the repository is provided by the organization responsible for maintaining the core implementation of PoT, and no rewards are offered for training the dummy model. Nodes' incentive during this period is limited to coinbase rewards and transaction fees. It is important to note that this situation should naturally resolve itself, as the cost of using the PoT network would decrease in the absence of active training requests.

Hiding the Dataset. A critical issue with the PoT protocol is the requirement for employers to make their datasets and architectures publicly available. This imposes a significant barrier, particularly for private enterprises, as it may lead them to refuse to use PoT altogether. Despite this, the core argument of PoT, which asserts the correctness of model training backed by independent votes, necessitates that nodes access the dataset. One potential solution is the utilization of fully homomorphic encryption [7,23], although current high execution times render this approach impractical. Alternatively, employing zero-knowledge systems such as those proposed in [17] could potentially replace the proposed way of computing the training secret as in Eq. 1. Nonetheless, it is worth noting that open-access datasets still constitute a significant portion of the datasets in use, especially among independent researchers and small to medium-sized companies, which are the primary target users of PoT services. Furthermore, it is common practice, even in large-budget projects, to begin training models on public datasets and later fine-tune them on private datasets. In such cases, the public portion of the training can be performed using the PoT.

Implementation. To show the feasibility of PoT we implemented a PoC implementation available at https://github.com/gwechta/proof-of-training.

4   Conclusion

In this paper, we introduced a novel approach for verifiable ML training in cloud-like environments by harnessing the consensus mechanism of a blockchain network. Our protocol leverages the intrinsic features of blockchain networks, thereby transforming the drawback of wasteful puzzle-solving into a PoT consensus mechanism, serving as a genuine alternative to PoW. We emphasize that our protocol can be run with almost any model and dataset, imposing only minimal restrictions. By distributing the verification process across the blockchain infrastructure, our approach ensures the utilization of dedicated nodes and enables training functionality with ample computational resources. Furthermore, we outlined a threat model and formalized malicious data manipulation attacks on ML models taking into account different forms of attacks as well as we have defined formal notion of describing adversary's advantage, for which our protocol serves as a solution.

References

1. Alfeld, S., Zhu, X., Barford, P.: Data poisoning attacks against autoregressive models. In: Proceedings of the AAAI Conference on Artificial Intelligence, vol. 30 (2016)

2. Arun, A., et al.: Verde: verification via refereed delegation for machine learning programs (2025). https://doi.org/10.48550/ARXIV.2502.19405
3. Ball, M., Rosen, A., Sabin, M., Vasudevan, P.N.: Proofs of useful work (2017). https://eprint.iacr.org/2017/203
4. Bentov, I., Lee, C., Mizrahi, A., Rosenfeld, M.: Proof of activity: extending bitcoin's proof of work via proof of stake (2014). https://eprint.iacr.org/2014/452
5. Biggio, B., Nelson, B., Laskov, P.: Poisoning attacks against support vector machines. arXiv preprint: arXiv:1206.6389 (2012)
6. Borrelli, F., Bemporad, A., Morari, M.: Predictive Control for Linear and Hybrid Systems. Cambridge University Press (2017)
7. Brand, M., Pradel, G.: Practical privacy-preserving machine learning using fully homomorphic encryption. Cryptology ePrint Archive, Paper 2023/1320 (2023). https://eprint.iacr.org/2023/1320
8. Chen, Y., Zhu, X.: Optimal attack against autoregressive models by manipulating the environment. In: Proceedings of the AAAI Conference on Artificial Intelligence, vol. 34, pp. 3545–3552 (2020)
9. Cina, A.E., et al.: Wild patterns reloaded: a survey of machine learning security against training data poisoning. ACM Comput. Surv. **55**(13s), 1–39 (2023)
10. Cohen, A., Hasidim, A., Koren, T., Lazic, N., Mansour, Y., Talwar, K.: Online linear quadratic control. In: International Conference on Machine Learning, pp. 1029–1038. PMLR (2018)
11. Dean, S., Mania, H., Matni, N., Recht, B., Tu, S.: On the sample complexity of the linear quadratic regulator. Found. Comput. Math., 1–47 (2019)
12. Dotan, M., Tochner, S.: Proofs of useless work – positive and negative results for wasteless mining systems (2020). https://arxiv.org/abs/2007.01046
13. Dua, D., et al.: UCI machine learning repository (2017)
14. Dunning, I., Huchette, J., Lubin, M.: JuMP: a modeling language for mathematical optimization. SIAM Rev. **59**(2), 295–320 (2017)
15. Fang, C., et al.: Proof-of-learning is currently more broken than you think (2022). https://doi.org/10.48550/ARXIV.2208.03567
16. Fitzi, M., Kiayias, A., Panagiotakos, G., Russell, A.: Ofelimos: combinatorial optimization via proof-of-useful-work a provably secure blockchain protocol (2021). https://eprint.iacr.org/2021/1379
17. Garg, S., Goel, A., Jha, S., Mahloujifar, S., Mahmoody, M., Policharla, G.V., Wang, M.: Experimenting with zero-knowledge proofs of training. Cryptology ePrint Archive, Paper 2023/1345 (2023). https://eprint.iacr.org/2023/1345
18. Goldblum, M., et al.: Dataset security for machine learning: data poisoning, backdoor attacks, and defenses. IEEE Trans. Pattern Anal. Mach. Intell. **45**(2), 1563–1580 (2023). https://doi.org/10.1109/TPAMI.2022.3162397
19. Gschossmann, I., van der Kraaij, A., Benoit, P.L., Rocher, E.: Mining the environment – is climate risk priced into crypto-assets? (2022). https://www.ecb.europa.eu/press/financial-stability-publications/macroprudential-bulletin/html/ecb.mpbu202207_3~d9614ea8e6.en.html
20. Jia, H., et al.: Proof-of-learning: definitions and practice (2021). https://arxiv.org/abs/2103.05633
21. King, S.: Primecoin: cryptocurrency with prime number proof-of-work (2013). https://primecoin.io/bin/primecoin-paper.pdf
22. King, S., Nadal, S.: PPCoin: peer-to-peer crypto-currency with proof-of-stake. Self-Published Paper (2012)
23. Lee, J.W., et al.: Privacy-preserving machine learning with fully homomorphic encryption for deep neural network (2021). https://arxiv.org/abs/2106.07229

24. Li, P.: Proof of training (PoT): harnessing crypto mining power for distributed AI training (2023). https://arxiv.org/abs/2307.07066

25. Lihu, A., Du, J., Barjaktarevic, I., Gerzanics, P., Harvilla, M.: A proof of useful work for artificial intelligence on the blockchain (2020). https://arxiv.org/abs/2001.09244

26. Liu, Y., Lan, Y., Li, B., Miao, C., Tian, Z.: Proof of learning (PoLe): empowering neural network training with consensus building on blockchains. Comput. Netw. **201**, 108594 (2021). https://doi.org/10.1016/j.comnet.2021.108594

27. Luccioni, A.S., Jernite, Y., Strubell, E.: Power hungry processing: watts driving the cost of AI deployment? (2023). https://arxiv.org/abs/2311.16863

28. Mei, S., Zhu, X.: Using machine teaching to identify optimal training-set attacks on machine learners. In: Proceedings of the AAAI Conference on Artificial Intelligence, vol. 29 (2015)

29. Nakamoto, S.: Bitcoin: a peer-to-peer electronic cash system (2008). https://bitcoin.org/bitcoin.pdf

30. Navarro, E., Standing, K.J., Dagher, G.G., Andersen, T.: Ensuring trustworthy neural network training via blockchain. In: 2023 IEEE 5th International Conference on Cognitive Machine Intelligence (CogMI), pp. 31–40 (2023). https://doi.org/10.1109/CogMI58952.2023.00015

31. Principe, J.C., Euliano, N.R., Lefebvre, W.C.: Neural and Adaptive Systems: Fundamentals Through Simulations. Wiley John + Sons, 1 edn. (1999)

32. Sevilla, J., Heim, L., Hobbhahn, M., Besiroglu, T., Ho, A., Villalobos, P.: Estimating training compute of deep learning models (2022). https://epochai.org/blog/estimating-training-compute. Accessed 08 Oct 2024

33. Song, C., Yi, Y., Zhou, T., Yang, J., Liu, L.: Undermining license plate recognition: a data poisoning attack, pp. 72–78 (2023)

34. Tschorsch, F., Scheuermann, B.: Bitcoin and beyond: a technical survey on decentralized digital currencies. IEEE Commun. Surv. Tutorials **18**(3), 2084–2123 (2016). https://doi.org/10.1109/comst.2016.2535718

35. Wang, W., et al.: A survey on consensus mechanisms and mining strategy management in blockchain networks. IEEE Access **7**, 22328–22370 (2019). https://doi.org/10.1109/ACCESS.2019.2896108

36. Álvarez, I.A., Gramlich, V., Sedlmeir, J.: Unsealing the secrets of blockchain consensus: A systematic comparison of the formal security of proof-of-work and proof-of-stake (2024). https://arxiv.org/abs/2401.14527

Multiple-Meta-Instance Selection. Combining the Properties of Many Instance-Selection Methods

Marcin Blachnik[✉][iD], Piotr Ciepliński[iD], and Daniel Dabrowski[iD]

Silesian University of Technology, Department of Industrial Informatics, 40-019 ul. Krasińskiego 8, Katowice, Poland
{marcin.blachnik,piotr.cieplinski,daniel.dabrowski}@polsl.pl

Abstract. This study presents two novel approaches for developing a multiple-meta-instance selection method, an advanced algorithm designed for efficient pruning of training sample in classification problems. The proposed meta-instance selection framework reformulates the traditional instance selection problem by introducing a meta-feature space, a problem-agnostic representation space. The transformation enables instance selection to be framed as a classification task in the meta-feature space, facilitating efficient computation with a time complexity of $O(nlog(n))$. A standard classification algorithm, such as Random Forest, can then be employed in the meta-feature space to determine the inclusion or exclusion of individual samples.

To enhance performance, we explore two strategies for combining multiple meta-instance selection algorithms: (1) constructing an ensemble of meta-classifiers and (2) concatenating many meta-sets. Experimental evaluations demonstrate that the meta-set concatenation approach surpasses both classical instance selection techniques and existing meta-instance selection methods. Moreover, the proposed algorithm significantly accelerates the instance selection process—achieving even by two or three orders of magnitude speed-up, depending on dataset size and the reference instance selection method.

Keywords: Instance selection · classification · knowledge distillation · data pruning

1 Introduction

As datasets used in machine learning applications continue to expand in size and complexity, managing, storing, and processing them efficiently is getting more challenging [19]. One of the approaches to tackle these problems is dataset pruning [13]. This is a group of methods used to reduce the size of a dataset while preserving its essential characteristics, making it a valuable tool for improving data management, model training, and overall system performance. More precisely it can be viewed as a process of selecting or constructing a subset of the most informative and representative data points from a larger dataset, with the goal of retaining the majority of the information and patterns present in the original data [15]. There are two main approaches including model-independent pruning and model-dependent pruning. The first group selects samples based on some external data characteristics [5], while the second one is also called dataset distillation [18] where the final model is used to select or construct new training samples. In both cases, the resulting dataset is typically much smaller. The reduction process involves identifying the most critical data points that capture the underlying structure and relationships within the data and removing redundant or noisy data points that do not contribute significantly to the task the model is built for.

One of the most commonly used techniques for dataset pruning is instance selection. Comprehensive reviews of instance selection methods can be found in [8] and [10]. This family of techniques was initially develped as a tool to improve the k-nearest neighbor (kNN) classifiers, but later it was effectively adopted also to other classification methods [3, 7].

However, instance selection suffers from significant computational complexity, as most algorithms iteratively evaluate the nearest-neighbor graph to prune redundant or noisy samples. To address this problem, a method called Meta Instance Selection (MetaIS) was proposed in [2]. This method tackles scalability issues by reframing the instance selection problem as a classification problem in a meta-feature space. In this approach, each sample is mapped into a new meta-feature space that describes the local properties of the nearest-neighbor graph, and a meta-classifier is used to determine whether a given sample should be retained or pruned. The meta-classifier is initially trained to emulate the behaviour of a specific instance selection algorithm. This is achieved by using the output of the classical instance selection method to label samples in the meta-feature space as either "to be kept" or "to be removed" and training a classifier (meta-classifier). Consequently, a single meta-classifier corresponds to a single instance selection algorithm.

In this work, we propose extending the capabilities of meta-instance selection by combining the properties of multiple instance selection methods into a single method which is called multiple-meta-instance selection (MMIS). This approach allows for further improvement of the properties of classical MetaIS solutions and achieving even higher performance without affecting execution time.

In the article, we discuss and empirically compare two methods of constructing MMIS. The first method is based of combining multiple meta-classifiers of MetaIS into a classifiers committee that combines the properties of individual instance selection methods into one system. The second approach is based on concatenating the meta-datasets obtained from the base instance selection algorithms into one large dataset, and consequently training a single meta-classifier on the combined meta-dataset. This assures gaining the knowledge from each base instance selection method.

The process of model evaluation allows for selecting the optimal combination of base-instance selection methods that complement each other. This allows for achieving superior performance compared to individual methods. Importantly, the gain in performance is achieved with minimal additional computational complexity.

The structure of this paper is as follows: first, we present an overview of meta-instance selection methods (Sect. 2). Next, in Sect. 3 the concept of multi meta instance selection is introduce. Section 4 describes setup of the experiments and results are presented in Sect. 5. Finally, the concluding section summarizes the findings and outlines potential directions for future research.

2   Meta Instance Selection

Meta-instance selection reformulates the problem of selection or rejection of a sample as a binary classification problem, where positive samples are labeled as "to keep" and negative samples are labeled as "to remove".

To make this process generic and applicable to any dataset, a common feature space is required in which the meta-classifier (a classifier responsible for assessing instance importance) operates—the so-called meta-feature space. In [2] it is suggested to use balanced random forest as a meta-classifier, and the meta-feature space is defined using local properties of the nearest neighbor graph (NNG), where each sample is characterized by statistics derived from the NNG. More details regarding the meta-feature space are provided in Subsect. 2.1. The basic concept is shown in Fig. 1.

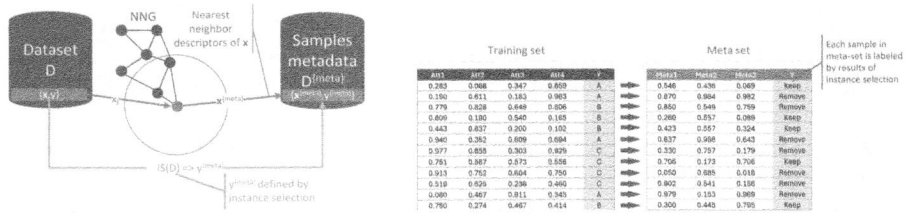

Fig. 1. The concept of MetaIS algorithm. The left figure shows the transformation process, where for the input dataset an NNG is constructed and properties of each vertex constitute the meta-feature space. Labeling of samples in the meta-set (marked in yellow) is applied only during the preparation of the meta-training set. The right figure shows the transformation results from regular to the meta-features space (Color figure online)

A schematic overview of the entire system is shown in Fig. 2, which is divided into two components: the training phase and the prediction/selection phase. The training phase illustrates the process of constructing the meta-classifier, while the prediction/selection phase represents the process of instance selection using the meta-classifier.

The training phase begins by applying a specific instance selection method, referred to as the reference method. This reference instance selection method is

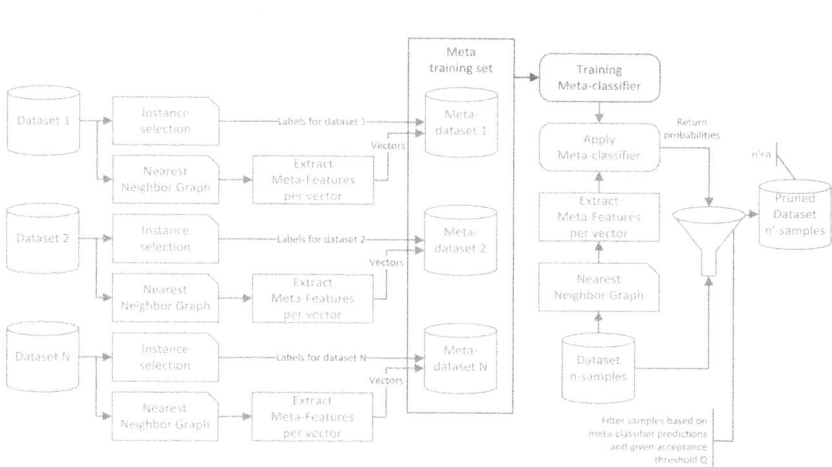

Fig. 2. The scheme of the MetaIS algorithm including preparation of the meta-set (green color), training meta-classifier (blue color), and application of MetaIS for pruning new dataset (orange color) (Color figure online)

executed on multiple datasets (the larger the number of datasets, the better the performance), and each sample in these datasets is labeled as either "to keep" or "to remove" based on the results of the reference method. Subsequently, for each sample in each dataset, meta-feature extraction is performed. The resulting pairs, represented as $< \mathbf{x}_{meta}, \{y_{Positive}(\text{to keep}), y_{Negative}(\text{to remove})\} >$, form the training set for the meta-classifier.

Following meta-feature extraction, the features for each sample in each dataset are normalized. This normalization step is crucial because the distances within individual datasets may vary significantly. After normalization, all meta-datasets are concatenated to create a unified training set for a binary classifier.

During the prediction phase, for a new dataset, meta-features are first extracted. Specifically, for each training sample ($\mathbf{x}, y$ in the dataset, its meta-features $\mathbf{x}_{meta}$ are computed. The meta-classifier is then applied to the samples represented using meta-features. Instead of producing a classical binary decision, the model typically outputs a probability score, which represents the importance of each instance rather than a definitive decision. Finally, based on these importance scores, the samples are ranked, and a selection is made according to a user-defined threshold Θ, where 0 indicates no instance selection and 1 indicates prunning the entire dataset, or based on preferences regarding the desired output size of the dataset.

2.1 Meta-Feature Space

As previously indicated, the meta-feature space is determined based on properties derived from the nearest neighbor graph (NNG). These meta-features are extracted from various attributes that were initially used in the reference instance selection methods. The details are provided in [2], and here, only the names and descriptions of the extracted features are presented:

- Average distance to k nearest neighbors from the same class: Among k nearest neighbors, find samples with the same class label as the query sample and get their average.
- Average distance to k nearest neighbors from the opposite class: Similar to the above, but calculate the average distance to the k nearest neighbors that belong to the opposite class of the query sample.
- Average distance to k nearest neighbors from any class: Compute the average distance to all k nearest neighbors, irrespective of class label.
- Minimum distance to samples from the same class: Determine the distance to the nearest neighbor that belongs to the same class as the query sample.
- Minimum distance to samples from the opposite class: Calculate the distance to the nearest neighbor that belongs to the opposite class of the query sample.
- Minimum distance to samples from any class: Determine the distance to the overall nearest neighbor, irrespective of class label.
- Number of samples from the same class among k nearest neighbors: Perform a vote among the k nearest neighbors to count how many belong to the same class as the query sample.
- Number of samples from the opposite class among k nearest neighbors: Similar to the above, but count how many of the k nearest neighbors belong to the opposite class.

When constructing the meta-feature space, these meta-features are computed for multiple values of k, specifically $k = \{3, 5, 9, 15, 23, 33\}$ as indicated in [2]. As demonstrated, the use of multiple values of k, combined with different types of meta-features, is critical for accurately characterizing the neighborhood of the query sample. This approach enables a more precise determination of the sample's importance.

3 Combining Multiple Instance Selection Methods

The Meta Instance Selection (MetaIS) method offers several advantages over traditional instance selection techniques, particularly in its ability to efficiently combine individual instance selection methods. While the concept of instance selection ensembles was initially introduced in [1] and later refined in [9], the MetaIS approach significantly enhances this process by enabling more efficient combinations of instance selection methods. Instead of running multiple computationally expensive instance selection algorithms, MetaIS leverages two efficient strategies: (1) an ensemble of meta-classifiers and (2) a meta-classifier trained on concatenated meta-datasets representing reference instance selection methods. A detailed explanation of these approaches is provided below.

Ensemble of Meta-Classifiers. This approach is based on the idea that each meta-classifier is trained to mimic a specific reference instance selection method for example HMN-EI, CCIS, etc. Since these reference methods differ in the subsets they select, the meta-datasets used for training meta-classifiers are distinct in terms of their labeling. Consequently, meta-classifiers trained on different reference methods exhibit variations in outputs (predictions) assuring diversity of the ensemble members. The challenge lies in determining which meta-classifiers should be combined to ensure they complement one another effectively and enhance overall performance.

Concatenation of Meta-Datasets. An alternative strategy for improving meta-classifier performance involves merging meta-datasets labeled according to specific reference instance selection methods into a single dataset. A single meta-classifier is then trained on this combined dataset. This approach results in a significantly larger training meta-dataset since it integrates subsets from multiple reference methods. However, the computational cost of training the meta-classifier is incurred only once, making this method practical and efficient in terms of training overhead.

Both approaches provide flexible and scalable solutions for combining multiple instance selection methods while maintaining computational efficiency, thereby addressing the key limitations of traditional instance selection techniques. Since the method combines multiple MetaIS methods into one system it is abbreviated as Multiple-Meta-Instance-Selection (MMIS).

4   Setup of the Experiments

The two proposed MMIS methods were evaluated empirically and validated on multiple datasets of varing size and domain. The details of the experiments are provided below.

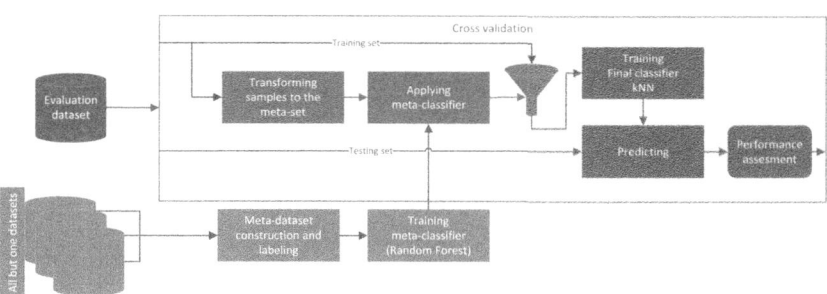

Fig. 3. The procedure used for performance assessment of the MetaIS and MMIS method. MetaIS/MMIS training is marked in green, brown is application of MMIS/MetaIS, and blue is standard cross-validation process (Color figure online)

4.1   Evaluation Procedure

The experiments were divided into two parts. In the first part, the leave-one-dataset-out methodology was employed to evaluate the performance of meta-instance selection. In the second part, the meta-classifier was tested on a single independent, large dataset containing around 500,000 samples. The leave-one-dataset-out method involves taking a collection of datasets (in this case, datasets with fewer than 100,000 samples), removing one dataset for testing, and using all remaining datasets to train the meta-classifier. Each dataset held out for testing was then pruned and the performance of the final classifier was assessed using 5-fold cross-validation. During the cross-validation procedure, the training data was first filtered using the instance selection method and subsequently evaluated using a 1NN classifier as shown in Fig. 3. The 1NN classifier is a standard practice in instance selection studies, and as meta-classifier the balanced random forest was used.

Table 1. Characteristics of the datasets used in the experiments

	Dataset	# samples	# attr.	# classes
1	abalone	4177	8	28
2	banana	5300	2	2
3	electricity	45312	8	2
4	letter	20000	16	26
5	magic	19020	10	2
6	nursery	12960	8	5
7	optdigits	5620	64	10
8	page-blocks	5472	10	5
9	penbased	10992	16	10
10	phoneme	5404	5	2
11	ring	7400	20	2
12	satimage	6435	36	6
13	shuttle	57999	9	7
14	spambase	4597	57	2
15	texture	5500	40	11
16	twonorm	7400	20	2
17	php89ntbG	488565	8	2

Since meta-instance selection outputs the probability of retaining a sample a curve representing the relation between reduction-rate and classification accuracy can be constructed using different Θ values. In particular, we used $\Theta = \{0.1, 0.2, 0.3, 0.4, 0.5, 0.6, 0.7, 0.8, 0.9\}$. Therefore, for measuring the overall performance the Area Under the Accuracy-Reduction Rate Curve ($AUARRC$)

was used, as described in [1]. This metric is calculated by determining the area defined by the polygon formed between the performance of the 1NN classifier (at a zero reduction rate) and the results achieved by the instance selection method at a reduction rate greater than zero ($\Theta > 0$).

4.2 Datasets Used in the Experiments

The experiments were conducted on 17 datasets of various sizes (Table 1). These datasets were obtained from the Keel Project repository [14], where they were preprocessed and provided in a format suitable for 5-fold cross-validation. While the repository contains additional, smaller datasets, only datasets with at least 4,000 samples were included in the study. This decision was made because removing redundant samples is not particularly useful for small datasets. In the experiments one larger datasets the *php89ntbG* dataset was obtained. It was obtained from the OpenML project [6].

4.3 Evaluation Parameters

In Sect. 3, it was noted that the effectiveness of combined meta-instance selection (MMIS) depends on the choice of reference instance selection methods. Five methods were used to label meta-set samples: *Edited Nearest Neighbor* ENN [16], Drop3 [17], *Interactive Case Filtering* ICF [4], *Hit Miss Network Editing* HMN-EI [11], and *Class Conditional Instance Selection* CCIS [12]. Each method has distinct behavior, so selected methods must complement each other to avoid performance degradation. Various combinations were tested (see Table 4, bottom rows) to find the optimal set. The five evaluated combinations were: CCIS, ICF, CCIS, ICF, Drop3, CCIS, ICF, Drop3, HMN-EI, CCIS, ICF, HMN-EI, and CCIS, ICF, HMN-EI, ENN.

The primary criterion for selecting the optimal combination was the performance of the individual methods, where CCIS and ICF consistently achieved the highest accuracy and compression rates. These two methods served as the foundation, with additional methods (HMN-EI, Drop3, ENN) incorporated sequentially in order of their individual performance.

The results for the individual models were obtained for $\Theta = \{0.1, 0.2, 0.3, 0.4, 0.5, 0.6, 0.7, 0.8, 0.9\}$, with performance assessment conducted using a 1NN classifier. For all reference instance selection methods (ENN, Drop3, CCIS, ICF and HMN-EI), the number of nearest neighbors was set to 3, as wherever parameter setup was required (3 is the default value suggested by the authors of particular methods). The experiments were carried out using our MetaIS library, implemented in Python and available at https://github.com/mblachnik/MetaIS. The reference instance selection methods were implemented in the RapidMiner Information Selection Extension https://github.com/mblachnik/infoSel, along with implementations from the Keel Project for selected algorithms (HMN-EI, CCIS).

The experiments consisted of two parts. In the first part, various combinations of reference methods were compared to identify the most effective combination. In the second part, the best-performing combination was compared against the individual reference instance selection methods.

5   Results

The first set of experiments was devoted to comparing the two approaches of creating MMIS, namely the meta-classifiers ensemble and concatenated meta-datasets. Within this comparison also the members of the ensemble including various IS models discussed in 4.3 were compared. The obtained results are presented in Table 2. The values represent the average AUARRC obtained using the given MMIS method for pruning training samples. For each dataset, the results are grouped according to the combination of ensemble members and according to the type of ensemble.

The last three rows of Table 2 summarize the obtained results. These are the mean AUARRC obtained by aggregating performances for each dataset, the average difference between the obtained results, and the p-value obtained using the Wilcoxon signed rank test obtained when comparing the two MMIS approaches. The average difference was calculated analogously to the Wilcoxon test, that is by $mean(left - right)$, where $left$ are the results of *models ensemble* and $right$ represent the column with *datasets ensemble*. A negative value of this indicator suggests that the *datasets ensemble* outperforms *models ensemble*, and a positive value indicates the opposite, that the *models ensemble* outperforms *datasets ensemble*.

The obtained results indicate that in all cases the mean difference is negative indicating that the *combined meta-datasets* performs better, and in all cases the difference is statistically significant, assuming $\alpha = 0.1$. When comparing the solutions among various members of the ensemble for the *combined meta-datasets* approach the best results are obtained by (CCIS, ICF, Drop3, HMN-EI), therefore this solution was used as a reference when comparing with the other methods.

Based on the results presented in Table 2, we conducted additional statistical analyses comparing the best-performing model (CCIS, ICF, Drop3, HMN-EI) with the remaining MMIS models from the *concatenated meta-dataset* families. These statistics are provided in Table 3. According to the results, the second-best performance is achieved by (CCIS, ICF, HMN-EI), followed by (CCIS, ICF, HMN-EI, ENN) in third place. While the differences among these three methods are not statistically significant, the positive values of the *mean differences* suggest that the (CCIS, ICF, Drop3, HMN-EI) based solution is the leading combination of reference instance selection methods. For the remaining approaches (CCIS, ICF) and (CCIS, ICF, Drop3), the differences in performance are statistically significant and worse then the best method.

A more detailed comparison of the differences is presented in Fig. 4, which illustrates the *accuracy-reduction rate* plots for four selected datasets. The results

Table 2. Results representing AUARRC performance measure obtained for two type of MMIS approaches - the meta-classifier based ensemble and concatenated meta-dataset - for various combination of base members. The last three rows summarize the obtained results. Statistically significant results are marked in bold ($\alpha = 0.05$)

	(ICF,CCIS)				(ICF,CCIS,Drop3)				(HMEI,ICF,CCIS)				(HMEI,ICF,CCIS,Drop3)				(HMEI,ENN,ICF,CCIS)			
	models ensemble		concatenated meta-datasets		models ensemble		concatenated meta-datasets		models ensemble		concatenated meta-datasets		models ensemble		concatenated meta-datasets		models ensemble		concatenated meta-datasets	
	mean	std	mean	std	mean	std	mean	std	mean	std	mean	std	mean	std	mean	std	mean	std	mean	std
php9gutbG	0.9282	0.0013	**0.9349**	0.0006	0.9061	0.0014	**0.9235**	0.0006	0.9440	0.0009	**0.9493**	0.0003	0.9205	0.0015	**0.9478**	0.0003	0.9426	0.0008	**0.9478**	0.0005
abalone	**0.2093**	0.0125	0.1897	0.0122	**0.2123**	0.0130	0.1938	0.0119	0.2228	0.0106	0.2183	0.0123	0.2205	0.0122	0.2153	0.0108	0.2222	0.0111	0.2152	0.0113
banana	0.8416	0.0111	0.8397	0.0086	0.8284	0.0104	0.8409	0.0124	0.8674	0.0078	**0.8796**	0.0042	0.8561	0.0069	**0.8745**	0.0073	0.8745	0.0052	**0.8836**	0.0050
letter	0.8015	0.0035	**0.8678**	0.0032	0.7686	0.0035	**0.8585**	0.0046	0.8178	0.0031	**0.8678**	0.0043	0.7887	0.0025	**0.8697**	0.0031	0.8108	0.0027	**0.8617**	0.0046
magic	0.7299	0.0044	**0.7716**	0.0070	0.6900	0.0038	**0.7585**	0.0064	0.8044	0.0086	0.8098	0.0033	0.7462	0.0032	**0.8054**	0.0043	0.8119	0.0067	0.8132	0.0041
nursery	0.8764	0.0106	0.8805	0.0093	0.8504	0.0173	**0.8767**	0.0124	0.8643	0.0093	0.8826	0.0123	0.8655	0.0100	**0.8862**	0.0145	0.8519	0.0107	**0.8859**	0.0140
optdigits	0.9241	0.0059	**0.9371**	0.0066	0.8748	0.0048	**0.9267**	0.0053	0.9243	0.0093	**0.9371**	0.0044	0.9087	0.0036	**0.9400**	0.0035	0.9135	0.0083	**0.9473**	0.0032
page-blocks	0.9072	0.0144	0.9092	0.0088	0.8649	0.0115	**0.9040**	0.0174	0.9480	0.0068	0.9494	0.0050	0.9174	0.0112	**0.9545**	0.0033	0.9440	0.0047	**0.9536**	0.0061
penbased	0.9773	0.0009	0.9774	0.0019	0.9584	0.0019	**0.9765**	0.0014	0.9700	0.0022	**0.9767**	0.0030	0.9682	0.0013	**0.9763**	0.0024	0.9605	0.0017	**0.9760**	0.0018
phoneme	0.8174	0.0076	**0.8288**	0.0090	0.7850	0.0055	**0.8203**	0.0088	0.8370	0.0099	0.8391	0.0089	0.8097	0.0072	**0.8361**	0.0118	0.8389	0.0051	0.8373	0.0094
ring	0.6840	0.0075	**0.6921**	0.0100	0.6738	0.0057	**0.7298**	0.0183	**0.7136**	0.0179	0.6348	0.0057	**0.7038**	0.0085	0.6780	0.0086	**0.6877**	0.0091	0.6142	0.0038
satimage	0.7944	0.0122	**0.8241**	0.0029	0.7406	0.0062	**0.8023**	0.0076	0.8650	0.0039	**0.8806**	0.0066	0.7890	0.0088	**0.8745**	0.0050	0.8622	0.0025	**0.8829**	0.0038
spambase	0.7158	0.0043	**0.7507**	0.0096	0.7198	0.0092	**0.7438**	0.0048	0.7907	0.0045	**0.8427**	0.0141	0.7338	0.0067	**0.8445**	0.0120	0.8105	0.0119	**0.8547**	0.0102
texture	0.9217	0.0075	**0.9396**	0.0070	0.8700	0.0073	**0.9321**	0.0052	0.9389	0.0036	**0.9471**	0.0047	0.9141	0.0058	**0.9489**	0.0074	0.9306	0.0032	**0.9485**	0.0065
twonorm	0.9357	0.0052	0.9379	0.0050	0.9081	0.0042	**0.9281**	0.0056	0.9511	0.0052	0.9502	0.0019	0.9314	0.0055	**0.9456**	0.0038	0.9533	0.0047	0.9508	0.0039
shuttle	**0.9964**	0.0007	0.9771	0.0081	**0.9924**	0.0018	0.9743	0.0091	0.9962	0.0009	0.9955	0.0002	0.9943	0.0017	0.9951	0.0006	**0.9952**	0.0007	0.9939	0.0009
electricity-normalized	0.6847	0.0044	**0.7623**	0.0026	0.6761	0.0039	**0.7610**	0.0025	0.7873	0.0031	**0.7946**	0.0024	0.7361	0.0040	**0.7963**	0.0009	**0.8013**	0.0032	0.7868	0.0031
Mean	0.8086		0.8247		0.7835		0.8206		0.8377		0.8444		0.8120		0.8464		0.8360		0.8443	
Mean diff.	−0.0162				−0.0371				−0.0067				−0.0344				−0.0084			
p-value	0.0129				0.0004				0.0129				0.0005				0.0887			

Table 3. Summary statistics - *mean difference* and *p-value* representing the comparison of the leading solution (see Table 2) DE(CCIS,ICF,Drop3,HMN-EI) method with the remaining solutions obtained using *concatenated datasets* approach. The statistics were obtained using data available in Table 2

	(CCIS, ICF,Drop3, HMN-EI)	(CCIS,ICF,HMN-EI)	(CCIS, ICF,HMN-EI,ENN)	(CCIS,ICF)	(CCIS,ICF,Drop3)
Mean diff.	ref.	0.0020	0.0021	0.0217	0.0258
p-value	ref.	0.8536	0.7563	0.0007	0.0026

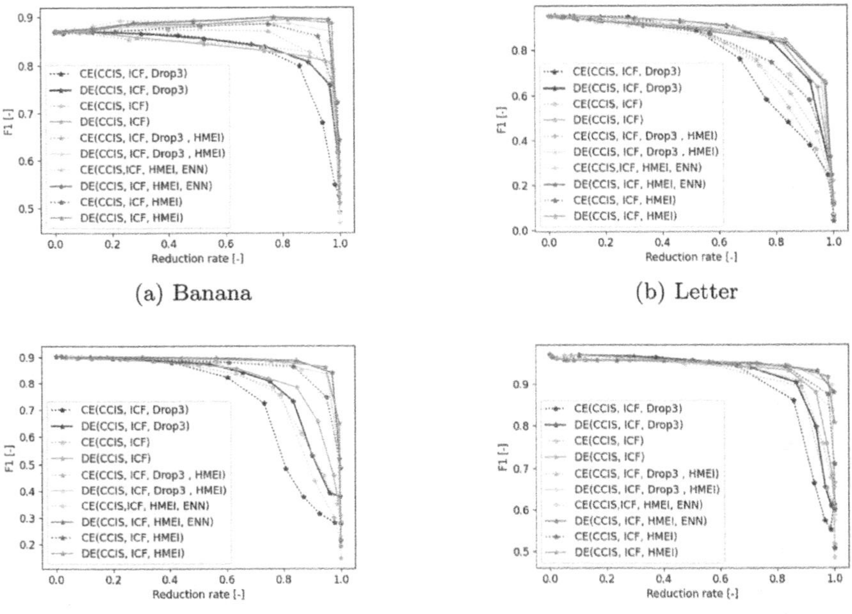

(a) Banana   (b) Letter

(c) Satimage   (d) Php89ntbG

Fig. 4. Comparison of the prediction performance vs reduction rate obtained for all evaluated MMIS methods on selected datasets. The CE( ) indicates classifiers ensemble and DE( ) indicates concatenated meta-datasets

indicate that, in all cases, the models highlighted in solid red, solid purple, and solid light green outperform the competitors. These models correspond to the *concatenated meta-dataset* approach based on concatenation of (CCIS, ICF, HMN-EI, ENN), (CCIS, ICF, HMN-EI), and (CCIS, ICF, Drop3, HMN-EI), respectively. Furthermore, in each case, the area under the *F1-reduction rate* curve is the largest, confirming their superior performance. Conversely, the approach based on (CCIS, ICF, Drop3) members yields the weakest results.

Next, we compared the best-performing MMIS model with the standard MetaIS models. The results, presented in Table 4, demonstrate that in all cases, the MMIS model is statistically significantly superior to each individual MetaIS model. The mean performance differences further highlight the advantages of the MMIS-based approach. A more detailed visualization is provided in Fig. 5, where *performance-reduction rate* plots are shown for four datasets, including the best MMIS method and the standard MetaIS models. The dominance of the MMIS approach is evident, as indicated by the dark blue curve, which consistently outperforms all other methods. Additionally, the MMIS method frequently surpasses the baseline reference instance selection methods, marked with an ×, which were originally used for labeling the meta-datasets in the training process.

Table 4. Results represent the standard MetaIS models and the best MMIS model. The last three rows summarize the results indicating the mean performance over all datasets, the mean differences in performance considering the MMIS model denoted as (CCIS, ICF, Drop3, HMN-EI) as a reference for the comparison and the p-value of the Wilcoxon signed-rank test

Dataset	DE(CCIS,ICF, Drop3,HMN-EI)	MetaHMN-EI	MetaENN	MetaCCIS	MetaICF	MetaDrop3
	mean ± std	mean ± std	mean ± std	mean ± std	mean ± std	mean ± std
banana	0.8745 ± 0.0073	0.6350 ± 0.0134	0.2839 ± 0.0055	0.8263 ± 0.0091	0.8610 ± 0.0076	0.8385 ± 0.0107
electricity-norm.	0.7963 ± 0.0009	0.7002 ± 0.0032	0.4727 ± 0.0030	0.6861 ± 0.0024	0.7401 ± 0.0034	0.7174 ± 0.0019
letter	0.8697 ± 0.0031	0.5845 ± 0.0033	0.3574 ± 0.0014	0.8611 ± 0.0025	0.7773 ± 0.0031	0.7381 ± 0.0062
magic	0.8054 ± 0.0043	0.5963 ± 0.0051	0.3526 ± 0.0034	0.7553 ± 0.0062	0.7009 ± 0.0036	0.6818 ± 0.0006
nursery	0.8862 ± 0.0145	0.7858 ± 0.0212	0.4021 ± 0.0305	0.8707 ± 0.0121	0.8817 ± 0.0125	0.8130 ± 0.0103
optdigits	0.9400 ± 0.0035	0.4241 ± 0.0053	0.1528 ± 0.0038	0.9427 ± 0.0057	0.9113 ± 0.0036	0.8549 ± 0.0049
page-blocks	0.9545 ± 0.0033	0.3769 ± 0.0047	0.1135 ± 0.0025	0.9369 ± 0.0093	0.9063 ± 0.0127	0.8563 ± 0.0189
penbased	0.9763 ± 0.0024	0.3319 ± 0.0070	0.0505 ± 0.0012	0.9687 ± 0.0011	0.9646 ± 0.0025	0.9591 ± 0.0038
phoneme	0.8361 ± 0.0118	0.6084 ± 0.0114	0.3822 ± 0.0014	0.8242 ± 0.0071	0.8033 ± 0.0084	0.7651 ± 0.0036
ring	0.6780 ± 0.0086	0.7857 ± 0.0095	0.3258 ± 0.0018	0.7492 ± 0.0084	0.6563 ± 0.0100	0.7133 ± 0.0035
satimage	0.8745 ± 0.0050	0.5032 ± 0.0041	0.2694 ± 0.0011	0.8491 ± 0.0041	0.7716 ± 0.0082	0.7386 ± 0.0041
shuttle	0.9951 ± 0.0006	0.5646 ± 0.0075	0.0060 ± 0.0003	0.9965 ± 0.0005	0.9922 ± 0.0012	0.9969 ± 0.0005
spambase	0.8445 ± 0.0120	0.6508 ± 0.0113	0.3416 ± 0.0112	0.7509 ± 0.0085	0.7479 ± 0.0087	0.7781 ± 0.0070
texture	0.9489 ± 0.0074	0.5360 ± 0.0072	0.1571 ± 0.0032	0.9539 ± 0.0043	0.8959 ± 0.0081	0.8529 ± 0.0125
twonorm	0.9456 ± 0.0038	0.5655 ± 0.0068	0.1954 ± 0.0044	0.9415 ± 0.0041	0.9208 ± 0.0039	0.9008 ± 0.0026
php89ntbG	0.9478 ± 0.0003	0.4754 ± 0.0032	0.1660 ± 0.0009	0.9456 ± 0.0005	0.9278 ± 0.0012	
Mean	**0.8858**	**0.5703**	**0.2518**	**0.8662**	**0.8412**	**0.8137**
Mean diff.		**0.3156**	**0.6340**	**0.0197**	**0.0447**	**0.0681**
p-value		**0.0002**	**0.0000**	**0.0182**	**0.0000**	**0.0004**

5.1   Execution Time Comparison

The final experiment demonstrates the efficiency of the proposed method in accelerating instance selection (Fig. 6). Speedup was calculated as the ratio of the execution time of base instance selection methods to that of MMIS (solid line) and MetaIS (dotted line). Results show that speedup scales with the logarithm of sample size or better, with values below 1 only for small datasets (4,0005,000 samples). The highest speedup—165×—was observed for Drop3.

This gain results from reduced time complexity, as MMIS requires only a single pass over the data. It includes meta-feature extraction in $O(n \log n)$ time

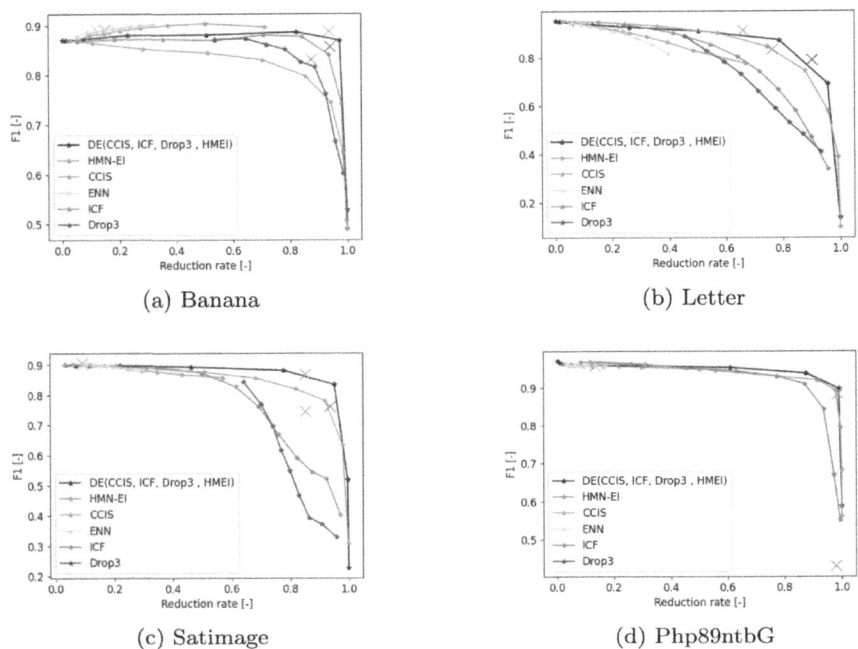

(a) Banana

(b) Letter

(c) Satimage

(d) Php89ntbG

Fig. 5. Comparison of the prediction performance vs reduction rate obtained for classical MetaIS methods and the best performing (CCIS, ICF, Drop3, HMN-EI) method based on concatenated meta-datasets approach. The DE( ) indicates concatenated meta-datasets

and classification using a balanced random forest in $O(n \log n^*)$. In the concatenated meta-dataset approach, execution time depends only on classifier prediction time $O(\log n^*)$, where n^* is tree depth. For large datasets, MMIS and MetaIS have similar runtimes due to shared meta-feature extraction and single classifier usage. Differences appear mainly for Drop3 and small datasets, where MetaIS built shallower trees, resulting in higher speedup for MMIS.

A difference between MetaIS and MMIS appears during the meta-classifier training. Here, for the *concatenated meta-datasets* approach used in MMIS the training dataset becomes larger since it is obtained by concatenation of multiple MetaIS datasets. But since the balanced random forest is used as a meta-classifier, the training time is acceptable because it scales by $O(k \cdot (n \log n))$ where k is the number of concatenated datasets. That is k- times slower than the MetaIS, but this process is conducted only once, and the meta-classifier can be used for any dataset. In the experiments, a single training took a couple of minutes.

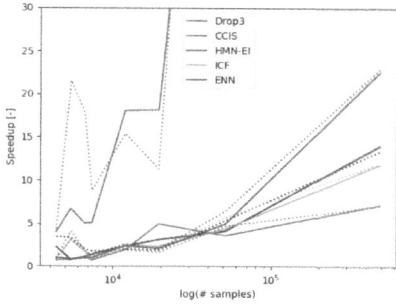

Fig. 6. Speed-up of the best proposed MMIS methods over the reference instance selection methods - marked with a solid line, and speed-up of MetaIS over the same reference instance selection, marked with a dotted line

6   Conclusions

This study introduced a novel approach to meta-instance selection, integrating multiple MetaIS methods into a unified framework called Multiple Meta Instance Selection (MMIS). Two strategies were explored: one leveraging an ensemble of meta-classifiers and another combining multiple individual methods by concatenating meta-training sets into a single dataset used to train meta-classifier. Experimental results demonstrate that the concatenated meta-dataset approach outperforms other approaches, both in terms of accuracy and reduction rate.

Furthermore, in the study we identified the most effective base-members. Among the tested methods, three demonstrated superior performance namely (CCIS, ICF, Drop3, HMN-EI), (CCIS, ICF, HMN-EI, ENN), and (CCIS, ICF, HMN-EI), with the (CCIS, ICF, Drop3, HMN-EI) yielding the best overall results.

In addition to improved selection accuracy, the proposed method significantly accelerates the instance selection process compared to baseline reference methods, offering substantial computational efficiency gains. These findings highlight the potential of meta-instance selection ensembles as a promising direction for optimizing data reduction in classification tasks. The proposed approach can be easily adapted also to other tasks such as support vector selection for the kernel-based methods, where instead of instance selection-based labeling support vectors can be used for labeling the meta-set.

Acknowledgments. The work was supported by the Excellence Initiative Research University at the Silesian University of Technology, project number 11/040/SDW/10-21-01 and the research project BK-227/RM4/2025 also funded by the Silesian University of Technology

Disclosure of Interests. The authors declare no relevant conflicts of interest.

References

1. Blachnik, M.: Ensembles of instance selection methods. a comparative study. Int. J. Appl. Math. Comput. Sci. **29**(1) (2019)
2. Blachnik, M., Ciepliński, P.: Meta-instance selection. instance selection as a classification problem with meta-features. arXiv preprint arXiv:2501.11526 (2025)
3. Blachnik, M., Kordos, M.: Comparison of instance selection and construction methods with various classifiers. Appl. Sci. **10**(11), 3933 (2020)
4. Brighton, H., Mellish, C.: Advances in instance selection for instance-based learning algorithms. Data Min. Knowl. Disc. **6**(2), 153–172 (2002)
5. Cunha, W., Viegas, F., França, C., Rosa, T., Rocha, L., Gonçalves, M.A.: A comparative survey of instance selection methods applied to non-neural and transformer-based text classification. ACM Comput. Surv. **55**(13s) (2023)
6. Feurer, M., et al.: Openml-python: an extensible python API for OpenML. J. Mach. Learn. Res. **22**(100), 1–5 (2021)
7. García, S., Luengo, J., Herrera, F.: Tutorial on practical tips of the most influential data preprocessing algorithms in data mining. Knowl.-Based Syst. **98**, 1–29 (2016)
8. García, S., Derrac, J., Cano, J.R., Herrera, F.: Prototype selection for nearest neighbor classification: taxonomy and empirical study. IEEE Trans. Pattern Anal. Mach. Intell. **34**(3), 417–435 (2012)
9. de Haro-García, A., Cerruela-García, G., García-Pedrajas, N.: Instance selection based on boosting for instance-based learners. Pattern Recogn. **96**, 106959 (2019)
10. Malhat, M., El Menshawy, M., Mousa, H., El Sisi, A.: A new approach for instance selection: algorithms, evaluation, and comparisons. Expert Syst. Appl. **149**, 113297 (2020)
11. Marchiori, E.: Hit miss networks with applications to instance selection. J. Mach. Learn. Res. **9**(Jun), 997–1017 (2008)
12. Marchiori, E.: Class conditional nearest neighbor for large margin instance selection. IEEE Trans. Pattern Anal. Mach. Intell. **32**(2), 364–370 (2010)
13. Sorscher, B., Geirhos, R., Shekhar, S., Ganguli, S., Morcos, A.: Beyond neural scaling laws: beating power law scaling via data pruning. Adv. Neural. Inf. Process. Syst. **35**, 19523–19536 (2022)
14. Triguero, I., et al.: Keel 3.0: an open source software for multi-stage analysis in data mining. Int. J. Comput. Intell. Syst. **10**(1), 1238–1249 (2017)
15. Wang, T., Zhu, J.Y., Torralba, A., Efros, A.A.: Dataset distillation. arXiv preprint arXiv:1811.10959 (2018)
16. Wilson, D.: Assymptotic properties of nearest neighbour rules using edited data. IEEE Trans Syst. Man, Cybern. **SMC-2**, 408–421 (1972)
17. Wilson, D., Martinez, T.: Reduction techniques for instance-based learning algorithms. ML **38**, 257–268 (2000)
18. Yu, R., Liu, S., Wang, X.: Dataset distillation: a comprehensive review. IEEE Trans. Pattern Anal. Mach. Intell. (2023)
19. Zha, D., Bhat, Z.P., Lai, K.H., Yang, F., Hu, X.: Data-centric AI: perspectives and challenges. In: Proceedings of the 2023 SIAM International Conference on Data Mining (SDM), pp. 945–948. SIAM (2023)

Using B-Spline Function Properties in the PIES Method to Handle Singularities in Boundary Value Problems

Agnieszka Bołtuć[(✉)] and Eugeniusz Zieniuk

Faculty of Computer Science, University of Bialystok, Bialystok, Poland
{a.boltuc,e.zieniuk}@uwb.edu.pl

Abstract. The paper presents the formulation of a parametric integral equation system (PIES) for 2D boundary value problems with singular solutions. The proposed approach combines B-spline basis functions with the PIES formalism to approximate the results in challenging problems such as notched plates or concentrated loads. The B-spline function with a specific degree and knots is used in approximating series instead of previously applied polynomials (such as Lagrange or Chebyshev). Increasing knot multiplicity to reduce continuity at some locations or increasing the number of knots in singular regions allows for more accurate results. The proposed approach is validated through selected problems. The results confirm that the B-spline-enhanced PIES approach can effectively solve singular problems with improved accuracy.

Keywords: Parametric integral equation system (PIES) · B-spline basis functions · Knots · Singular solutions

1 Introduction

The parametric integral equation system (PIES) [1–4] is one of the methods that is used to solve boundary value problems. The authors of the paper develop it as an alternative to known numerical methods such as the finite element method (FEM) [5–7] and the boundary element method (BEM) [7–9]. The main difference between PIES and the above methods is that PIES does not require classical discretization. It uses parametric curves [10,11] to model the boundary. Each segment of the boundary is defined globally by a single curve. The boundary segments are created naturally on the basis of the shape. Polygonal shapes have segments that coincide with each side, whereas in the case of curved shapes, the way they are divided depends on their complexity. In both cases, modeling is limited to assigning a small number of control points. Such defined shapes are analytically incorporated into the mathematical formalism of PIES. Therefore, each modification of the shape results in a modification of the integral equation. The authors have used mainly the Bezier and NURBS curves for this purpose.

© The Author(s), under exclusive license to Springer Nature Switzerland AG 2025
M. H. Lees et al. (Eds.): ICCS 2025, LNCS 15904, pp. 237–248, 2025.
https://doi.org/10.1007/978-3-031-97629-2_17

Using the appropriate number of curves of the appropriate degree ensures the accuracy of shape modeling. The accuracy of the solutions is independently controlled and determined by entirely different factors. PIES solutions are approximated using series with a specific number of terms and with selected basis functions. Changing both of these parameters affects the accuracy. So far, the authors have analyzed various numbers of series terms and various polynomials (such as Chebyshev, Legendre, Lagrange) as basis functions [1–4]. However, when solutions are singular, which is common in boundary value problems, such an approach may give unsatisfactory results. Using a high-degree polynomial in such situations is necessary, which may lead to numerical instability (Runge's phenomenon). Moreover, these polynomials are global over the range of approximation, so they cannot accurately take the local singularity into account. It would be appropriate to consider functions for which accuracy can be improved only in the singularity regions without affecting the entire segment. Such functions are B-spline, described by degree and knots. The use of these functions for the interpolation or approximation of solutions instead of polynomials is known [12]. Their properties allow for practical application in complex functions with high fluctuations. Therefore, it was decided to use them, among other basis functions, in the efficient PIES for demanding singularity problems.

B-spline functions are also used in other methods to solve boundary value problems, such as isogeometric FEM and BEM [13,14] or meshless methods [15]. However, the idea behind these methods differs from that of PIES. The lack of elements is a common feature of meshless methods and PIES. However, the way geometry is represented is entirely different. Meshless methods use scattered nodes in the domain. Lack of structured connectivity can cause higher computational costs and problems with boundary condition implementation. PIES uses a parametric boundary representation. Both approaches can apply B-spline functions to approximate solutions. On the other hand, isogeometric methods are known for using B-spline functions for both modeling and approximation of solutions. In PIES, it is also possible, but these two stages are separated, and various methods can be applied independently. Moreover, all the above-mentioned approaches (besides PIES) still divide the domain/boundary into some cells/elements to integrate numerically.

This paper proposes the use of B-spline functions to approximate solutions in PIES. This allows for more accurate results for some singular cases, still without changing the way of modeling. The shape is defined using Bezier curves to emphasize the independence of shape approximation from solution approximation. Some examples are considered. They are solved using PIES with the proposed B-spline functions and previously applied Chebyshev polynomials. It is also investigated how B-spline properties (degree and knots) changed by, e.g., knot multiplicity, clamped knots or adaptive knot placement can affect the accuracy of solutions. Two examples analyzed confirm the accuracy of the proposed approach.

2 Parametric Integral Equation System (PIES)

The parametric integral equation system (PIES) is presented in the example of 2D elastic problems without body forces [3]

$$0.5\boldsymbol{u}_l(\bar{s}) = \sum_{j=1}^{n} \int_{s_{j-1}}^{s_j} \left\{ \mathbf{U}_{lj}^*(\bar{s}, s)\boldsymbol{p}_j(s) - \mathbf{P}_{lj}^*(\bar{s}, s)\boldsymbol{u}_j(s) \right\} J_j(s)ds, \qquad (1)$$

where $l, j = 1, 2, ..., n$, $s_{l-1} \le \bar{s} \le s_l$, $s_{j-1} \le s \le s_j$. $J_j(s)$ is a Jacobian of transformation to the parametric reference system and n is the number of segments in the considered boundary. The parametric boundary functions $\boldsymbol{p}_j(s)$ and $\boldsymbol{u}_j(s)$ are known or searched on the particular segments of the boundary.

The boundary fundamental solution $\mathbf{U}_{lj}^*(\bar{s}, s)$ from (1) is presented in the following form [3]

$$\mathbf{U}_{lj}^*(\bar{s}, s) = -\frac{1}{8\pi(1-\nu)\mu} \begin{bmatrix} (3 - 4\nu)ln(\eta) - \frac{\eta_1^2}{\eta^2} & -\frac{\eta_1\eta_2}{\eta^2} \\ -\frac{\eta_1\eta_2}{\eta^2} & (3 - 4\nu)ln(\eta) - \frac{\eta_2^2}{\eta^2} \end{bmatrix}, \qquad (2)$$

where $\eta_1 = \Gamma_j^{(1)}(s) - \Gamma_l^{(1)}(\bar{s})$, $\eta_2 = \Gamma_j^{(2)}(s) - \Gamma_l^{(2)}(\bar{s})$, $\eta = \left[\eta_1^2 + \eta_2^2\right]^{0.5}$, ν is Poisson's ratio and μ is the shear modulus.

The second integrand $\mathbf{P}_{lj}^*(\bar{s}, s)$ from (1) is also presented in the matrix form by [3]

$$\mathbf{P}_{lj}^*(\bar{s}, s) = -\frac{1}{4\pi(1 - \nu)\eta} \begin{bmatrix} P_{11} & P_{12} \\ P_{21} & P_{22} \end{bmatrix}, \qquad (3)$$

where

$$P_{11} = \left\{ (1 - 2\nu) + 2\frac{\eta_1^2}{\eta^2} \right\} \frac{\partial\eta}{\partial n}, \quad P_{22} = \left\{ (1 - 2\nu) + 2\frac{\eta_2^2}{\eta^2} \right\} \frac{\partial\eta}{\partial n},$$

$$P_{21} = P_{12} = \left\{ 2\frac{\eta_1\eta_2}{\eta^2} \frac{\partial\eta}{\partial n} - (1 - 2\nu) \left[\frac{\eta_1}{\eta} n_2(s) + \frac{\eta_2}{\eta} n_1(s) \right] \right\},$$

and $\frac{\partial\eta}{\partial n} = \frac{\partial\eta_1}{\partial n}n_1(s) + \frac{\partial\eta_2}{\partial n}n_2(s)$, while $n_1(s)$ and $n_2(s)$ are direction cosines of the external normal to jth segment of the boundary.

The shape of the boundary is directly defined in the kernels (2) and (3) by parametric curves $\boldsymbol{\Gamma}$ of arbitrary degree [10,11]. The following section shows that they can efficiently model and modify various boundary shapes.

The form of PIES for other types of problems like potential, acoustic or elastoplastic can be found, among others, in [1,2,4].

3 Modeling of 2D Geometry in PIES

As mentioned in the introduction, the boundary in PIES is modeled entirely by curves. The most popular, among others, are the Bezier, B-spline, or NURBS curves [10,11]. These curves can be polynomials of any degree. The most commonly used polynomials are first-degree to model polygonal domains and third-degree to model curved domains. The main advantage of parametric curves is that their definition requires a small number of control points. In polygonal problems, only corner points can be posed.

Parametric curves are analytically incorporated into the PIES formalism (1). As a result, the PIES is defined on a straight line in a parametric reference system for an arbitrary shape. The shape is modeled and modified by the control points mentioned above. This process is independent of approximating the boundary functions (solutions). For this reason, in this paper, Bezier curves were used to model the shape of the boundary instead of NURBS to emphasize the independence of the shape approximation from the approximation of the solutions. The use of NURBS curves for some shapes would probably additionally improve the accuracy of the solutions. An example of modeling the polygonal and curved domain in PIES using Bezier curves is shown in Fig. 1.

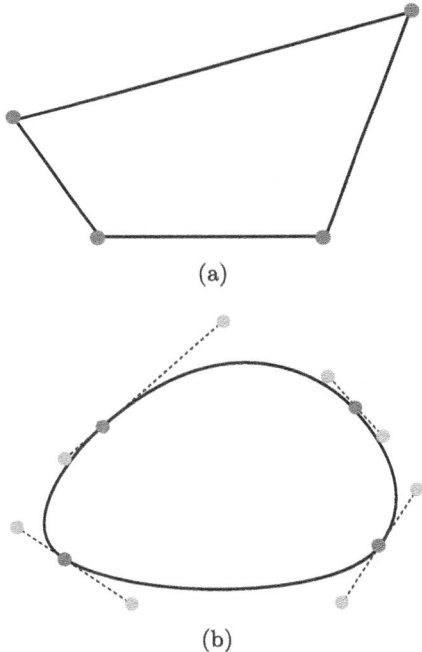

(a)

(b)

Fig. 1. The way of modeling in PIES: a) polygonal shape, b) curved shape

Both shapes presented in Fig. 1 consist of four segments. The polygonal shape (Fig. 1a) is built from linear segments defined by first-degree Bezier curves, while the curved one (Fig. 1b) is defined by third-degree curves. Both are modeled using only control points.

4   Approximation of PIES

The PIES solution is reduced to finding the unknown functions $u_j(s)$ and $p_j(s)$ from (1) on individual boundary segments. They are approximated by the following series [3]

$$u_j(s) = \sum_{k=0}^{M-1} u_j^{(k)} f_j^{(k)}(s), \quad p_j(s) = \sum_{k=0}^{M-1} p_j^{(k)} f_j^{(k)}(s), \tag{4}$$

where $u_j^{(k)}$, $p_j^{(k)}$ are searched coefficients, M is the number of coefficients on segment j, $f_j^{(k)}(s)$ are arbitrary basis functions and $j = 1, 2, ..., n$.

So far, Legendre, Chebyshev or Lagrange polynomials have been used as basis functions [3,4]. However, modeling singular solutions (e.g. near cracks, notches, corners) requires functions that effectively deal with local discontinuities and rapidly changing values. The polynomials mentioned above are global, so it is impossible to consider the local singularity. An increase in their degree can help, but also lead to numerical instability. Therefore, the authors decided to use B-spline basis functions that provide more local control, allowing for a more accurate capture of strong gradients near singularities.

A B-spline basis function of degree q is defined recursively using the Cox-de Boor recurrence formula starting with a piecewise constant polynomial ($q = 0$) [16]

$$N_{i,0}(v) = \begin{cases} 1 & if \quad v_i \leq v < v_{i+1}, \\ 0 & otherwise, \end{cases} \tag{5}$$

where v_i are knots. They are sorted in non-decreasing order in vector V, and their role is partitioning the parameter space. The number of knots is $M + q + 1$.

For $q = 1, 2, 3, ...$ B-splines are defined by [16]

$$N_{i,q}(v) = \frac{v - v_i}{v_{i+q} - v_i} N_{i,q-1}(v) + \frac{v_{i+q+1} - v}{v_{i+q+1} - v_{i+1}} N_{i+1,q-1}(v). \tag{6}$$

B-spline functions have properties that affect the accuracy of modeling singular solutions, such as their degree and the arrangement of knots (their local intensification or multiplication).

When knots are closely spaced around a certain point, the B-spline basis functions are sharper and more localized around it. This increases the resolution and flexibility of the basis functions in that area and allows to capture more detail where the solution has high gradients. In other words, a local refinement is preserved.

Increasing the multiplicity of a knot (i.e. repeating the same knot value) changes the behavior of the basis functions by reducing the continuity at that knot. This is important when solving problems with the physical discontinuities, e.g. concentrated loads.

The B-spline functions with $q = 2$ and different knot vectors (uniform, concentrated and duplicated) are shown in Fig. 2.

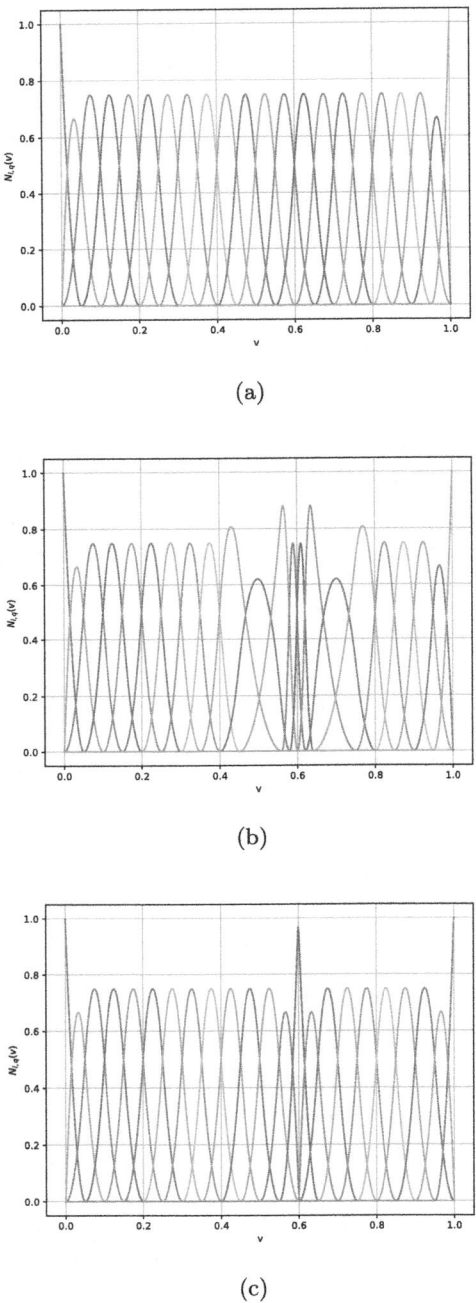

(a)

(b)

(c)

Fig. 2. B-spline basis functions for $q = 2$ and knot vector: a) uniform, b) concentrated around point 0.6, c) duplicated at point 0.6

Finally, the PIES method is solved using the collocation method [17]. The arrangement of collocation points is also essential. However, the authors have studied the influence of collocation points' arrangement and number (M) on the solutions in the previous research. Therefore, in this paper, both parameters are fixed, and the analysis focuses on the properties of the B-spline function.

5   Numerical Examples

5.1   Example 1 - Half-Space with a Concentrated Boundary Condition

The first example concerns a half-space with a concentrated Neumann boundary condition given by the Dirac function. The problem is modeled by the Laplace equation and is presented in Fig. 3.

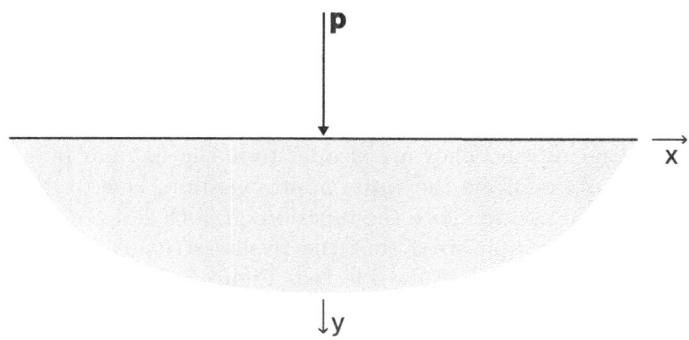

Fig. 3. A half-space with a concentrated boundary condition

As a result of applying PIES presented in [1] to the problem being solved, an analytical solution on the boundary is obtained

$$u(x) = \frac{1}{\pi} ln(\frac{1}{\sqrt{x^2}}). \tag{7}$$

Then, PIES was solved numerically for the case where the half-space is bounded by a large region modeled using Bezier curves of the first degree. Chebyshev polynomials and B-spline functions were used to approximate the solutions using (4) with $M = 10$. B-spline function properties are $q = 2$ and $\boldsymbol{V} = \{-0.5, -0.5, -0.5, -0.15, -0.013, 0, 0, 0, 0.013, 0.15, 0.5, 0.5, 0.5\}$. Due to the expected singularity of the solution at point $x = 0$, the middle knot has been duplicated to reduce continuity. The obtained results in a certain region around the singularity are compared with the analytical solution (7) and presented in Fig. 4.

As shown in Fig. 4, solutions obtained by PIES with B-spline functions are much more accurate than those generated using Chebyshev polynomials. It can

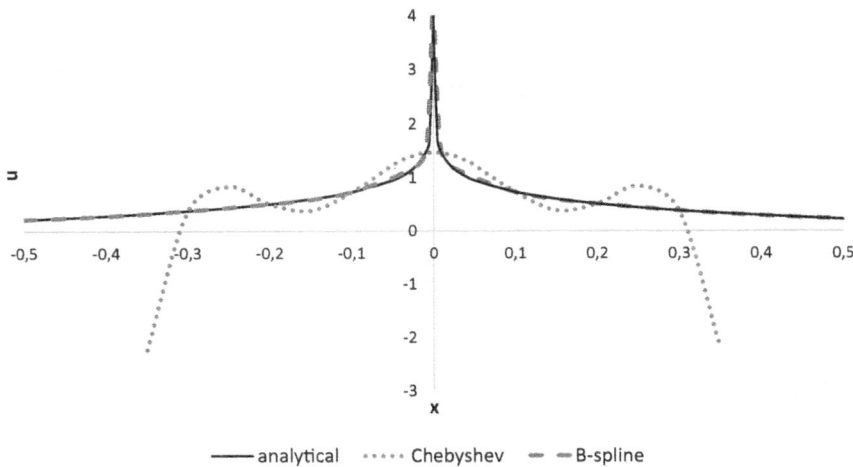

Fig. 4. Comparison of solutions u on the boundary

be seen that when polynomials are used to approximate a function with a singularity, oscillations appear. They are also far from the singular point because a singularity in one place affects the entire approximation. Due to its local structure, the B-spline can approximate the function on both sides of the singularity well, but it should be emphasized that the results strongly depend on knots. Therefore, they should be carefully selected. Figure 5 presents solutions using various knot vectors $V1, V2, V3$. The knot arrangement is given at the bottom of the figure (the knots' color corresponds to the solution's color). It should also be remembered that some knots are duplicated, which may not be visible in the figure. $q + 1$ knots from the beginning and end of the approximation interval force the function to pass through the first and last control points. The middle knot duplication reduces continuity at that location to enable modeling problems with concentrated boundary conditions.

The solutions presented in Fig. 5 show that accurate considering the singularity requires concentrating knots in its vicinity, not only the multiplication of the middle knot. This is visible in the vector $V3$ (green), where the knot at point $x = 0$ is multiplicated, but the neighboring knots are moved slightly away from the center. This results in a much lower accuracy of solutions exactly at the singular point (lower peak). On the other hand, the densification only in the center, while shifting the remaining knots to the ends (vector $V2$, blue) causes disturbances in the broader range around the singular point and at the ends compared to the $V3$ case.

5.2   Example 2 Elastic Plate with a Circular Hole

The second example concerns the elastic plate with the circular hole subjected to the tensile load p at its ends (Fig. 6). Plane stress conditions are assumed

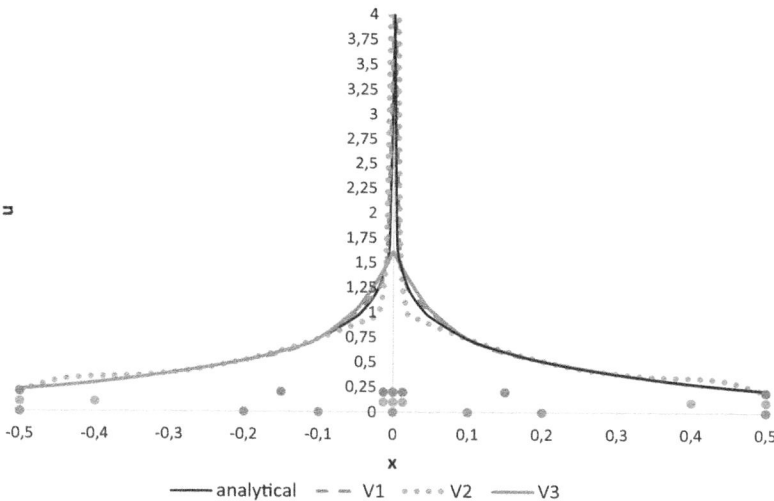

Fig. 5. Solutions u depending on various knot vectors (Color figure online)

with the following material properties $E = 200 \times 10^9 MPa$ and $v = 0.33$. Each segment is modeled by the Bezier curve (of the first or third degree).

Once again, for approximation of boundary solutions, two different basis functions (Chebyshev and B-spline) are applied in (4) with $M = 7$. Various q degrees and knots are analyzed for the B-spline function. The results for stresses σ_y at the part of the bottom boundary segment (around the hole) are obtained and compared with the existing analytical solution [18]

$$\sigma_y = \frac{p}{2}(2 + \frac{1}{r^2} + 3\frac{1}{r^4}), \tag{8}$$

where r is one of the polar coordinates and $r^2 = x^2 + y^2$. The result of the comparison together with the knots arrangement is presented in Fig. 7.

As seen in Fig. 7, the degree of the B-spline function plays an essential role in the accuracy of the obtained solutions. The best fit to the analytical results is obtained for $q = 4$, but it should be emphasized that even the solution for $q = 2$ is much more accurate than that obtained using Chebyshev polynomials. The average error of the solutions obtained using the B-spline function on the considered part of the segment decreased from 10.34% (for $q = 1$) to 2.27% ($q = 4$). For comparison, this error for Chebyshev polynomials is 11.89%.

Analyzing the knots arrangement, it is visible that the desired distribution is their more significant number in the singular region, which occurs near the hole. Once again, $q + 1$ extreme knots are duplicated, which ensures that the function starts and ends precisely at the ends of the boundary segment.

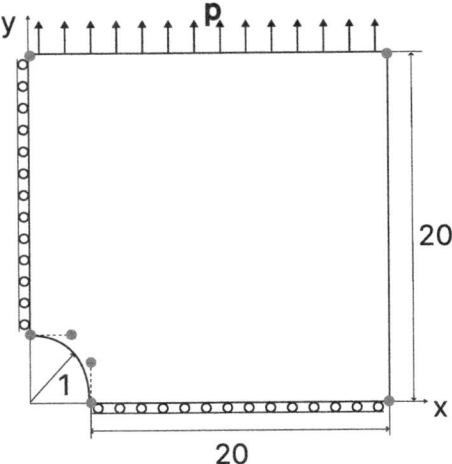

Fig. 6. The plate with a circular hole

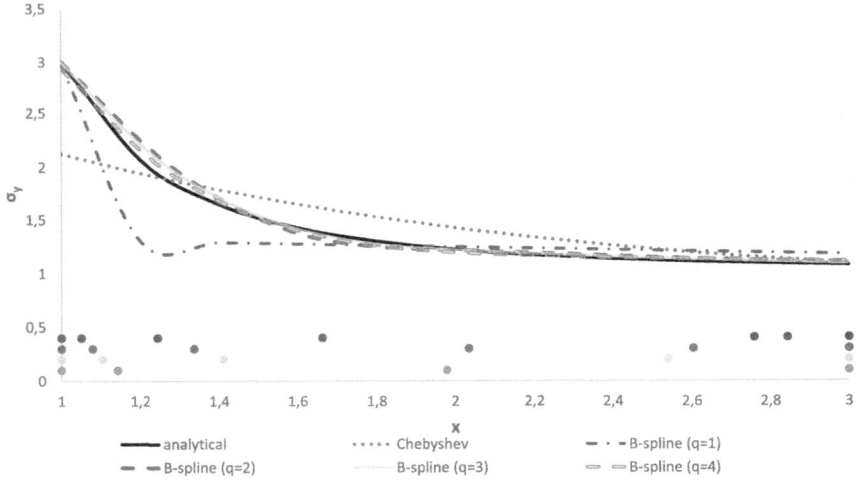

Fig. 7. Solutions σ_y for various degrees of B-spline functions

6   Conclusions

This paper has derived and tested the approach for solving boundary problems with singular solutions. The PIES method combines the B-spline functions as basis functions in the approximation series. They require determining the degree and defining knots, which allows controlling the accuracy of the singular solutions by placing knots adaptively or increasing knot multiplicity.

Two examples were considered in the paper. The first one concerned a half-space with a boundary condition concentrated at a single point modeled by the Laplace equation. This example was chosen because of the possibility of obtaining an exact solution for comparison purposes. The second one is a typical elastic example, where we deal with a plate with a hole subjected to the tensile load. In both cases, the solutions are singular.

The results obtained from the numerical analysis show that using the B-spline as basis functions significantly increases the accuracy of modeling singular solutions. With an appropriately selected degree and knots, the errors of the solutions are minimized. The biggest challenge is the selection of appropriate knots, because their arrangement has a significant impact on the final results. It would be worth looking for optimal arrangements of knots using any optimization algorithm. Some metaheuristics can be used to optimize the form of the knot vector, but this requires a suitable objective function. The most straightforward would be to minimize the difference between the analytical and numerical solutions. However, developing an approach for cases where the exact solution does not exist would be worth trying. Moreover, in this paper, some PIES properties are fixed (collocation points number and arrangement). Still, it would also be worth examining what relation between those properties and knots gives the best results. Again, the optimal relationship using optimization algorithms can be found. Finally, comparing the results obtained by PIES with other methods that use B-spline functions to approximate solutions also would be interesting.

Disclosure of Interests. The authors have no competing interests to declare that are relevant to the content of this article.

References

1. Zieniuk, E.: Bézier curves in the modification of boundary integral equations (BIE) for potential boundary-values problems. Int. J. Solids Struct. **40**(9), 2301–2320 (2003)
2. Zieniuk, E., Boltuc, A.: Bézier curves in the modeling of boundary geometry for 2D boundary problems defined by Helmholtz equation. J. Comput. Acoust. **14**(3), 353–367 (2006)
3. Zieniuk, E., Boltuc, A.: Non-element method of solving 2D boundary problems defined on polygonal domains modeled by Navier equation. Int. J. Solids Struct. **43**(25–26), 7939–7958 (2006)
4. Bołtuć, A.: Parametric integral equation system (PIES) for 2D elastoplastic analysis. Eng. Anal. Boundary Elem. **69**, 21–31 (2016)
5. Zienkiewicz, O.C.: The Finite Element Methods. McGraw-Hill, London (1977)
6. Liu, G.R., Quek, S.S.: The Finite Element Method: A Practical Course. Butterworth Heinemann, Oxford (2003)
7. Ameen, M.: Computational elasticity. Alpha Science International Ltd, Harrow (2005)
8. Aliabadi, M.H.: The Boundary Element Method, vol. 2. Applications in Solids and Structures. Wiley, Chichester (2002)
9. Gao, X.W., Davies, T.G.: Boundary Element Programming in Mechanics. Cambridge University Press, Cambridge (2002)

10. Farin, G.: Curves and surfaces for CAGD: A practical Guide. Morgan Kaufmann Publishers, San Francisco (2002)
11. Salomon, D.: Curves and Surfaces for Computer Graphics. Springer, New York (2006)
12. Idais, H., Yasin, M., Pasadas, M., González, P.: Optimal knots allocation in the cubic and bicubic spline interpolation problems. Math. Comput. Simul. **164**, 131–145 (2019)
13. Cottrell, J.A., Hughes, T., Bazilevs, Y.: Isogeometric Analysis: Toward Integration of CAD and FEA. Wiley, Hoboken (2009)
14. Simpson, R.N., Bordas, S., Trevelyan, J., Rabczuk, T.: A two-dimensional isogeometric boundary element method for elastostatic analysis. Comput. Methods Appl. Mech. Eng. **209–212**, 87–100 (2012)
15. Liu, G.R.: Meshfree Methods: Moving Beyond The Finite Element Method. CRC Press, Boca Raton (2009)
16. Piegl, L., Tiller, W.: The NURBS Book. Springer, Germany (1997)
17. Gottlieb, D., Orszag, S.A.: Numerical Analysis of Spectral Methods. SIAM, Philadelphia (1977)
18. Timoshenko, S.P., Goodier, J.N.: Theory of Elasticity. McGraw-Hill, Tokyo (1970)

cuTeBool: Fast and Scalable Boolean Matrix Factorization on GPUs Using Tensor Cores

Andrea Beyer[✉][iD], Valentin Henkys[✉][iD], Robin Kobus[iD], Stefan Kramer[iD], and Bertil Schmidt[iD]

Institute of Computer Science, Johannes Gutenberg University Mainz, Mainz, Germany
abeyer@students.uni-mainz.de, henkys@uni-mainz.de

Abstract. Boolean matrix factorization aims to represent binary data as a product of two factor matrices, in order to uncover the underlying structure of the data and find a compressed representation. However, finding the factors of a given ground truth is computationally hard and calls for fast implementations that accomplish a good approximation in reasonable time. We present cuTeBool, a novel parallel algorithm that exploits Tensor Cores on CUDA-enabled GPUs for fast matrix operations based on a randomized approach. Our comprehensive performance evaluation shows that it produces approximate factorization competitive to other state-of-the-art tools within vastly reduced runtime for a variety of input matrices. Moreover, our algorithm is the only available method that scales well with the size of the ground truth and is able to factorize matrices that are at least one order-of-magnitude larger than all competitors. We further analyze algorithmic parameters allowing us to find a trade-off between performance and reconstruction quality.

Keywords: GPUs · Matrix Factorization · Parallel Computing

1 Introduction

Matrix factorization is an important technique in the field of unsupervised data mining and compression. It aims to find a decomposition of a ground truth matrix $\mathbf{C}$ into a product of n matrices $\mathbf{A_i}$

$$\mathbf{C} = \prod_{i=1}^{n} \mathbf{A_i}.$$

In most practical applications, $n \in \{2, 3\}$, like in *Singular Value Decomposition* with $n = 3$ and *Non-Negative Matrix Factorization* with $n = 2$.

Associative data can be naturally expressed in terms of Boolean matrices over $\mathcal{B}^{m \times n}$ with entries in $\mathcal{B} = \{0, 1\}$ instead of real-valued matrices. Rows and columns of such matrices are then interpreted as *objects* and *attributes*: an entry

© The Author(s), under exclusive license to Springer Nature Switzerland AG 2025
M. H. Lees et al. (Eds.): ICCS 2025, LNCS 15904, pp. 249–264, 2025.
https://doi.org/10.1007/978-3-031-97629-2_18

M_{ij} is set to 1, if object i has attribute j. Typical examples include whether a user listened to a song, read a book, or knows another person in a social network. Boolean matrices spawned by real-world applications are often sparse, i.e., the number of objects and attributes is large compared to the average number of attributes that are assigned to a single object.

In this context, **Boolean Matrix Factorization (BMF)** is defined as: Find $\mathbf{A} \in \mathcal{B}^{m \times k}$ and $\mathbf{B} \in \mathcal{B}^{k \times n}$ for a given ground truth $\mathbf{C} \in \mathcal{B}^{m \times n}$ such that

$$\mathbf{C} = \mathbf{A} \circ \mathbf{B}.$$

Here, $\circ$ denotes matrix multiplication over the semi-ring $(\{0, 1\}, \vee, \wedge)$. Since finding an exact representation is NP-hard [10,12], we often aim to find an approximation with $k \ll m$ and $k \ll n$. This results in a compressed representation of the ground truth and can be used for noise reduction, among others.

A lot of work has gone into optimizing BMF for non-parallel processors [2,9, 19]. To improve associated runtimes some libraries started to use parallelism on multi-core CPUs [6] and many-core e Graphics Processing Units (GPUs) [4,5]. With the rise of Machine Learning (ML), another type of compute unit has now emerged: the neural processing unit (NPU). It is optimized for Matrix Multiply-Accumulate (MMA) operations, which are frequently used in ML tasks. Our work focuses on NVIDIA's implementation, called Tensor Cores (TCs), available on most modern CUDA-enabled GPUs.

While typically designed for floating-point arithmetic, TCs now also feature support for Boolean matrix operations, making them a suitable candidate platform for BMF. In this work, we present cuTeBool – the first TC-enabled parallel BMF algorithm achieving high efficiency on modern GPUs.

Our detailed contributions consist of:

1. A novel BMF implementation utilizing TCs on GPUs, providing average speedups of $2\times$, $20\times$, $38\times$, $20\times$ compared to other limited-rank competitors cuBool, Panpal, Primp, MEBF, respectively, while maintaining better F_1-scores across all tested datasets.
2. Support for ground truth matrices that are at least one order-of-magnitude bigger than all competitors.
3. Automatic hyperparameter tuning, providing a good trade-off between reconstruction quality and performance.
4. Open-source implementation at https://gitlab.rlp.net/pararch/cutebool.

The remainder of the paper is structured as follows. We provide necessary background in Sect. 2, followed by Sect. 3 discussing related work. Algorithmic methods are introduced in Sect. 4. Parallelization details are described in Sect. 5. Performance is evaluated in Sect. 6. Finally, Sect. 7 concludes.

2   Background

2.1   Boolean Matrix Factorization

Formally, BMF operates within the Boolean semi-ring $(\{0,1\}, \vee, \wedge)$. Boolean Matrix Multiplication (BMM, denoted as $\circ$) is then defined as

$$C_{i,j} = \bigvee_{l=1}^{k} A_{i,l} \wedge B_{l,j}. \tag{1}$$

We denote the set of Boolean $x \times y$ matrices as $\mathcal{B}^{x \times y}$. BMF is defined as the task of finding two factor matrices $\mathbf{A} \in \mathcal{B}^{m \times k}$ and $\mathbf{B} \in \mathcal{B}^{k \times n}$ for a given ground truth $\mathbf{C} \in \mathcal{B}^{m \times n}$, such that

$$\mathbf{C} = (\mathbf{A} \circ \mathbf{B}). \tag{2}$$

We distinguish between the *decomposition rank*, which is the dimension k of a given factorization, and the *Boolean rank* of the ground truth: the smallest k, for which Eq. (2) can be satisfied.

Since finding exact solutions is NP-hard [10,12], low-rank approximations are typically preferred as a means to cancel out noise and compress data. The underlying notion is that the relations present in the data are influenced by a relatively small set of unknown factors. Hence, a factorization is commonly understood as a pair of basis vectors and coefficients that produce a close approximation and reveal structure of the underlying data.

In this work, we approach *approximate BMF* with a rank $k \leq 128$. The goal here is to find $\mathbf{A}, \mathbf{B}$ with a given k that minimize the *reconstruction error*:

$$E_{rec} = |\mathbf{C} - (\mathbf{A} \circ \mathbf{B})| = fp + fn. \tag{3}$$

fp and fn denote the number of false positives and false negatives, respectively. In some cases, false negatives and false positives may vary in significance for the reconstruction. This can be captured by the *weighted reconstruction error*:

$$E_{rec}^{w} = w \cdot fn + fp. \tag{4}$$

For $w > 1$, this implies that false negatives are penalized harder than false positives. We use *recall* (R), *precision* (P), and the *F-measure* (F_1) as metrics to assess the quality of approximations. They are expressed in terms of *true/false positives* (tp, fp) and *negatives* (tn, fn) as follows.

$$R = \frac{tp}{tp + fn}, P = \frac{tp}{tp + fp}, F_1 = 2\frac{P \cdot R}{P + R} = \frac{2tp}{2tp + fp + fn} \tag{5}$$

2.2   CUDA and Tensor Cores

CUDA-enabled GPUs feature multiple Streaming Multiprocessors (SMs) that can be programmed using a large number of threads. Threads are partitioned into *thread blocks* and *warps*. Each warp consists of 32 threads, each thread block in turn hosts up to 32 warps. Threads within the same thread block can access a common, but limited shared memory with reduced latency compared to global memory, which is accessible to all threads. Within a warp, threads benefit from data exchange comparable to register-speed through warp shuffles. The CUDA language extension for C++ distinguishes between host and device code. The host (CPU) can launch kernels on the device (GPU).

Working at warp-level, TCs are optimized to perform fast matrix multiplication. This is achieved through the MMA operation $\mathbf{W} = \mathbf{X} \cdot \mathbf{Y} + \mathbf{Z}$. $\mathbf{X}$ and $\mathbf{Y}$ are referred to as *factors*, $\mathbf{Z}$ and $\mathbf{W}$ as *accumulators*. Large matrices are partitioned into so-called *fragments* of fixed size, that can be understood as submatrices. Each warp can compute one fragment of the result by iterating over fragments in the factors, using the accumulator for subresults. This allows for efficient parallelization of arbitrary sized products.

Each thread only holds a subset of entries for each fragment, so that the warp collectively holds all entries of the considered submatrix. The choice of fragment dimensions and data type for the multiplication is restrained by the special layout required by TCs [14].

Newer TCs also provide support for 1-bit data types with a `m8n8k128` fragment type. That is, factor fragments of dimensions 128×8 and 8×128, and accumulator fragments of size 8×8 are supported. Since we aim for a low factorization rank, we do not extend the factors beyond 128 bit in their common dimension, so that the factorization rank either coincides with or is lower than the side length of a single fragment. Note that TCs do not directly perform BMM as defined in Eq. (1), but compute a sum instead of the logical OR:

$$w_{ij} = \texttt{POPC}\left(\mathbf{X}[i] \,\&\, \mathbf{Y}[j]\right) + z_{ij} = \sum_{l=1}^{k} \left(x_{il} \wedge y_{lj}\right) + z_{ij} \tag{6}$$

The obtained value can then be transformed using $w_{ij} \leftarrow w_{ij} \neq 0$ to acquire the desired result for OR-based BMM.

3   Related Work

In an early attempt, Miettinen et al. adapted the *Asso*-algorithm [9] to BMF by first generating a number of candidate vectors and greedily selecting k such vectors to form a factor that minimizes the reconstruction error. Dynamic BMF [8] provides an online algorithm that incorporates recent changes of the ground truth into an existing factorization. Moreover, Miettinen and Vreeken worked on the problem of the Minimum Description Length (MDL) for BMF [10]. Here, the error of BMF is encoded in a separate matrix $\mathbf{E}$, so that $(\mathbf{A} \circ \mathbf{B}) \oplus \mathbf{E}$ is an exact representation of the ground truth.

More recently, Wan et al. [19] presented median expansion for BMF (*MEBF*). This approach identifies patterns by permuting the rows and columns of the original matrix, such that positive entries accumulate in the top right. The best rank-1 approximation of the resulting matrix is then added to the factors of the ground truth.

Belohlavek and Vychodil [2] connected BMF to formal concept analysis by proposing the *GreCon* and *GreConD* algorithms. Both algorithms greedily add candidate rows to the factors and pursue a from-below approach, i.e., no false positives are allowed. GreConD differs from GreCon by only considering a subset of formal concepts which increases performance but yields comparable results. These algorithms have since been extensively studied and optimized. Trnecka and Vyjidacek revisited GreCon and presented *GreCon2* [17] – speeding up the algorithm without a loss in quality. In [1], *GreConD+* is introduced, which deviates from the from-below approach taken by GreConD. Here, accepted factors are expanded, allowing false positives if the overall reconstruction error benefits from the expansion. Trnecka and Krajča developed *ParaGreConD* [6], a performance optimized version of the GreConD algorithm, by using multiple CPU-threads to evaluate candidates. Note, that their implementation approaches BMF as a coverage problem in the sense that it extends the factors until a desired coverage factor is reached. This crucially differs from our approach, which tries to minimize the reconstruction error for a fixed factorization rank. As ParaGreConD is the only GreCon modification that explores parallelization, it is most relevant to the work presented in this paper.

Similarly, Outrata and Trnecka [15] note that many previous BMF algorithms work with greedy heuristics, sequentially improving the result. They introduce a general parallelization scheme for such algorithms, that computes multiple locally optimal subresults in parallel, and returns the best found approximation. Here, the main benefit is the improved quality of the factorization, compared to single-core execution.

Hess et al. implemented the *PALTiling* framework [4], featuring the algorithms *Panpal* and *Primp*. Both algorithms are iterative in nature, but use NVIDIA's cuBLAS library to benefit from GPU-acceleration for individual operations. PALTiling attempts to decompose the ground truth into densely populated tiles, allowing factors with values in $[0, 1]$ for computation, and enforcing a binary result by applying a threshold to the individual factor values. The main difference between the algorithms is the use of different cost measures. Panpal uses a L_1 regularization, following the example of Panda [7], whilst Primp adapts the cost function of Krimp [18], originally used in the context of MDL.

Also recognizing the inherent capability for massive parallelization of matrix multiplication, *cuBool* [5] harnesses GPUs to compute a factorization. Here, an initial factorization is guessed and iteratively improved by evaluating random bit-flips, until an error threshold or time limit is reached. Multiple warps of threads are used to explore different updates in parallel. Our work presented in this paper builds upon the basic idea of cuBool [5], but extends it to gain even higher speeds by making it amenable to TCs.

4 Algorithmic Method

Our overall algorithmic approach is based on local search: initially, random factor matrices are generated and iteratively improved. Within each iteration, we update factors **A** and **B** successively, aiming to minimize the reconstruction error. The algorithm terminates, when it was unable to improve the reconstruction for a certain number of iterations, indicating a local minimum.

Updates are generated by introducing minor changes to factor values by flipping randomly selected bits. Subsequently, the resulting weighted reconstruction error is calculated. Updates that improve the error are written to memory.

When generating updates, single bit-flips are preferred over multiple flips within the same row. This has two main causes. First, flipping many bits at once makes it impossible to determine which of the individual flips are beneficial, as only the update as a whole can be kept or discarded. This potentially keeps non-beneficial bit-flips or discards beneficial ones. Second, performing many bit-flips tends to overpopulate the factors, leading to dense results and low precision.

Since updates in one factor may influence the error calculation for updates in the other factor, each factor has to be updated successively. As the error calculation of different values in the same factor can be done independently, cuTeBool can harness GPUs to compute multiple updates on the same factor in parallel. Our novel parallelization scheme using TCs is described in detail in Sect. 5.2 and 5.3.

A common issue with the randomized approach is that the algorithm may produce an empty reconstruction. Especially on sparse datasets the initial reconstruction will contain a relatively high number of false positives. This incentivizes discarding true positives along with false positives to reduce the reconstruction error. As a result, the algorithm may produce a zero reconstruction with high probability, losing all information about the data.

To avoid this problem, we apply a weight w to the reconstruction error (see Eq. (4)) that penalizes false negatives more than false positives in an attempt to achieve a higher recall. I.e., a true positive is only discarded if at least w false positives are discarded along with it. This weight is gradually decreased over many iterations. At a weight of 1, false negatives and false positives are penalized equally. Initial weight and reduction factor can be customized as command line arguments. A higher weight typically favors higher recall at the expense of precision. During execution, we monitor the (weighted) improvement achieved by each update to detect local minima. We terminate, when it got stuck for a certain number of iterations without finding an update, that would improve the error. Alternatively, it can also be set to run for a fixed amount of iterations. Due to the randomized nature of this approach, the algorithm may get stuck in a suboptimal minimum. To increase the chance of finding a good solution, we suggest to spawn multiple instances with different seeds, as discussed in Sect. 5.

5   Parallelization Scheme

The goal of our implementation scheme is to minimize CPU-GPU communication and let the GPU handle most of the computation. Therefore, after reading the input data, the ground truth is moved to the GPU and the factors are initialized directly on the GPU. Subsequently, our main optimization loop starts.

In this loop we iteratively update the factors by a few random bits, check the new errors and decide whether to keep the new bits or not. Since each update of a factor has to be performed independently from the other factor, we use two distinct kernels: One for column and one for row updates, where each has to wait for the previous update to be finished.

We support the initialization of multiple instances to be computed at once, so that available GPU resources can be fully utilized. Each instance differs in the random seed used to initialize the factors and to determine the bits flipped in each iteration. This variety leads to possibly faster convergence and better results than exploring a single seed. Only when the break condition is reached, we choose the instance with the lowest error and move the computed factors back to the CPU for possible storage. This overall workflow is shown in Algorithm 1.

5.1   GPU Data Layout

While our factor and ground truth matrices are read frequently, they generally do not fit into the fast CUDA shared memory. Consequently, we use an efficient data layout for global memory to allow for efficient coalesced memory accesses as proposed by cuBool [5], seen in Fig. 1.

Our focus is on a decomposition rank of 128 or less, due to the perfect fit into the Tensor Cores' $8 \times 8 \times 128$ fragment dimensions for 1-bit data types. To allow the threads to perform coalesced data accesses, we store our factors in a row-major fashion for $\mathbf{A}$ and column-wise for $\mathbf{B}$, respectively, using the 4×32-bit wide integer type `uint4`. For a decomposition rank lower than 128, we pad the matrices with zeros to reach the fragment size.

For the ground truth matrix another storage layout is needed, since it has to be read row- and column-wise, depending on which error we want to compute. We compute rows and columns in batches of 32 to better accommodate the required layout for TCs. Thus, we divide the ground truth into multiple 32×32 submatrices. If necessary, the ground truth's dimensions are padded with zeros to a multiple of 32. Each submatrix can then be read in a coalesced way.

5.2   Update Kernels

While there are two distinct kernels for row and column updates, both work analogously. Therefore, we explain the column kernel as example. A general overview over its workflow is shown in Algorithm 2.

In the column kernel, the selected random bits of the factor $\mathbf{B}$ are changed and the resulting error is calculated. If the changes reduce the error, they are kept, else discarded. Since these steps are the most time consuming part of the program, a lot of care has been taken to make them as efficient as possible.

Each thread block is assigned a batch of 32 consecutive columns of factor $\mathbf{B}$. These 32 columns are further sub-divided into 4 groups of 8 columns corresponding to the width of the chosen TC fragments. Each batch is processed by one GPU thread block, which we spawn with 32 warps. Thus 8 warps cooperate

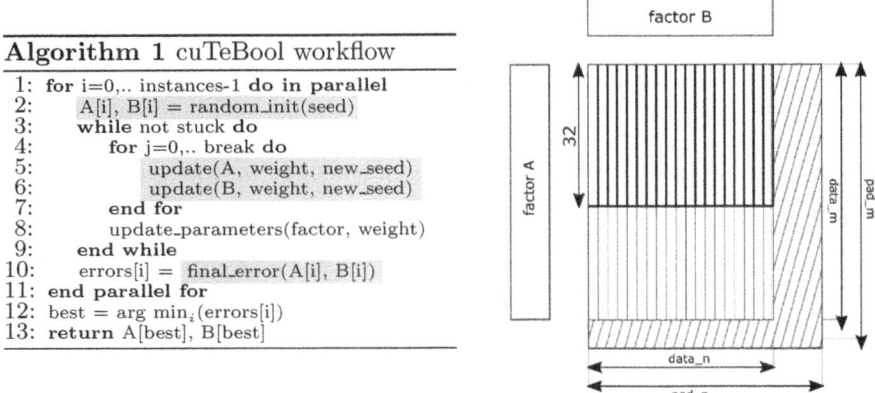

Algorithm 1 cuTeBool workflow

```
 1: for i=0,.. instances-1 do in parallel
 2:     A[i], B[i] = random_init(seed)
 3:     while not stuck do
 4:         for j=0,.. break do
 5:             update(A, weight, new_seed)
 6:             update(B, weight, new_seed)
 7:         end for
 8:         update_parameters(factor, weight)
 9:     end while
10:     errors[i] = final_error(A[i], B[i])
11: end parallel for
12: best = arg min_i(errors[i])
13: return A[best], B[best]
```

Alg. 1: Most work is done on the GPU.

Fig. 1. Data layout of the ground truth.

on the same columns in a warp group. The values of **A** are needed by multiple threads. Hence, we can benefit from coalesced reads and the lower latency of shared memory, by splitting **A** into chunks of 1024 rows, that are read in a coalesced way and temporarily stored in shared memory. This warp cooperation scheme is illustrated in Fig. 2. Each of the warp groups update random bits of their respective columns and then calculate the old and updated errors, splitting the rows of **A** equally across the collaborating warps.

5.3   Error Computation on Tensor Cores

In our implementation, we address TCs using Parallel Thread Execution (PTX) instructions, which are used as an intermediate code representation for the NVIDIA CUDA compiler (NVCC). PTX allows for direct control over the registers participating in operations on Tensor Cores giving a transparent data layout. This has the benefit, that data does not need to be written to and read from memory between individual TC computations.

To decide if our updated columns lead to an improvement, we need to compute the original error and the updated (weighted) error. Here, each warp group is responsible for 8 columns of the ground truth matrix. Using this distribution, each thread loads its respective column of **B** and its updated version into registers and iteratively works through the rows of the current chunk of **A**. For each thread, the **B** values remain constant throughout kernel execution, removing the need to read **B** multiple times. Once all rows of the current chunk of **A** are exhausted, the block collectively loads the next chunk into shared memory and proceeds iterating over **A**.

Within each 8 × 8 tile of the result matrix, each thread of one warp is assigned two horizontally adjacent elements. This corresponds to a transposed submatrix, compared to how the ground truth is stored. To avoid extra memory accesses

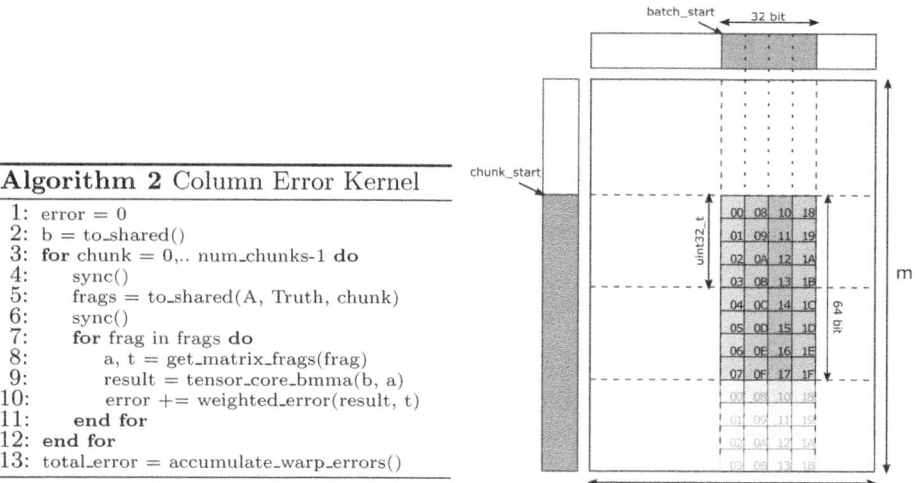

Algorithm 2 Column Error Kernel

```
 1: error = 0
 2: b = to_shared()
 3: for chunk = 0,.. num_chunks-1 do
 4:     sync()
 5:     frags = to_shared(A, Truth, chunk)
 6:     sync()
 7:     for frag in frags do
 8:         a, t = get_matrix_frags(frag)
 9:         result = tensor_core_bmma(b, a)
10:         error += weighted_error(result, t)
11:     end for
12: end for
13: total_error = accumulate_warp_errors()
```

Alg. 2: Note that the algorithm for computing the row-wise errors only differs in memory access and data exchange patterns.

Fig. 2. Warp distribution: each group of 8 consecutive warps work together on an 8-bit wide column. Warps are numbered in hexadecimal format.

or computational overhead, we pass the input for each BMMA operation in a reversed order, which will produce the transposed result $\mathbf{B}^T\mathbf{A}^T = (\mathbf{AB})^T$. After each MMA, the threads compare their results to the ground truth and update the reconstruction error accordingly.

Afterwards all partial errors for each column are accumulated. This process is performed in three steps:

1. Gather the partial errors from collaborating threads within each warp using warp-shuffles.
2. Employ shared memory to exchange these column-specific errors with cooperating warps.
3. Execute another warp shuffle to accumulate the errors that have been received in the previous step.

These steps have to be performed both for the updated errors, as well as the original one. If the total updated error for a column is lower than the previous version, the change is kept and written to memory.

5.4   Automatic Hyperparameter Selection

Our program depends on a number of hyperparameters influencing cuTeBool's behavior and the quality of results. We automatically select the best parameters for a good trade-off between reconstruction quality and efficiency, based on a grid search on multiple datasets. The outcome on one representative smaller dataset, namely AOffice (see Table 1), is illustrated in Fig. 3.

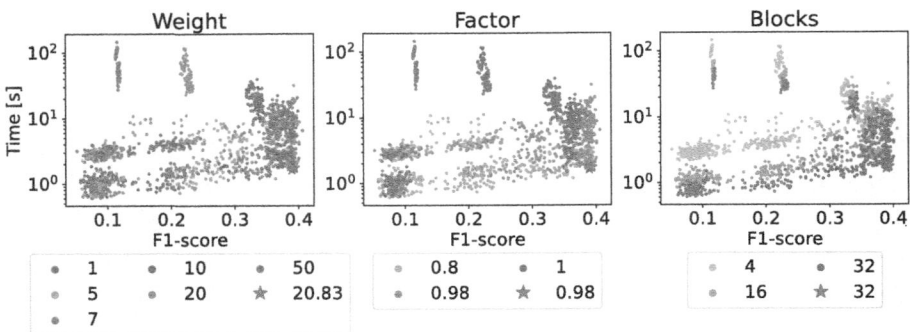

Fig. 3. Effect of hyperparameters on runtime (y-axis) and F_1-score (x-axis) of the AOffice dataset. Bottom right is better. Each point represents one run. Automatically set parameters are indicated as a star.

While the starting weight, shown in the left most plot, is dataset dependent, there is a general trade-off noticeable between reconstruction quality and runtime. Additionally, we noticed that, if the weight is high, the algorithm tends to find more true positives on sparser datasets early on. Thus, we implemented the following measure, clamping the results to avoid extremes:

$$\text{weight} = \max \left\{ 3, \ \min \left\{ \frac{\text{\#entries in ground truth}}{\min\{m, n\}}, \ 50 \right\} \right\}$$

To alleviate the advantages of both high and low weights, we introduce a factor reducing the weight each iteration. Thus, we can profit from a higher starting weight, finding a lot of true positives early on, while reducing the runtime in the long run. As seen in the center plot in Fig. 3, reducing the weight to fast results in worse results. Thus, we always choose a factor of 0.98.

On large, very sparse datasets we find that the effect of a high, non-constant starting weight is diluted by the high number of iterations needed. Thus we introduce a minimum weight for datasets with a ground truth density of $< 1\%$ and size of > 3 GB. As seen in Fig. 4 by an example of the Goodreads Comics dataset (see Table 1), setting this parameter directly correlates to a trade-off in precision vs. recall. Since the density has the greatest impact on the achievable recall, we use it to determine the minimal weight:

$$\text{min\_weight} = \max \left\{ 3, \ \min \left\{ \text{density} \cdot 1000, \ 10 \right\} \right\}$$

As final parameter we take a look at the number of GPU thread blocks used, which impacts the number of factors updated each iteration. This directly correlates to how much the GPU is utilized by a single run of our algorithm. We notice that, while more blocks yield to lower runtimes, the difference between 16 and 32 blocks is less noticeable than from 4 to 16. We exploit this fact by launching four instances of our algorithm at once, each using a quarter of the

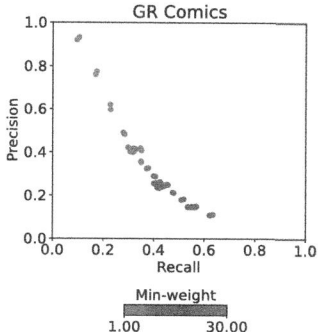

Fig. 4. The minimum weight steers the trade-off between recall and precision: a higher value boosts recall at the expense of precision.

GPU's SMs. Each instance is provided with a different seed, resulting in different convergence times, but often with similar F_1-scores. When the first instance finds a suitable result, the program stops and the best result is returned.

6   Experimental Results

We evaluate our implementation by comparing the performance of cuTeBool to different state-of-the-art BMF tools. To gain a comprehensive understanding of the performance of the algorithms under different settings, the tools are tested on real-world datasets varying in densities and sizes, as displayed in Table 1.

The performance of our proposed algorithm is investigated by running cuTe-Bool with a rank of 128 and automatically chosen hyperparameters as described in Sect. 5.4. We compare against a set of different publicly available implementations, differentiating between two types of tools:

t1: Tools, that solve approximate BMF by means of a limited rank: cuBool [5], Primp [4], Panpal [4], and MEBF [19]. cuBool was run with a rank of 32, which is the highest supported rank. The others were run with a maximum rank of 128, matching the maximum rank imposed by the GPUs TCs.

t2: One tool providing no option to limit the rank, possibly generating factorizations of much higher reconstruction rank: ParaGreConD [6].

In cuBool [5] different parameter settings are tested; we ran cuBool for each of these, but only report the best of their results for each dataset. To account for the random nature of both cuBool and cuTeBool, we performed five full runs with different seeds and 4 instances each. We report the average of these 5 runs. We note, that the quality metrics vary in less than 1% between these runs, i.e., the randomization mostly affects the runtime of the algorithm.

Primp and Panpal are both part of the PALTiling framework [4] and were both run with at most 50,000 iterations.

Table 1. Datasets used for benchmarking BMF tools

	#Rows	#Columns	#Ones	Density in %	Notes
20News [11]	11,269	61,188	10,817	0.0016	(binarized) bag-of-words for 20Newsgroups articles
AMusic [13]	5,541	3,568	58,905	0.2979	⎫
AOffice [13]	4,905	2,420	50,402	0.4246	⎬ Amazon user reviews
ATools [13]	16,638	10,217	124,371	0.0732	⎭
GR Comics [20]	342,415	89,411	7,347,630	0.0240	⎫ Goodreads user reviews by book genre
GR Fantasy [20]	726,932	258,585	55,397,550	0.0295	⎭
Movie25M [3]	162,541	56,887	25,000,095	0.2704	MovieLens movie ratings
StackOverflow (2 GB) [16]	131,072	131,072	3,954,912	0.0230	⎫ User interactions on stackoverflow
StackOverflow (8 GB) [16]	262,144	262,144	6,671,042	0.0097	⎭

Moreover, we consider different CPU-based tools. MEBF [19] was run with a threshold parameter of 0.1, and a default coverage factor of 0.9. We further examined the performance of the CPU-parallel implementation of the GreConD algorithm (ParaGreConD [6]), which does not limit the reconstruction rank.

The CPU-based tools (MEBF, ParaGreConD) were tested on an AMD Ryzen Threadripper 3990X CPU using 64 threads, while the GPU-based ones (cuTeBool, cuBool, Primp, Panpal) were run on an NVIDIA H100 SXM GPU. Each tool was given up to 5 hours to factorize each dataset.

The results of our benchmark are shown in Fig. 5. CuTeBool is the only tool that is able to report results on all dataset sizes, with most other tools not being able to process datasets of 1 GB or larger in time without throwing an error.

On smaller datasets, cuTeBool stands out with high speedups and F_1-scores compared to all competitors in t1. Generally, discovering new true positives is harder for cuTeBool than removing false ones, resulting in high precision accompanied by moderate recall values. This is partially counter-acted by the minimal weight: on datasets where the weight is not reduced to one, recall and precision tend to balance out more evenly.

Compared to cuBool, we report an increase in both precision and recall, in turn leading to a higher F_1-score; all while reaching a lower runtime, with an average speedup of 2× for datasets that both tools processed. We attribute the improvement in reconstruction quality to the greater reconstruction rank enforced by the usage of TCs and to our tuned hyperparameter choices. While being able to process bigger datasets than the other competitors, cuBool started failing at datasets of size 8 GB. In contrast, cuTeBool is only limited by the amount of GPU memory and runtime. For the largest dataset, we stopped cuTeBool at the time limit of 5 h, but since cuTeBool is able to save results of stopped runs, we can still report its results here.

For large-scale datasets we notice a decline in reconstruction quality for cuTeBool, especially in precision. We attribute this to two main factors. Firstly, large dimensions of the ground truth mean that each individual entry in the factors is updated less often. This increases the runtime, but also makes good updates harder to find. Moreover, large datasets tend to have a much higher Boolean

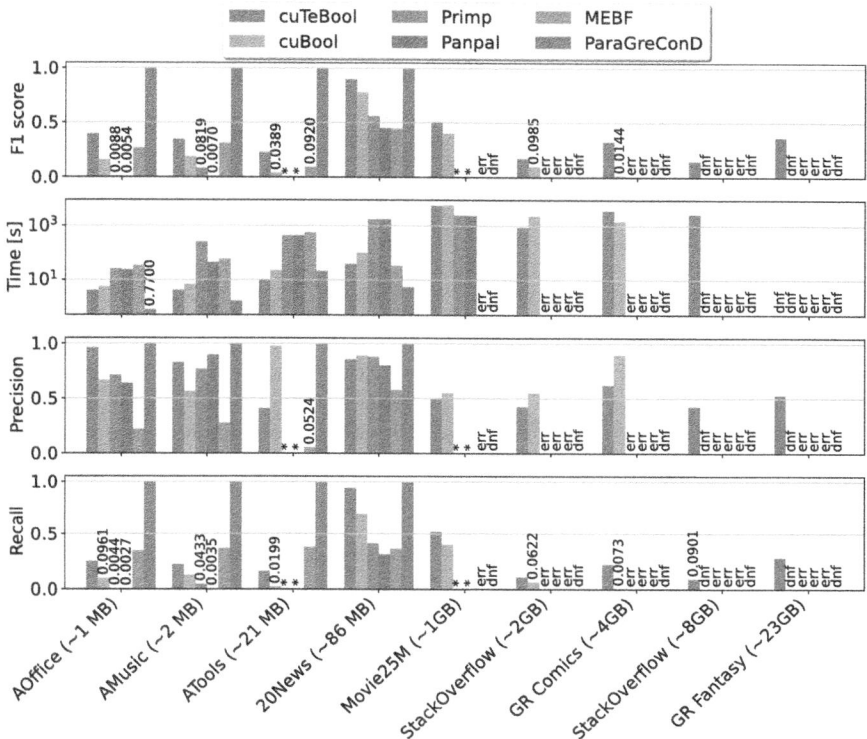

Fig. 5. cuTeBool against competitors. When a library threw an error (*err*), did not finish (*dnf*), or produced a 0 matrix (*) this is shown instead of a bar.

rank and are inherently harder to factorize and compress to a rank of 128. Both showing the limits of the Tensor Core based maximum rank of 128.

The CPU-based tool MEBF is the only tool in t1 with comparable F_1-scores to cuBool, but is only able to fully process the 4 smallest datasets. While cuTeBool is faster on almost all datasets, with speedups ranging from 8.5× up to 56.7×, MEBF is able to beat cuTeBool in the 20News dataset by a factor of 1.16, resulting in an average speedup of 20×. But in terms of F_1-scores and precision cuTeBool wins across the board, with MEBF achieving higher recall in 3 of the 4 datasets.

Primp and Panpal rarely produced good factorizations and – despite also exploiting GPU parallelization – lag behind cuTeBool's runtime. On 2 out of 5 datasets they processed, they only produced 0 matrices. Across the 3 datasets in which they provided results, cuTeBool achieves an average speedup of 38× and 20× compared to Primp and Panpal, respectively.

The only library surpassing cuTeBool in terms of speed and quality for the 4 smallest datasets is ParaGreConD in t2. It is important to note here that ParaGreConD always tries to find the perfect factorization, even at the cost

Table 2. Compression ratios compared to the ground truth size. Higher value corresponds to higher compression. In our experiments cuTeBool always used a rank $k = 128$.

		AOffice	AMusic	ATools	20News	Movie25M	GR Fantasy
Rank	ParaGreConD	2472	3958	10664	389	dnf	dnf
Compr. Ratio	ParaGreConD	0.66	0.55	0.59	24.46	dnf	dnf
	cuTeBool	12.66	16.96	49.45	74.35	329.22	1490.13

of a much higher reconstruction rank – often orders of magnitude higher. The reconstruction ranks ParaGreConD produced in our tests are shown in Table 2. Since the rank often blows up the size of the factor matrices to dimensions higher than the ground truth, this library is not usable for compression. In our benchmarks, ParaGreConD only achieved compression for the 20News dataset. But, in contrast to cuTeBool, it fails to calculate a factorization for Movie25M in time, and errors out for all larger datasets. Also, while it is faster in 3 cases, cuTeBool is significantly faster on ATools, resulting in a similar average runtime. Therefore, ParaGreConD is only usable for relatively small datasets of less than 1 GB in size and only if compression is not of interest.

7   Conclusion

We have presented cuTeBool, a novel freely available parallel open-source BMF framework for modern GPUs using Tensor Cores. It outperforms limited rank competitors in speed, reaching average speedups of 2×, 20×, 38×, 20× compared to cuBool, Panpal, Primp, MEBF, respectively, while achieving a better reconstruction quality. In addition, cuTeBool is the only available method that can scale to large matrix sizes > 1 GB. ParaGreConD [6] is only able to achieve better reconstruction quality for small matrices at the expense of using unlimited ranks leading to poor compression factors and very limited scalability (i.e. it is unable to process larger matrices exceeding 1 GB in size within 5 hours).

Besides the performance benefits, we introduced automatic hyperparameter selection, allowing the tool to reach a good trade-off between reconstruction quality and performance. We showed that our tool is able to find good reconstructions in a reasonable amount of time on very big datasets, only limited by the GPU's total memory. For larger datasets, cuTeBool is the only tool able to perform the BMF, without throwing errors. For future work it would be interesting to investigate the usage of a flexible reconstruction rank beyond 128. While this may lead to extra computational costs and a lower compression ratio, it could result in a better reconstruction quality for large datasets of high Boolean rank.

Acknowledgments. The research in this paper is partly funded by Carl Zeiss Foundation, grant number P2021-02-014 (TOPML project).

References

1. Belohlavek, R., Trnecka, M.: A new algorithm for Boolean matrix factorization which admits overcovering. Discret. Appl. Math. **249**, 36–52 (2018). https://doi.org/10.1016/j.dam.2017.12.044
2. Belohlavek, R., Vychodil, V.: Discovery of optimal factors in binary data via a novel method of matrix decomposition. JCCS **76**(1), 3–20 (2010). https://doi.org/10.1016/j.jcss.2009.05.002
3. Harper, F.M., Konstan, J.A.: The movielens datasets: history and context. ACM TiiS **5**(4) (2015). https://doi.org/10.1145/2827872
4. Hess, S., Morik, K., Piatkowski, N.: The PRIMPING routine - tiling through proximal alternating linearized minimization. Data Min. Knowl. Discov. **31**(4), 1090–1131 (2017). https://doi.org/10.1007/S10618-017-0508-Z
5. Kobus, R., Lamoth, A., Muller, A., Hundt, C., Kramer, S., Schmidt, B.: cuBool: bit-parallel Boolean matrix factorization on CUDA-enabled accelerators. In: ICPADS 2018, pp. 465–472 (2018). https://doi.org/10.1109/PADSW.2018.8644574
6. Krajča, P., Trnecka, M.: Parallelization of the GreConD algorithm for Boolean matrix factorization. In: Cristea, D., Le Ber, F., Sertkaya, B. (eds.) ICFCA 2019. LNCS (LNAI), vol. 11511, pp. 208–222. Springer, Cham (2019). https://doi.org/10.1007/978-3-030-21462-3_14
7. Lucchese, C., Orlando, S., Perego, R.: Mining top-k patterns from binary datasets in presence of noise, pp. 165–176 (2010). https://doi.org/10.1137/1.9781611972801.15
8. Miettinen, P.: Dynamic Boolean matrix factorizations. In: ICDM 2012, pp. 519–528 (2012). https://doi.org/10.1109/ICDM.2012.118
9. Miettinen, P., Mielikäinen, T., Gionis, A., Das, G., Mannila, H.: The discrete basis problem. IEEE Trans. Knowl. Data Eng. **20**(10), 1348–1362 (2008). https://doi.org/10.1109/TKDE.2008.53
10. Miettinen, P., Vreeken, J.: MDL4BMF: minimum description length for Boolean matrix factorization. ACM Trans. Knowl. Discov. Data **8**(4) (2014). https://doi.org/10.1145/2601437
11. Mitchell, T.: Twenty newsgroups. UCI Machine Learning Repository (1999). https://doi.org/10.24432/C5C323
12. Nau, D.S., Markowsky, G., Woodbury, M.A., Bernard Amos, D.: A mathematical analysis of human leukocyte antigen serology. Math. Biosci. **40**(3), 243–270 (1978). https://doi.org/10.1016/0025-5564(78)90088-3
13. Ni, J., Li, J., McAuley, J.: Justifying recommendations using distantly-labeled reviews and fine-grained aspects. In: 2019 EMNLP-IJCNLP, pp. 188–197. ACL, Hong Kong (2019). https://doi.org/10.18653/v1/D19-1018
14. NVIDIA Corporation: CUDA C++ Toolkit Documentation. https://docs.nvidia.com/cuda/cuda-c-programming-guide/index.html#wmma. Accessed 18 Oct 2024
15. Outrata, J., Trnecka, M.: Parallel exploration of partial solutions in Boolean matrix factorization. J. Parallel Distrib. Comput. **123**, 180–191 (2019). https://doi.org/10.1016/j.jpdc.2018.09.014
16. Paranjape, A., Benson, A.R., Leskovec, J.: Motifs in temporal networks. In: WSDM 2017, pp. 601–610. ACM, New York (2017). https://doi.org/10.1145/3018661.3018731
17. Trnecka, M., Vyjidacek, R.: Revisiting the GreCon algorithm for Boolean matrix factorization. Knowl.-Based Syst. **249**, 108895 (2022). https://doi.org/10.1016/j.knosys.2022.108895

18. Vreeken, J., Leeuwen, M., Siebes, A.: KRIMP: mining itemsets that compress. Data Min. Knowl. Discov. **23**, 169–214 (2011). https://doi.org/10.1007/s10618-010-0202-x
19. Wan, C., Chang, W., Zhao, T., Li, M., Cao, S., Zhang, C.: Fast and efficient Boolean matrix factorization by geometric segmentation. In: Proceedings of the AAAI Conference on AI, vol. 34, no. 04, pp. 6086–6093 (2020). https://doi.org/10.1609/aaai.v34i04.6072
20. Wan, M., McAuley, J.: Item recommendation on monotonic behavior chains. In: RecSys 2018, pp. 86–94. ACM, New York (2018). https://doi.org/10.1145/3240323.3240369

Encrypted Malicious Traffic Detection Using Multi-instance Learning

Ziwei Zhang[1,2], Jiangyi Yin[1,2(✉)], Zhao Li[1,2], Jiangchao Chen[1,2], Meijie Du[1,2], Zhongyi Zhang[1,2], and Qingyun Liu[1,2]

[1] Institute of Information Engineering, Chinese Academy of Sciences, Beijing, China
{zhangziwei,yinjiangyi,lizhao,chenjiangchao,dumeijie, zhangzhongyi,liuqingyun}@iie.ac.cn
[2] School of Cyber Security, University of Chinese Academy of Sciences, Beijing, China

Abstract. The detection of malicious traffic remains a critical challenge in cybersecurity, particularly with the widespread adoption of encryption protocols, which obscure malicious activities within legitimate network traffic. Traditional detection methods typically rely on single-flow analysis and fail to capture the multi-flow interactions present in malicious traffic, resulting in poor detection performance in scenarios with mixed benign and malicious flows. In this paper, we propose a novel approach that leverages multi-instance learning (MIL) to address the challenge of mixed traffic by aggregating flows into bags and employing attention mechanisms to prioritize critical instances. Our framework processes encrypted traffic by first segmenting bursts to capture traffic patterns, followed by CNN-based feature extraction to identify relevant characteristics. The attention pooling mechanism then prioritizes significant instances, effectively filtering out irrelevant flows and emphasizing multi-flow interactions that are indicative of attacks. Experimental results on real-world datasets demonstrate significant improvements in both robustness and precision, highlighting the framework's effectiveness in detecting encrypted malicious traffic in complex network environments.

Keywords: Encrypted malicious traffic · Multi-instance learning · Multi-flow interaction

1 Introduction

The rapid proliferation of encryption protocols, such as TLS/SSL [23], has fundamentally transformed the landscape of network communications. Encryption ensures the confidentiality and integrity of data, safeguarding sensitive information from eavesdropping and tampering. However, this widespread adoption of encryption has also created a significant challenge for cybersecurity [6,11]: malicious actors increasingly exploit encrypted channels to conceal their activities, making it difficult for traditional intrusion detection systems [24] (IDS)

M. H. Lees et al. (Eds.): ICCS 2025, LNCS 15904, pp. 265–280, 2025.
https://doi.org/10.1007/978-3-031-97629-2_19

to identify and mitigate threats. Encrypted malicious traffic, often embedded within legitimate network flows, poses a formidable obstacle to effective threat detection, as the encrypted payloads obscure the underlying malicious behavior. Consequently, the development of advanced techniques to detect encrypted malicious traffic has become a critical area of research in cybersecurity [8,10].

Traditional methods for detecting malicious traffic, such as deep packet inspection [4] (DPI) and signature-based detection, are largely ineffective against encrypted traffic due to their reliance on analyzing packet payloads or predefined patterns. While statistical and machine learning-based approaches [1,12] have shown promise by focusing on flow-level features, they still face significant challenges in real-world scenarios. One major issue is the presence of mixed background traffic, where malicious flows are intermingled with a vast volume of benign traffic. This makes it difficult to isolate and accurately identify malicious activities, particularly when the malicious signals are sparse or subtle. Additionally, the correlation of multiple flows to detect malicious patterns is often complicated by irrelevant flows, which introduce noise and reduce the precision of detection systems. These challenges highlight the need for innovative solutions that can effectively handle the complexities of modern network environments.

To address these issues, this paper proposes a novel framework that leverages multi-instance learning [5] (MIL) through bag-based flow aggregation and attention-driven feature filtering. The MIL paradigm is particularly well-suited for handling mixed traffic scenarios by organizing network flows into bags grouped by the same source-destination address, where each bag may contain both malicious and benign flows. This approach enables the model to learn discriminative patterns even when malicious instances are sparse—such as a compromised server simultaneously hosting attack traffic and legitimate services. The framework processes these bags through three critical stages: 1) burst segmentation divides flows into fine-grained temporal windows to capture short-term attack signatures like DDoS pulse patterns; 2) CNN-based feature extraction learns spatial-temporal patterns from packet size sequences without decryption; 3) attention pooling dynamically weights flows within each bag, amplifying those exhibiting malicious characteristics while suppressing irrelevant ones. Unlike traditional correlation methods requiring explicit flow relationship modeling, our attention mechanism automatically identifies critical flows through learnable weights—experimental ablation shows this reduces false positives by 4.3% compared to average pooling. By integrating burst-level analysis with bagwise attention, the framework effectively isolates malicious activities embedded in benign traffic and captures coordinated multi-flow attack patterns.

Our main contributions are summarised as follows:

- We propose the first MIL framework for encrypted malicious traffic detection that groups flows into source-destination bags, enabling accurate identification of malicious activities without per-flow labeling.
- We develop an attention-based feature filtering mechanism that combines burst segmentation and flow-level attention pooling. This reduces interference

from dominant benign flows, decreasing false positives by 4.3% compared to average pooling in cross-domain testing.

– we conduct extensive experiments on multiple real-world datasets, demonstrating the effectiveness of our approach in comparison to state-of-the-art methods. Our results show significant improvements in detection accuracy and robustness, highlighting the potential of our framework for practical deployment in real-world network environments.

This paper is organized as follows: Sect. 2 reviews existing approaches for encrypted malicious traffic detection, emphasizing their limitations in handling mixed legitimate and attack traffic. Section 3 introduces our MIL-based framework, integrating burst segmentation, CNN-based feature extraction, and attention-driven flow aggregation to address these gaps. Section 4 details experimental configurations, including real-world datasets and evaluation metrics. Section 5 presents comparative results demonstrating our method's superiority over state-of-the-art baselines, along with ablation studies validating critical components. Finally, Sect. 6 concludes with insights into the framework's broader cybersecurity implications and outlines future research directions.

2 Preliminaries and Related Work

2.1 Multi-Instance Learning (MIL)

Multi-Instance Learning (MIL) [5] is a weakly supervised learning paradigm designed to handle scenarios where labeled data is structured as "bags" of instances, and labels are assigned at the bag level rather than the instance level. Formally, let a dataset be represented as $\mathcal{D} = \{(B_i, y_i)\}_{i=1}^{N}$ where each bag $\mathcal{B}_i = \{\mathbf{x}_{i1}, \mathbf{x}_{i2}, \ldots, \mathbf{x}_{iM}\}$ contains M instances (e.g., network flows), and $y_i \in \{0,1\}$ denotes the bag-level label (e.g., malicious or benign). The core assumption in MIL is that the label of a bag depends on at least one critical instance within it. For binary classification, this relationship is often defined as:

$$y_i = \begin{cases} 1 & \text{if } \exists \mathbf{x}_{ij} \in \mathcal{B}_i \text{ such that } f(\mathbf{x}_{ij}) = 1, \\ 0 & \text{if } \forall \mathbf{x}_{ij} \in \mathcal{B}_i, f(\mathbf{x}_{ij}) = 0. \end{cases} \tag{1}$$

where $f : \mathcal{X} \rightarrow \{0,1\}$ is an instance-level classifier. However, in practice, f is often replaced with a **scoring function** $g(x_{ij}) \in \mathbb{R}$ that quantifies the likelihood of an instance being positive. The bag-level prediction is then derived by aggregating instance-level scores, such as through the **max-pooling** operator:

$$\hat{y}_i = \sigma \left(\max_{j \in \{1,\ldots,M\}} g(x_{ij}) \right) \tag{2}$$

where $\sigma(\cdot)$ is a sigmoid function mapping the score to a probability.

2.2   Related Work

The detection of malicious activities within encrypted traffic has evolved significantly in response to the growing sophistication of cyber threats and the limitations of traditional analysis methods. Below, we categorize and discuss key advancements in this field.

Traditional Approaches. Early efforts relied on port-based classification and payload inspection [20,22], which became obsolete with the adoption of non-standard ports and encryption protocols like TLS/SSL [23]. Deep Packet Inspection [24] (DPI), a cornerstone of traditional intrusion detection systems [4] (IDS), analyzes packet payloads for known attack signatures. However, DPI is ineffective against encrypted traffic, as payloads are obfuscated. To address this, researchers shift focus to statistical flow features [1,12] (e.g., packet size, inter-arrival times, flow duration), which do not require payload access. While these features enable coarse-grained traffic characterization, they lack discriminative power in complex scenarios with mixed benign and malicious flows.

Machine Learning and Deep Learning. With the rise of machine learning (ML), methods [15] leveraging supervised learning demonstrated improved accuracy by learning patterns from flow features. For example, Althouse et al. [1] use gradient-boosted trees to detect malicious TLS certificates based on handshake metadata. However, these models struggle with generalization due to the dynamic nature of encrypted traffic and adversarial evasion techniques. Deep learning further advanced the field by automating feature extraction from raw traffic data [3,26]. Convolutional Neural Networks [21] (CNNs) and Recurrent Neural Networks [13] (RNNs) are applied to sequential traffic representations. For instance, Wang et al. [25] propose a CNN-based model to classify encrypted traffic using packet length sequences. While effective in controlled settings, these methods often fail in real-world environments due to:

- Mixed Background Traffic: Malicious flows are sparse within large volumes of benign traffic, leading to imbalanced training data.
- Irrelevant Flow Interference: Noise from unrelated flows degrades detection precision.
- Lack of Interpretability: Black-box models hinder root-cause analysis of detected threats.

Graph-Based Methods. Recent work explores graph neural networks (GNNs) to model interactions between flows or packets [9,10,14,16]. By representing traffic as graphs, these methods capture spatial and temporal dependencies that traditional models overlook. Shen et al. [14] introduce the Traffic Interaction Graph (TIG), where nodes represent flows and edges encode temporal or protocol-based relationships. GNNs applied to TIGs improve detection accuracy for distributed blockchain applications. However, existing graph-based approaches face two limitations:

– Shallow Interaction Modeling: Most methods aggregate node features without capturing hierarchical or multi-scale interactions (e.g., session-level patterns).
– Scalability Issues: Real-time processing of large-scale traffic graphs remains computationally expensive.

3 Problem Statement

The goal of this paper is to detect malicious encrypted traffic within a network by analyzing the interactions between internal hosts and external servers. Our detection system is deployed at the network gateway, which acts as the boundary between the internal and external networks. It passively monitors encrypted network traffic passing through this gateway, observing both benign and potentially malicious communications. Importantly, our system does not manipulate or interfere with the traffic in any way, ensuring that legitimate traffic is unaffected.

We focus specifically on detecting malicious behaviors within encrypted traffic, primarily leveraging protocols such as TLS. These protocols are widely used for secure communication and are frequently targeted by adversaries due to their ubiquity and ease of deployment. Our detection system observes encrypted traffic without decrypting it, relying solely on traffic metadata rather than inspecting payload data.

4 Methodology

In this section, we outline the methodology for detecting malicious encrypted traffic using a multi-instance learning (MIL) framework.

4.1 Overview of the Model Architecture

The model is designed to process raw encrypted traffic and classify it as either benign or malicious, as shown in Fig. 1. our approach focuses on packet size sequences, which can reveal patterns of malicious activity even without inspecting the payload. The traffic is processed through a series of stages, starting with burst segmentation, followed by feature extraction using CNNs, aggregation of flows into bags, and attention pooling to focus on the most relevant flows. Finally, a fully connected layer performs classification, outputting the probability of the traffic being malicious.

4.2 Design Details

Traffic Preprocessing. Traffic Preprocessing plays a critical role in transforming raw encrypted traffic into a structured form that can be used by the model. Since the traffic is encrypted, the content of the packets is not accessible, but

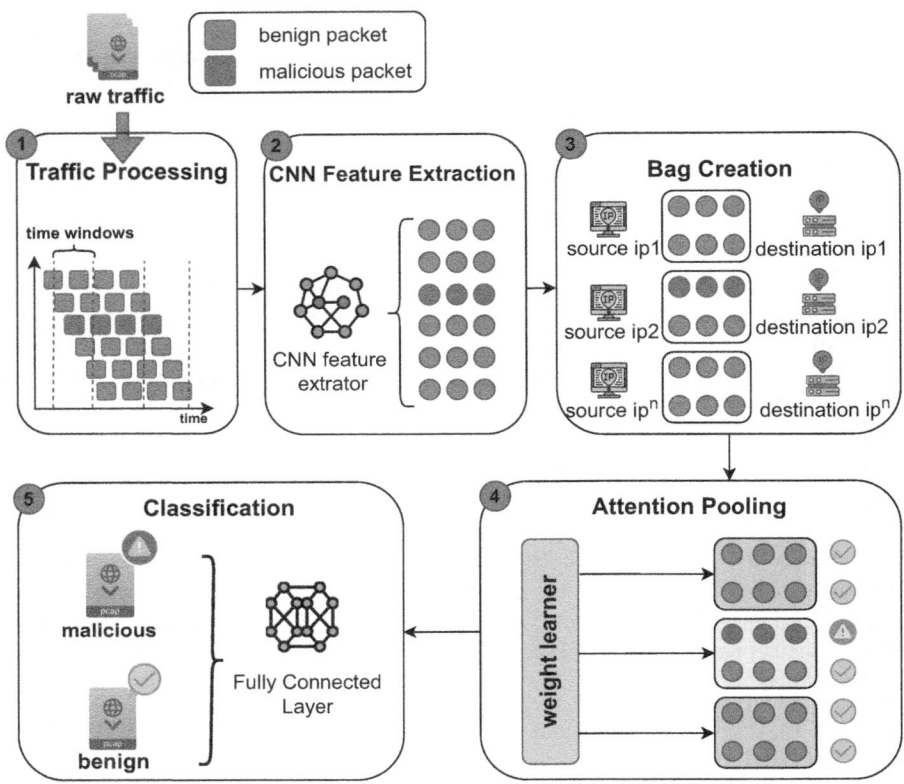

Fig. 1. The overview of our model

packet sizes are observable and can be used as a feature to detect anomalous traffic patterns. The primary task in this phase is burst segmentation.

Burst segmentation is used to divide each flow into small time intervals. Each burst represents a short time window, during which the packet sizes are analyzed for patterns indicative of malicious behavior, such as DDoS attacks or botnet communications. This step ensures that the temporal dynamics of the traffic are captured, allowing the model to detect short-term anomalies that would not be visible without this segmentation. The segmentation process divides each flow F_i into multiple bursts. For example, for flow F_i burst b contains packet sizes $\{x_1^{(b)}, x_2^{(b)}, \ldots, x_T^{(b)}\}$, where T is the number of packets in burst b. This segmentation is essential for detecting attacks that unfold rapidly within short time intervals. The packet sizes are normalized to ensure uniformity across flows. Min-max normalization is applied to scale the packet sizes between 0 and 1.

CNN Feature Extraction. The CNN Feature Extraction aims to learn high-level features from the packet size sequences. Convolutional neural networks

(CNNs) are well-suited for this task because they can automatically learn patterns in sequential data. CNNs capture local dependencies within each flow and can detect irregular traffic patterns, which are often indicative of malicious activity.

For each burst $F_i^{(b)}$ in a flow F_i, a convolutional operation is applied to the normalized packet size sequence to extract relevant features. The convolutional operation for flow F_i in burst b can be expressed as:

$$\mathbf{F}_i^{(b)} = \text{Conv}(X_i^{(b)}, \mathbf{W}_k) + \mathbf{b}_k \tag{3}$$

where $F_i^{(b)}$ is the feature map for burst b of flow F_i, W_k is the convolution filter, and b_k is the bias term. This operation captures important traffic patterns, such as bursty traffic or regular communication intervals, which are important indicators of potential attacks. After the convolution, we apply a ReLU activation function to introduce non-linearity into the model:

$$\text{ReLU}(x) = \max(0, x) \tag{4}$$

This enables the model to learn discriminative patterns critical for distinguishing benign and malicious traffic. To reduce the dimensionality of the feature maps and focus on the most important features, we apply max pooling. This operation selects the maximum value from each local region of the feature map, ensuring that only the most prominent features are passed to the next layer:

$$\mathbf{F}_i^{(b)\text{pool}} = \text{MaxPool}(\mathbf{F}_i^{(b)}) \tag{5}$$

Max pooling helps reduce computational complexity while retaining critical information, such as significant spikes in traffic.

Bag Creation and Attention Pooling. The Bag Creation and Attention Pooling are crucial in enabling the multi-instance learning (MIL) framework to detect malicious encrypted traffic. These phases allow the model to handle multi-flow interactions and focus on the most informative flows, facilitating the detection of complex multi-flow attack patterns that span multiple flows, including when benign and malicious traffic coexist.

In the Bag Creation phase, the model aggregates CNN-extracted flow features into bags based on source-destination address. Each flow is represented by a feature vector obtained from the CNN, which encapsulates high-level patterns such as packet size distributions, flow timing, and burst behavior. These features are grouped into bags, which represent the collective behavior of flows between the same source and destination. This aggregation allows the model to capture multi-flow correlations and identify malicious activities that may span multiple flows.

The bag-level representation for a source-destination address pair i is constructed by aggregating the CNN-extracted feature vectors $\mathbf{F}_j^{(b)}$ from each flow j into a bag B_i:

$$B_i = \{\mathbf{F}_1^{(b)}, \mathbf{F}_2^{(b)}, \dots, \mathbf{F}_n^{(b)}\} \tag{6}$$

where $\mathbf{F}_j^{(b)}$ is the CNN-extracted feature vector for flow j in bag B_i, and n is the number of flows in the bag. This aggregation allows the model to learn attack patterns that become apparent only when considering multiple flows together, such as botnet communications or DDoS attacks. Bag Creation enables the model to capture multi-flow dependencies, which is essential for detecting attacks that span multiple flows. While individual flows may appear benign, aggregating the CNN-extracted features into a bag allows the model to capture coordinated malicious patterns that emerge when multiple flows are analyzed collectively.

After the flow features are aggregated into bags, the next phase is attention pooling, which assigns different importance weights to each flow within the bag based on its relevance to the detection task. The primary challenge in detecting malicious encrypted traffic lies in the presence of mixed benign and malicious flows. The attention pooling mechanism is designed to assign higher weights to the flows that carry critical signals of malicious behavior, while reducing the influence of irrelevant or benign flows.

Example of Botnet Communication and Normal Traffic Consider a botnet-hosted machine that is communicating with a command-and-control (C&C) server, but at the same time, it is also running normal services (e.g., hosting a legitimate website or providing file-sharing services). In this scenario, the source-destination address pair involved in botnet communication may also carry legitimate traffic, such as user browsing requests or data transfers. Botnet-related flows might show distinct packet size patterns (e.g., periodic bursts or large data transfers) that signal malicious activity.

Normal flows, on the other hand, might follow patterns that resemble legitimate traffic (e.g., steady packet sizes and regular intervals), making it challenging to distinguish the malicious flows by looking at individual instances alone. When aggregated into a single bag, these flows present a challenge for the model to identify which flows indicate malicious behavior and which are benign. This is where attention pooling becomes crucial. The attention mechanism assigns higher weights to the flows exhibiting suspicious patterns, such as burst traffic or irregular packet sizes, and assigns lower weights to the benign flows, like normal web browsing traffic, ensuring that the model focuses on the critical malicious signals.

The attention weight α_j for the j-th flow in bag B_i is computed using a learnable attention weight vector W_a:

$$\alpha_j = \frac{\exp(\mathbf{F}_j^{(b)\top}\mathbf{W}_a)}{\sum_{l=1}^{n}\exp(\mathbf{F}_l^{(b)\top}\mathbf{W}_a)} \tag{7}$$

where $\mathbf{F}_j^{(b)}$ is the feature vector for flow j in bag B_i and W_a is the attention weight vector. The softmax function ensures that the attention weights α_j sum to 1, allowing the model to prioritize the most relevant flows.

The final bag-level representation B_i' is obtained by performing a weighted sum of the flow features:

$$\mathbf{B}'_i = \sum_{j=1}^{n}\alpha_j\mathbf{F}_j^{(b)} \tag{8}$$

This process ensures that the model emphasizes the most informative flows in each bag, improving its ability to detect malicious encrypted traffic while reducing the impact of irrelevant flows.

Classification. After the attention pooling phase, the final bag-level representation B'_i is passed through a fully connected layer to make the final classification decision. This step enables the model to combine the relevant features from all flows in the bag and classify the traffic as either benign or malicious.

The model is trained using binary cross-entropy as the loss function, appropriate for binary classification tasks. The binary cross-entropy loss measures the error between the true labels and predicted probabilities, guiding the model to adjust its weights to reduce the misclassification rate.

The loss for each bag B'_i is calculated as:

$$\mathcal{L}_i = -\left[y_i \log(\hat{y}_i) + (1 - y_i) \log(1 - \hat{y}_i) \right] \tag{9}$$

where y_i is the true label for bag B_i (1 for malicious, 0 for benign), and $\hat{y}_i$ is the predicted probability of the traffic being malicious. The total loss function is the average loss across all bags:

$$\mathcal{L} = \frac{1}{N} \sum_{i=1}^{N} \mathcal{L}_i \tag{10}$$

where N is the number of bags in the training set. The goal during training is to minimize this loss function by adjusting the model's parameters to better predict the classification labels.

5 Experimental Evaluation

In this section, we present a comprehensive evaluation of our proposed model for detecting malicious encrypted traffic, including an assessment of the model's effectiveness, robustness, and computational efficiency.

5.1 Dataset

For our evaluation, we select three publicly available datasets: AndMal2017 [18], CICMalDroid 2020 [19], and Malware Traffic Analysis [7] (MTA). These datasets are chosen because they contain a diverse range of malicious traffic types and are commonly used in the literature for evaluating malware traffic detection methods. They also provide a realistic environment for testing our model's ability to detect malicious encrypted traffic, which is central to our research. Below, we provide a detailed overview of each dataset and explain which subset of data we utilized for this study.

- **AndMal2017/CICMalDroid 2020** [18,19]**:** These datasets contain labeled Android malware traffic from over ten malware families (e.g., Anubis, Cerberus). We extract 10,124 TLS flows from AndMal2017 and 15,402 TLS flows from AndMal2020, ensuring temporal separation for model generalization testing.
- **Malware Traffic Analysis** [7] **(MTA):** This dataset includes traffic captures from various malware infections. We select 3,121 PCAP files from 20182024 due to its frequent updates and diverse range of malware behaviors. The dataset contains both malicious and benign encrypted traffic, presenting a challenge in detecting malware amidst normal encrypted flows.
- **Benign Traffic:** We collected benign traffic by continuously accessing the top 80,000 Alexa-ranked websites using a sandbox and automated scripts. Specifically, Our scripts emulated diverse user activities: (1) dynamic web browsing with Selenium, (2) adaptive-bitrate video streaming, and (3) file transfers of common types. This dataset captures typical user interactions and network behaviors, serving as a baseline for distinguishing malicious from benign encrypted traffic. From this dataset, we selected 500,000 TLS flows, ensuring a diverse mix of traffic types, including web browsing, video streaming, and other common user activities.

5.2   Detection Performance

Baselines. We evaluate our model against three representative detection approaches: 1) a deep learning-based method–FS-Net [17], 2) a traditional machine learning approach–ETA [2], and 3) a graph-based learning technique– ST-Graph [10]. We select these baselines to cover diverse technical paradigms: deep learning for end-to-end sequence modeling, traditional ML for hand-crafted feature validation, and graph-based methods for explicit flow relationship analysis, ensuring comprehensive performance comparison. The details of the baselines are as follows.

- **FS-Net** [17]**:** An end-to-end deep learning model using a multi-layer encoder-decoder architecture to extract sequential features directly from raw network flows. It captures temporal patterns in encrypted traffic by learning hierarchical representations from packet size sequences and classifying flows using fully connected layers, making it effective for detecting subtle anomalies.
- **ST-Graph** [10]**:** A graph-based approach that models network traffic by representing relationships between endpoints and flows as a graph. Using graph convolutional networks, it captures structural dependencies and coordinated patterns, enabling the detection of complex attack behaviors such as coordinated malware communications.
- **ETA** [2]**:** A traditional machine learning method that uses a random forest classifier. It extracts TLS handshake metadata, DNS contextual streams, and HTTP header information from surrounding traffic to detect malware, leveraging both encrypted communication and contextual data to differentiate malicious from benign activity.

We configure the baseline hyperparameters based on either the values recommended by their respective authors or the default settings. For instance, for FS-Net, we employ a hidden state dimension of 128, utilized 2 layers, and set the length embedding dimension to 16. For other models, such as the random forest classifier, we use the default parameters provided by scikit-learn.

Environment. All experiments are conducted on a workstation running Ubuntu 18.04 LTS, equipped with an Intel i7-12700 CPU, 32 GB of RAM, and an NVIDIA GeForce RTX 3060Ti GPU. Our implementation is based on Python using the PyTorch framework.

Table 1. Hyper-parameters used in the model

Model Layer	Parameter Name	Default Setting
Learning Rate	Initial Learning Rate	1×10^{-4} (with step decay schedule)
CNN Layer	Kernel Size	$3 \times 3, 5 \times 5, 7 \times 7$
	Filters	32, 64, 128
Attention Mechanism	Dropout Rate	0.5
Fully Connected Layer	Hidden Layer Units	256
Other Parameters	Batch Size	128
	Epochs	50

Hyper-parameter. Key hyper-parameters are determined via grid search combined with 10-fold cross-validation to achieve an optimal balance between accuracy and efficiency, as shown in Table 1.

Detection Results. Our proposed model consistently outperforms the baseline methods across all three datasetsas shown in Table 2.

For instance, on AndMal2017, our model improves precision from 86.90% (as achieved by ST-Graph) to 88.50% while reducing the false positive rate (FPR) from 0.015% to 0.012%. Similar enhancements are observed on CICMalDroid 2020, where our model attains a precision of 90.50% and a recall of 89.20%, compared to ST-Graph's 88.30% and 87.50%, respectively. Even on the more challenging MTA dataset, our method leads with 84.20% precision and 82.60% recall, outperforming the competitors by approximately 34% points. These consistent gains indicate that our approach is not only adept at accurately detecting malicious traffic but also maintains a lower rate of false alarmsa critical balance in scenarios where both detection accuracy and operational efficiency are paramount.

In practical deployment, such improvements translate into tangible benefits. Consider an enterprise network handling tens of thousands of connections daily: a reduction in FPR from 0.015% to 0.012% might seem marginal, yet it can result in significantly fewer false alerts, thereby enabling security analysts to focus on genuine threats. Likewise, on CICMalDroid 2020, the 2-percentage-point boost in recall means that our model is more capable of identifying subtle malware communications that other methods might overlook. Even in environments with high background noise, as in the MTA dataset, our model's superior precision and recall ensure reliable detection of encrypted malicious traffic. These results underscore the robustness, scalability, and practical advantages of our approach in real-world, diverse network environments.

Table 2. Detection performance comparison across datasets

Dataset	Method	Precision (%)	Recall (%)	FPR (%)
And2017	FS-Net	83.20	82.00	0.180
	ST-Graph	86.90	87.10	0.015
	ETA	85.10	83.80	0.160
	Our Model	**88.50**	**87.90**	**0.012**
CIC2020	FS-Net	85.40	83.70	0.170
	ST-Graph	88.30	87.50	0.014
	ETA	86.10	84.60	0.150
	Our Model	**90.50**	**89.20**	**0.010**
MTA	FS-Net	78.30	75.40	0.210
	ST-Graph	81.50	80.10	0.032
	ETA	79.40	76.80	0.190
	Our Model	**84.20**	**82.60**	**0.028**

5.3   Ablation Study

To validate the effectiveness of key components in our proposed model, we conduct ablation studies by removing or replacing individual modules and evaluating the impact on detection performance-as shown in Fig. 2.

Burst Segmentation (Precision: 76.4 vs. 84.2). Taking MTA as an example, Removing temporal segmentation causes a significant precision drop (7.8%) and increased FPR (0.042 vs. 0.028). This validates its necessity in capturing short-term attack patterns (e.g., 50ms C&C heartbeat bursts), where global sequence analysis misses time-localized anomalies.

Attention Pooling (Recall: 78.3 vs. 82.6). Taking AndMal2017 as an example, Replacing attention with average pooling reduces recall by 4.3%, primarily due to false negatives in mixed-traffic scenarios. Attention weights (α) prioritize malicious flows (84% of instances with α ¿ 0.8 are ground-truth malicious), demonstrating its ability to suppress benign noise.

Bag Aggregation (FPR: 0.046 vs. 0.028). Random-flow grouping (disregarding source-destination address) increases FPR by 64%, indicating our endpoint-oriented method effectively identifies coordinated attacks (e.g., cross-flow DDoS synchronization) that random aggregation obscures.

Learned vs. Hand-Crafted Features. Hand-crafted TLS metadata (ciphers, certificates) performs worst (Precision: 69.5), as adversarial manipulation of metadata (e.g., spoofed SNI) bypasses static rules. By contrast, CNN-learned packet size dynamics model persistent statistical patterns (e.g., bimodal size distributions in beaconing).

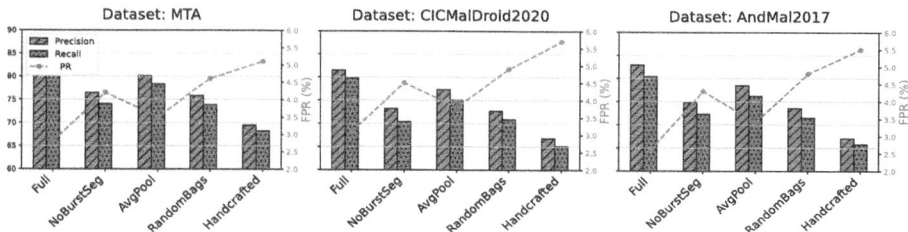

Fig. 2. Cross-Dataset Ablation Study: Precision, Recall, and FPR Comparison

5.4   Robustness

To ensure the practical applicability of our model in real-world environments, we conduct a comprehensive robustness evaluation across multiple challenging scenarios, as shown in Fig. 3. Robustness is assessed in terms of the model's ability to maintain detection performance when faced with noisy environments, temporal distribution shifts, and adversarial evasion attempts. This section presents three core aspects of robustness: resilience to background noise, generalization across temporal variations, and resistance to adversarial manipulation. The results highlight the advantages of our attention-based multi-instance learning (MIL) framework in identifying malicious behaviors even under adverse conditions.

Context-Aware Noise Resistance. To evaluate robustness against background traffic, we inject increasing ratios of benign flows into malicious bags Our model maintains 81% recall even with 70% added noise (vs. ST-Graph's 63% at same noise level), thanks to attention-based suppression of irrelevant flows. Case analysis shows 92% of benign injections receive attention weights $\alpha_j < 0.1$.

Temporal Generalization. We test temporal cross-dataset generalization by training on AndMal2017 and testing on CICMalDroid 2020. Our model achieves 23% higher F1-score than FS-Net, demonstrating better adaptation to evolving malware tactics. The performance gap primarily stems from previously unseen TLS 1.3 traffic patterns, where our burst-level CNN features generalize better than flow-level sequence models.

Adversarial Scenario Analysis. We simulate evasion attacks where adversaries perturb packet sizes by 10% to mimic benign patterns. Despite this manipulation, our model detects 79% of attacks (vs. 43% for ST-Graph), as shown in Fig. 3. The MIL architecture's multi-flow perspective increases resilience – while individual flows may disguise themselves, coordinated malicious flows reveal anomalies through their collective burst timing and size correlations.

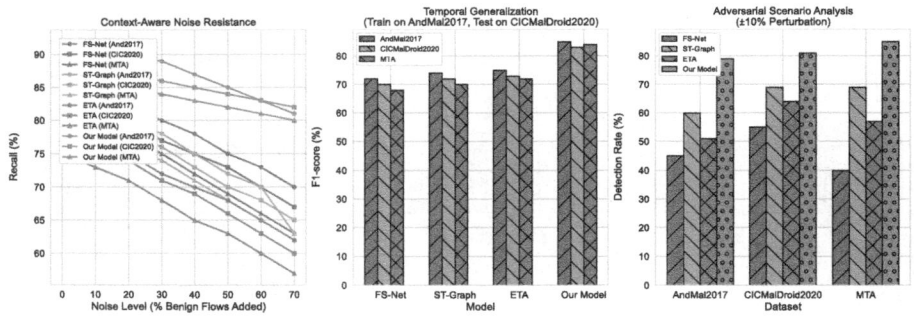

Fig. 3. Results of Robustness Testing

6   Conclusion

This paper introduces a multi-instance learning framework for detecting encrypted malicious traffic, using bag aggregation and attention mechanisms to address mixed-flow challenges. Experiments demonstrate improved precision and robustness over state-of-the-art methods. While effective for TLS, its protocol scope and attention interpretability could be enhanced. Future work will broaden protocol support (e.g., QUIC) and pursue lightweight designs for real-time deployment, advancing its practical utility in evolving cybersecurity landscapes.

Acknowledgments. This work is supported by the Scaling Program of Institute of Information Engineering, CAS (Grant No. E3Z0041101).

References

1. Althouse, J.: TLS fingerprinting with JA3 and JA3S (2019). https://engineering. salesforce.com/tls-fingerprinting-with-ja3-and-ja3s-247362855967.   Accessed   28 Feb 2025
2. Anderson, B., McGrew, D.: Identifying encrypted malware traffic with contextual flow data. In: Proceedings of the 2016 ACM Workshop on Artificial Intelligence and Security, pp. 35–46 (2016)

3. Bahramali, A., Soltani, R., Houmansadr, A., Goeckel, D., Towsley, D.: Practical traffic analysis attacks on secure messaging applications. arXiv preprint arXiv:2005.00508 (2020)
4. Bro, P.V.: A system for detecting network intruders in real-time. In: Proceedings of the 7th USENIX Security Symposium (1998)
5. Carbonneau, M.A., Cheplygina, V., Granger, E., Gagnon, G.: Multiple instance learning: a survey of problem characteristics and applications. Pattern Recogn. **77**, 329–353 (2018)
6. Cimpanu, C.: NOPEN is the equation group's backdoor for Unix systems (2022)
7. Duncan, B.: Malware-traffic-analysis.net (2024). https://malware-traffic-analysis.net. Accessed 28 Feb 2025
8. Fu, C., Li, Q., Shen, M., Xu, K.: Detecting tunneled flooding traffic via deep semantic analysis of packet length patterns. In: Proceedings of the 2024 on ACM SIGSAC Conference on Computer and Communications Security, pp. 3659–3673 (2024)
9. Fu, C., Li, Q., Xu, K.: Detecting unknown encrypted malicious traffic in real time via flow interaction graph analysis. arXiv preprint arXiv:2301.13686 (2023)
10. Fu, Z., et al.: Encrypted malware traffic detection via graph-based network analysis. In: Proceedings of the 25th International Symposium on Research in Attacks, Intrusions and Defenses, pp. 495–509 (2022)
11. Gallagher, S.: Nearly half of malware now use TLS to conceal communications. Sophos news (2021)
12. Gezer, A., Warner, G., Wilson, C., Shrestha, P.: A flow-based approach for TrickBot banking trojan detection. Comput. Secur. **84**, 179–192 (2019)
13. He, Y., Huang, P., Hong, W., Luo, Q., Li, L., Tsui, K.L.: In-depth insights into the application of recurrent neural networks (RNNs) in traffic prediction: a comprehensive review. Algorithms **17**(9), 398 (2024)
14. Hong, Y., Li, Q., Yang, Y., Shen, M.: Graph based encrypted malicious traffic detection with hybrid analysis of multi-view features. Inf. Sci. **644**, 119229 (2023)
15. Jiang, X., Zhang, H.R., Zhou, Y.: Multi-granularity abnormal traffic detection based on multi-instance learning. IEEE Trans. Netw. Serv. Manage. **21**(2), 1467–1477 (2023)
16. Li, W., Zhang, X.Y., Bao, H., Shi, H., Wang, Q.: ProGraph: robust network traffic identification with graph propagation. IEEE/ACM Trans. Netw. **31**(3), 1385–1399 (2022)
17. Liu, C., He, L., Xiong, G., Cao, Z., Li, Z.: FS-net: a flow sequence network for encrypted traffic classification. In: IEEE INFOCOM 2019-IEEE Conference on Computer Communications, pp. 1171–1179. IEEE (2019)
18. Mahdavifar, S., Alhadidi, D., Ghorbani, A.A.: Effective and efficient hybrid android malware classification using pseudo-label stacked auto-encoder. J. Netw. Syst. Manage. **30**(1), 22 (2022)
19. Mahdavifar, S., Kadir, A.F.A., Fatemi, R., Alhadidi, D., Ghorbani, A.A.: Dynamic android malware category classification using semi-supervised deep learning. In: 2020 IEEE International Conference on Dependable, Autonomic and Secure Computing, International Conference on Pervasive Intelligence and Computing, International Conference on Cloud and Big Data Computing, International Conference on Cyber Science and Technology Congress (DASC/PiCom/CBDCom/CyberSciTech), pp. 515–522. IEEE (2020)
20. Mimura, M., Otsubo, Y., Tanaka, H., Tanaka, H.: A practical experiment of the http-based rat detection method in proxy server logs. In: 2017 12th Asia Joint Conference on Information Security (AsiaJCIS), pp. 31–37. IEEE (2017)

21. Nasr, M., Bahramali, A., Houmansadr, A.: DeepCorr: strong flow correlation attacks on tor using deep learning. In: Proceedings of the 2018 ACM SIGSAC Conference on Computer and Communications Security, pp. 1962–1976 (2018)
22. Nelms, T., Perdisci, R., Ahamad, M.: {ExecScent}: mining for new {C&C} domains in live networks with adaptive control protocol templates. In: 22nd USENIX Security Symposium (USENIX Security 13), pp. 589–604 (2013)
23. Rescorla, E.: The transport layer security (TLS) protocol version 1.3. Technical report, Internet Engineering Task Force (IETF) (2018)
24. Roesch, M., et al.: Snort: lightweight intrusion detection for networks. In: Lisa, pp. 229–238 (1999)
25. Wang, G., Gu, Y.: Multi-task scenario encrypted traffic classification and parameter analysis. Sensors **24**(10), 3078 (2024)
26. Zhang, Y., Chen, X., Jin, L., Wang, X., Guo, D.: Network intrusion detection: based on deep hierarchical network and original flow data. IEEE Access **7**, 37004–37016 (2019)

Dimensionality Reduction in Product
of Metric Spaces

Aleksander Denisiuk[(✉)] [iD]

University of Warmia and Mazury in Olsztyn, ul. Słoneczna 54,
10-710 Olsztyn, Poland
denisiuk@matman.uwm.edu.pl

Abstract. The purpose of the article is to develop a new dimensionality
reduction algorithm for data that are described by many features of
different nature. A method of feature selection is based on a new concept
of metrical importance of the features. The concept of feature importance
is based on metrical properties of data and is inspired by the principle
component analysis. Numerical experiments confirm the effectiveness of
the method and certain accordance of it with other concepts of feature
importance.

Keywords: dimensionality reduction · feature selection · feature
importance · explainable machine learning · metric learning · weighted
metric · classification

1 Introduction

Dimensionality reduction of the data space while retaining as much information as possible is important problem in the data analysis. Beside other things,
it reduces computational complexity of various algorithms, mitigates the curse
of dimensionality, and thus has many applications in clustering, classification,
visualization, and compression of high-dimensional data (see, for instance, the
survey [20]).

The purpose of this article is further development of the dimensionality reduction method from [8] proposed for categorical data. That method was inspired
by the classical linear PCA feature selection. Namely, it was shown in [8] that
PCA has the following metrical interpretation. Consider the affine transform that
minimizes the total squared inner-class distance. It turns out that the major feature is scaled with the minimal multiplier, the minor feature—with the maximal
one, like at Fig. 1. This interpretation was transferred to the categorical data
space with the Hamming metic, and problem of feature selection was reduced to
certain linear programming problem. See [8] for more details.

In this article the same interpretation of the PCA leads to formulation of
appropriate non-linear optimization problem in product of metrical spaces. The
solution of the optimization problem allows to calculate multipliers that are

© The Author(s), under exclusive license to Springer Nature Switzerland AG 2025
M. H. Lees et al. (Eds.): ICCS 2025, LNCS 15904, pp. 281–293, 2025.
https://doi.org/10.1007/978-3-031-97629-2_20

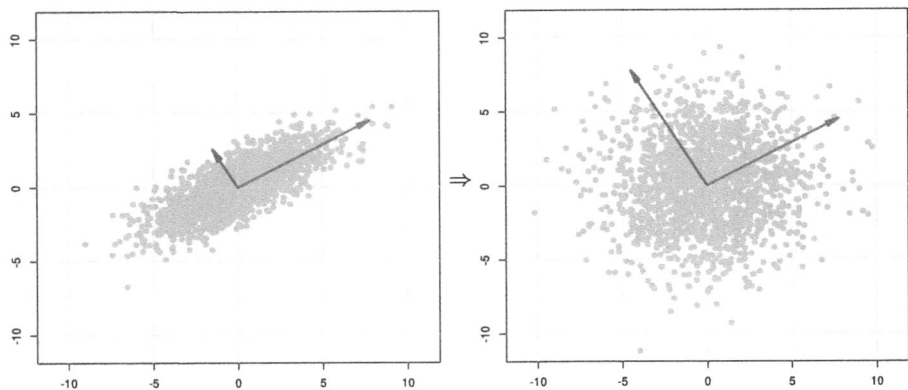

Fig. 1. Scaling that minimizes total relative inner-class squared distance.

interpreted as the feature importances. In what follows it called the *metrical importances*. In such a way, dimensionality reduction consists of dropping less important features first.

To prove the concept, numeric experiments on three datasets are performed. The data features were discarded in order of growing importances and the F_1 Score of classification was measured. Beside new introduced metrical importances other known feature importances were also considered: random forest mean decrease in accuracy, mean decrease in Gini index, the "standard errors" of the permutation-based importance measure, and the Shapley values. In all cases the average value of feature importance with respect to all classes was considered.

Experiments show that despite the importance values of individual features for different methods are different, all the methods show similar efficiency. It should also be mentioned that the metrical importance is much simpler to calculate. In fact, the metrical importance is calculated with two explicit formulas: (1) and (5). The computational time for the Shapley values increases exponentially with the number of features [15, Chapter 17]. Calculation of random forest related features importance involves building of decision trees and grows significally with number of features [5].

The rest of the paper is organized as follows. In Sect. 2 there is a short survey of the basic related works. Section 3 contains the proposed algorithm for metrical importances calculation. The performed numerical experiments are discussed in the Sect. 4. Finally, some concluding remarks are given in the Sect. 5.

2   Related Works

The key notion of new algorithm is the metrical importance of a feature. Importances of data features are of active study in recent years in the framework of explainable machine learning. One can find a comprehensive review in recently published book [15]. Here only some concepts will be mentioned. Three importances related to the random forest classifier: mean decrease in accuracy, the mean decrease in Gini index, and the "standard errors" of the permutation-based importance measure [5]. The last one concept is based on the Shapley values that were introduced in [19] for the game theory. In the article [21] the authors suggest to interpret the Shapley values as a contribution of individual feature to data classification. Metric-based importance comes from the principle components analysis and is direct continuation of the work [8].

All the mentioned concepts of feature importance are used in numerical experiments in Sect. 4.

Another field of machine learning that concerns this work is the metric learning. The current state of the metric learning can be found in surveys [4] and [12]. Most of methods concern data with pure numerical features. Non-numerical features are often embedded into continuous space. Some papers develop methods of metric learning for structured data: graphs with the graph-editing metric or text strings with the Levenshtein distance [3,17]. The proposed in this article approach can be used to data with mixed features of any nature and metric.

Determining the weights assigned to individual features of mixed data was recently used in context of supervised and unsupervised machine learning respectively in articles [7] and [9].

3   Metrical Importance of the Features and Algorithm for Dimensionality Reduction

Assume that the dataset $\mathbf{X}$ of M instances is given. Let each instance $x \in \mathbf{X}$ has n features of different kind, $x = (x_1, \ldots, x_n)$. Suppose that each feature x_i, $i = 1, \ldots, n$ is equipped with an appropriate distance $\mathrm{dist}_i(\cdot, \cdot)$ that measures dissimilarity of the data.

The product distance on $\mathbf{X}$ in defined in the following way:

$$\mathrm{dist}^2(x, y) = \sum_{i=1}^{n} \mathrm{dist}_i^2(x_i, y_i).$$

Introduce the weights vector $u = (u_1, \ldots, u_n) \in \mathbb{R}^n$, where $u_i \geq 0$ corresponds to a *scaling* of the feature i for $i = 1, \ldots, n$. The weighted distance is defined as follows:

$$\mathrm{dist}_u^2(x, y) = \sum_{i=1}^{n} u_i^2 \, \mathrm{dist}_i^2(x_i, y_i).$$

The last assumption is that the dataset is divided into c classes, $\mathbf{X} = C_1 \cup \cdots \cup C_c$.

The total inner-class squared distance is

$$G(u) = \frac{1}{M^2} \sum_{k=1}^{c} \sum_{x,y \in C_k} \mathrm{dist}_u^2(x,y) = \frac{1}{M^2} \sum_{k=1}^{c} \sum_{x,y \in C_k} \sum_{i=1}^{n} u_i^2 \, \mathrm{dist}_i^2(x_i, y_i)$$

$$= \sum_{i=1}^{n} u_i^2 \left(\frac{1}{M^2} \sum_{k=1}^{c} \sum_{x,y \in C_k} \mathrm{dist}_i^2(x_i, y_i) \right) = \sum_{i=1}^{n} u_i^2 z_i,$$

where

$$z_i = \frac{1}{M^2} \sum_{k=1}^{c} \sum_{x,y \in C_k} \mathrm{dist}_i^2(x_i, y_i). \tag{1}$$

Consider the following constraint minimization problem:

$$\begin{cases} G(u) = \sum_{i=1}^{n} u_i^2 z_i \to \min, \\ \frac{1}{n} \sum_{i=1}^{n} u_i = 1, \quad u_i \geq 0 \text{ for } i = 1, \ldots, n. \end{cases} \tag{2}$$

Definition 1. *The metrical importance I_m of the feature $i = 1, \ldots, n$ equals*

$$I_m(i) = 1/u_i^0,$$

where $u^0 = (u_1^0, \ldots, u_n^0)$ is the solution of the minimization problem (2).

Remark 1. Consider multivariate normally distributed data with distribution function

$$f(x) = \frac{\exp\left(-\frac{1}{2} x^T \Sigma^{-1} x\right)}{\sqrt{(2\pi)^n \det \Sigma}},$$

where Σ is diagonal matrix, $\Sigma = \mathrm{diag}(\lambda_1, \ldots, \lambda_n)$, $\Sigma^{-1} = \mathrm{diag}(\lambda_1^{-1}, \ldots, \lambda_n^{-1})$, $\det \Sigma = \lambda_1 \cdots \lambda_n$, and $\lambda_i > 0$ for $i = 1, \ldots, n$. One can show [8] that solution of the problem (2) is proportional to vector $\lambda = (\lambda_1^{-1}, \ldots, \lambda_n^{-1})$. In such a way features with greater variance have greater metrical importance. This is in accordance with the principle components analysis (see, for instance [1]).

Remark 2 (Data normalization). Since different features may change at different ranges, as is the case of most metric-based algorithms, data normalization must be performed. Numerical features can be rescaled such that the mean value will be 0, and variance 1. In this case the mean squared distance $\mathrm{dist}^2(x,y) = (x-y)^2$ is equal to 2. So, for non-numerical features the corresponding distance should be scaled with multiplier to make the mean squared distance also be equal to 2. Specifically, for categorical feature of cardinality k the standard Hamming distance

$$\mathrm{dist}_h(x,y) = \mathrm{diff}(x,y) = \begin{cases} 1, & x \neq y, \\ 0, & x = y, \end{cases}$$

should be multiplied by $\sqrt{2k/(k-1)}$.

To solve the problem (2) one can use the Lagrange multipliers method. Corresponding Lagrange function is

$$L(u, \Lambda) = \sum_{i=1}^{n} u_i z_i - \Lambda \left(\sum_{i=1}^{n} u_i - n \right). \tag{3}$$

Differentiating (3) with respect to u_i and equalizing the result to zero, one obtains

$$u_i = \frac{\Lambda}{2z_i}, \quad i = 1, \ldots, n. \tag{4}$$

Differentiation with respect to Λ implies

$$\Lambda = \frac{2n}{\sum_{i=1}^{n} z_i^{-1}}.$$

Substituting this to (4) one gets

$$u_i = \frac{1}{z_i} \left(\frac{1}{n} \sum_{i=1}^{n} z_i^{-1} \right)^{-1}, \quad i = 1, \ldots, n.$$

Finally, for metrical importance (definition 1) one have

$$I_m(i) = z_i \left(\frac{1}{n} \sum_{i=1}^{n} z_i^{-1} \right), \quad i = 1, \ldots, n. \tag{5}$$

The above considerations are summarized in the Algorithm 3.1.

Algorithm 3.1. Reduction of m dimensions

Require: the dimension of dataset $\mathbf{X}$ is n, $n > m$
Ensure: the dimension of dataset $\mathbf{X}$ is $n - m$
  compute coefficients z_i with the formula (1)
  compute importances with the formula (5)
  $s \leftarrow 0$
  while $s < m$ **do**
    discard the feature with the less importance
    $s \leftarrow s + 1$
  end while

---

4   Numerical Experiments

To illustrate the concept a few R scripts have been created. The code is available as a project on Gitlab at https://gitlab.com/adenisiuk/l2.

The purpose of the tests is to show that the introduces in this article Algorithm 3.1 allows to reduce dimensionality while retaining as much information as possible. To do this, consider the classification problem. The data first were classified with complete set of features ordered by decreasing importance. Then less important features one by one were discarded and F_1Score of classification was measured. This order is called the *pca* order (red line in the figures).

The method was also considered with other importances mentioned in the Sect. 2. The order that corresponds to the Shapley values is referred as the *shapley* order, the green line on the pictures. The order that corresponds to the mean decrease of accuracy in random forest classifier is plotted in blue color and is referred as the *rf* order. The order related to the decrease in Gini index is referred as the *gini* order (magenta line). And the order generated by the "standard errors" of the permutation-based importance measure for random forest classifier has cyan color and is called as the *sd* order.

The Shapley values were calculated with the `fastshar` R package [10]. The random forest classifier was used as the user-specified prediction wrapper.

Three importances related to the random forest classifier were calculated with the R implementation [13].

The reverse orders for every importance were also tested. They are referred correspondingly as *acp*, *yelpahs*, *fr*, *inig*, *ds*. Related lines in the figures has the same color, but the dashed style.

Implementations of three classifiers: random forest, SVM and XGBoost in R were used in experiments: [6,13,14]. For the random forest classifier an average result for 100 tests is presented.

In all the above classifiers the standard implementation settings were used.

3 datasets from the UCI Machine Learning Repository [2] were considered: the Australian Credit Approval [18], the Bank Marketing [16] and the Heart Disease [11]. These datasets contain numerical and categorical data. The standard difference metric for numerical and categorical parts, normalized according to Remark 2 was used in experiments.

Note that dataset features were rearranged for the implementation: continuous features are the first, and categorical features are placed at the end of the feature list.

All the tested datasets were split into train (80%) and test (20%) parts.

The F_1Score measure with respect to the first (having a smaller amount of records) class was used to estimate the classification rate.

The considered datasets have only two classes, but the method can be used to dataset with greater number of classes as well.

For each dataset the importances with respect to five considered concepts, corresponding orders of the features discarding and performance of classification after discarding the features were calculated. The importances were calculated with full datasets, and then were scaled to sum to 100%.

One can see that despite the importances and orders are different, in all the tests direct orders have similar performance and give the minimal information loss, while reverse orders give greater lost.

4.1   Australian Credit Approval

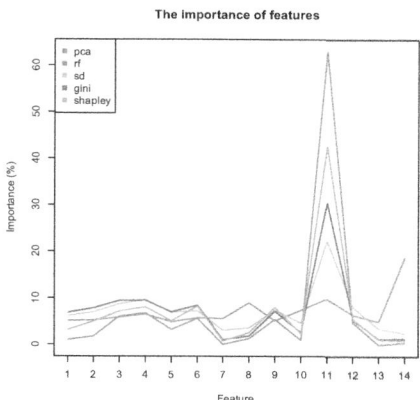

Fig. 2. Importances of features for the Australian Credit Approval dataset.

Table 1. Orders of feature discarding for the Australian Credit Approval dataset

Feature discarding order														
pca	5	13	1	9	2	7	6	3	12	4	10	8	11	14
rf	13	7	14	10	1	8	2	5	12	9	6	3	4	11
sd	14	7	13	8	10	1	5	2	6	9	12	3	4	11
gini	7	13	14	8	10	12	1	5	9	2	6	3	4	11
shapley	7	14	13	10	8	1	2	5	12	3	9	4	6	11

The Australian Credit Approval [18] dataset has 6 continuous, 8 nominal attributes, 690 records, and 2 decision categories.

The importances of individual features and the orders of feature discarding are presented in the Fig. 2 and the Table 1. The F_1Score for selected classifiers and different orders of the feature discarding are presented in the Fig. 3.

One can see that all the algorithms marked the feature 11 as very important.

Large difference can be observed in the most important feature. The metric based algorithm marked 14 as most important, while the other algorithms marked 11th feature. The reason probably is that these features are strongly dependent. χ^2 dependency test produced p-value of 0.0006133. Dependency in some sens means that these features contain similar information. So, it should have similar importance. This guess is confirmed if one observe the *acp* order: after discarding the feature 14 and keeping 11, information lost is low.

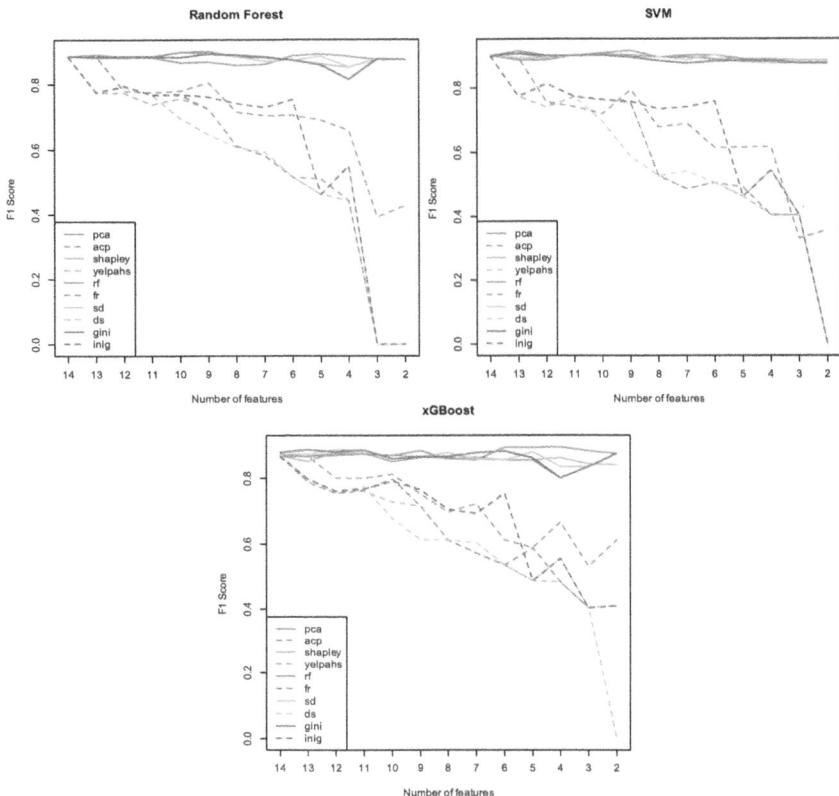

Fig. 3. Classification accuracy for the Australian Credit Approval dataset.

Other difference one can see analysing the less important feature. The new algorithm mark the feature 5, according to the *rf* order it is the feature 13, *sd* marked the feature 14, rest of algorithms suggest the feature 7. Despite of all the mentioned differences, the performance of classification after the less important feature discarding is almost identical. That can be corollary of the fact that the levels of importance of less significant features are generally low and approximately the same.

4.2   Bank Marketing

The Bank Marketing dataset [16] has 7 continuous, 9 nominal attributes, 4521 records, 2 decision categories.

The importances of individual features and the orders of feature discarding are presented in the Fig. 4 and the Table 2. The F_1 Score for selected classifiers and different orders of the feature discarding are presented in the Fig. 5.

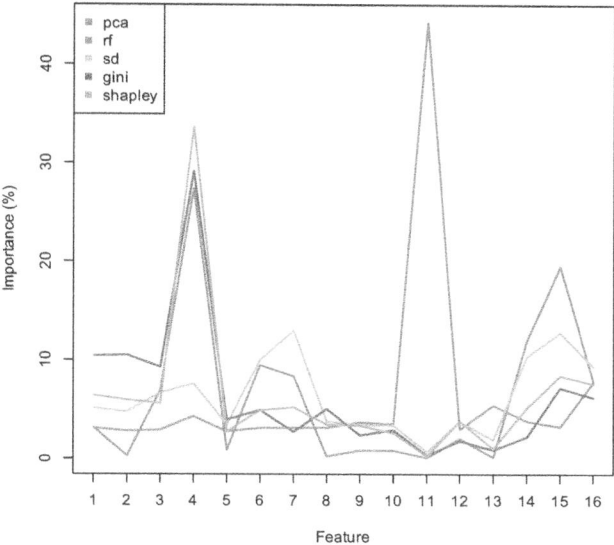

Fig. 4. Importances of features discarding for the Bank Marketing dataset.

Table 2. Orders of feature discarding for the Bank Marketing dataset

Feature discarding order																
pca	5	2	3	12	1	6	7	8	15	10	9	14	4	13	16	11
rf	11	13	8	2	10	9	5	12	1	3	16	7	6	14	15	4
sd	11	13	10	9	5	12	8	2	1	3	4	16	6	14	15	7
gini	11	13	12	14	9	7	10	5	6	8	16	15	3	1	2	4
shapley	11	13	10	5	9	8	12	6	14	7	3	2	1	16	15	4

Again one can observe a difference in the features order, but similar performance for all direct orders. For the random forest classifier the metrical importance order even overperforms the others. Indeed, most of algorithms marked the feature 11 as less important. But dropping this feature in random forest classifier experiments caused significant lost of F_1 Score.

Note, however, that generally performance of classification for this dataset is poor.

4.3   Heart Disease

The Heart Disease dataset [11] has 5 continuous and 8 nominal attributes, 270 records, 2 decision categories.

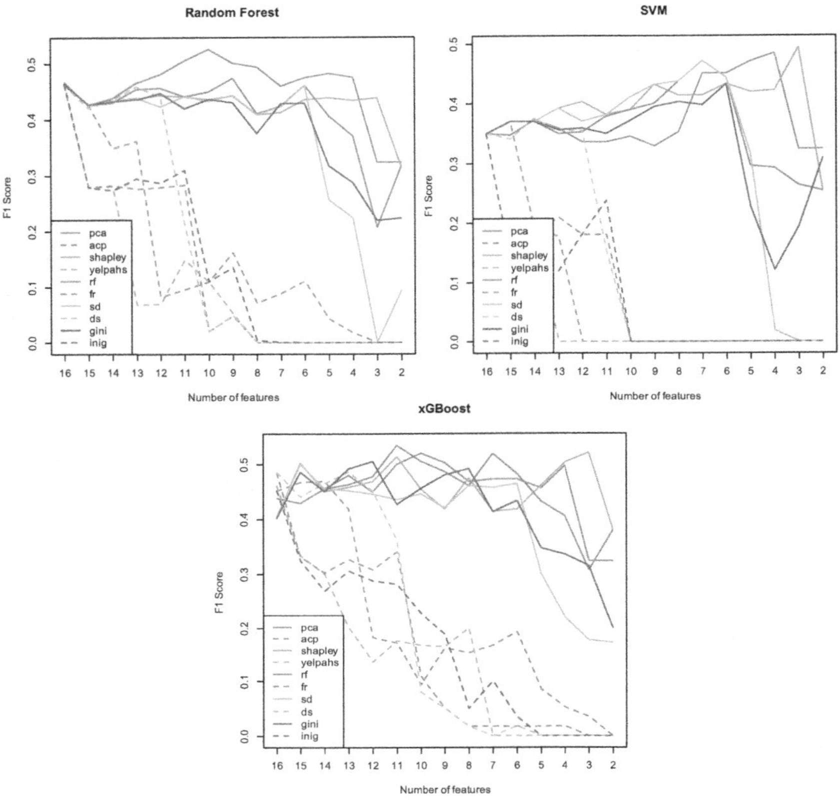

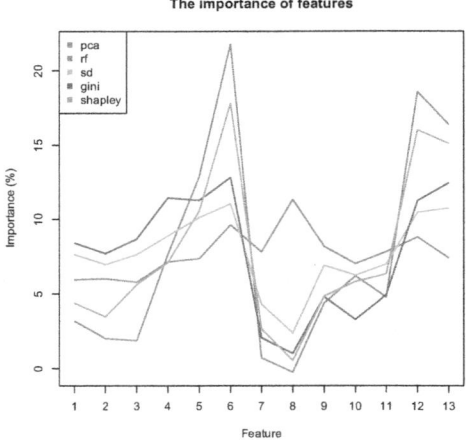

Fig. 5. Classification accuracy for the Bank Marketing dataset.

Fig. 6. Importances of features for the Heart Disease dataset.

Table 3. Orders of feature discarding for the Heart Disease dataset

Feature discarding order													
pca	3	1	2	10	4	5	13	7	11	9	12	6	8
rf	8	7	3	2	1	9	11	10	4	5	13	12	6
sd	8	7	10	9	2	11	3	1	4	5	12	13	6
gini	8	7	10	9	11	2	1	3	12	5	4	13	6
shapley	8	7	2	1	9	3	10	11	4	5	13	12	6

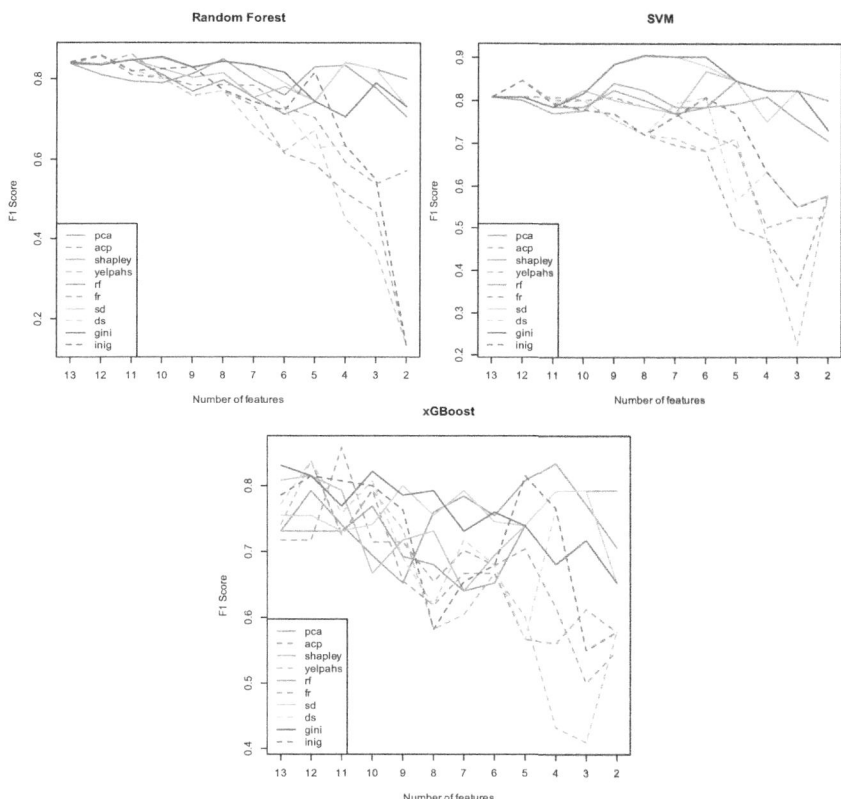

Fig. 7. Classification accuracy for the Heart Disease dataset.

The importances of individual features and the orders of feature discarding are presented in the Fig. 6 and the Table 3. The F_1 Score for selected classifiers and different orders of the feature discarding are presented in the Fig. 7.

For this dataset one can observe, like in case of the Australian Credit Approval dataset, that the 6th feature was marked as very important for all the algorithms. But the *pca* marked 8th feature as the most important. And, again, observing reverse order tests one can see that dropping the 6th feature at the first step does not cause the lost of F_1 Score, but even results in F_1 Score gain.

The following difference between the metrical importance and other concepts should be mentioned. Namely, the metrical importance stems from the metrical properties of the data and is independent of any classifier. The remaining concepts are closely related to the random forest classifier. This difference has two consequences. On the one hand, the *pca* ordering produces more stable results across different classifiers (this can be seen especially in the Australian Credit Approval dataset). On the other hand, if the metric used to compare data does not reflect its structure, the classification results may be unsatisfactory.

5   Conclusion and Future Work

In this article a new concept of metrical importance of data features is proposed. The notion is inspired by the classical PCA, but can be applied to any mixed data with appropriate metric defined on individuals features.

A simple algorithm of dimensionality reduction based on the new notion was developed: discard less important features first.

Numerical experiments with three classifiers were performed: random forest, SVM, XGBoost. Other concepts of feature importance were also considered: three related to the random forest classifier and the Shapley values (see the Sect. 2 for more details.

Experiments show that for considered datasets the new metrical importance produces dimensionality reduction algorithm of efficiency that is close to the above-mentioned knows concepts of feature importance, while the metrical importance has much smaller computational complexity.

So, one can suggest to use the metrical importance as a new instrument in analysis of mixed data.

The concept of metric importance deserves further development in the context of explainable machine learning. It would be interested to analyse it with real data set.

References

1. Afifi, A., May, S., Donatello, R., Clark, V.: Practical Multivariate Analysis. Chapman & Hall/CRC Texts in Statistical Science Series. CRC Press, Taylor & Francis Group (2019). https://books.google.pl/books?id=AUyrswEACAAJ
2. Asuncion, A., Newman, D.J.: UCI machine learning repository (2007). http://www.ics.uci.edu/~mlearn/MLRepository.html
3. Bellet, A., Habrard, A., Sebban, M.: Good edit similarity learning by loss minimization. Mach. Learn. **89**, 5–35 (2012). https://doi.org/10.1007/s10994-012-5293-8
4. Bellet, A., Habrard, A., Sebban, M.: Metric Learning. Springer, Cham (2015). https://doi.org/10.1007/978-3-031-01572-4

5. Breiman, L.: Random forests. Mach. Learn. **45**, 5–32 (2001)
6. Chen, T., et al.: XGBoost: extreme gradient boosting (2023). https://CRAN.R-project.org/package=xgboost. R package version 1.7.6.1
7. Denisiuk, A.: Weighted hamming metric and kNN classification of nominal-continuous data. In: Mikyška, J., de Mulatier, C., Paszynski, M., Krzhizhanovskaya, V.V., Dongarra, J.J., Sloot, P.M. (eds.) Computational Science - ICCS 2023, pp. 306–313. Springer, Cham (2023)
8. Denisiuk, A.: PCA dimensionality reduction for categorical data. In: Franco, L., de Mulatier, C., Paszynski, M., Krzhizhanovskaya, V.V., Dongarra, J.J., Sloot, P. (eds.) Computational Science - ICCS 2024, pp. 179–186. Springer, Cham (2024)
9. Denisiuk, A., Grabowski, M.: Embedding of the hamming space into a sphere with weighted quadrance metric and c-means clustering of nominal-continuous data. Intell. Data Anal. **22**(6), 1297001314 (2018). https://doi.org/10.3233/IDA-173645
10. Greenwell, B.: fastshap: fast approximate shapley values (2024). https://github.com/bgreenwell/fastshap, https://bgreenwell.github.io/fastshap/. R package version 0.1.1
11. Janosi, A., Steinbrunn, W., Pfisterer, M., Detrano, R.: Heart disease. UCI Machine Learning Repository (1989). https://doi.org/10.24432/C52P4X
12. Kulis, B.: Metric learning: a survey. Found. Trends® Mach. Learn. **5**(4), 287–364 (2013). https://doi.org/10.1561/2200000019
13. Liaw, A., Wiener, M.: Classification and regression by randomforest. R News **2**(3), 18–22 (2002)
14. Meyer, D., Dimitriadou, E., Hornik, K., Weingessel, A., Leisch, F.: e1071: misc functions of the department of statistics, probability theory group (formerly: E1071). TU Wien (2022). R package version 1.7-12
15. Molnar, C.: Interpretable Machine Learning, 2 edn. (2022). https://christophm.github.io/interpretable-ml-book
16. Moro, S., Rita, P., Cortez, P.: Bank marketing. UCI Machine Learning Repository (2014). https://doi.org/10.24432/C5K306
17. Neuhaus, M., Bunke, H.: Automatic learning of cost functions for graph edit distance. Inf. Sci. **177**(1), 239–247 (2007). https://doi.org/10.1016/j.ins.2006.02.013
18. Quinlan, R.: Statlog (Australian credit approval). UCI Machine Learning Repository (1987). https://doi.org/10.24432/C59012
19. Shapley, L.S., et al.: A value for n-person games (1953)
20. Van Der Maaten, L., Postma, E., Van den Herik, J.: Dimensionality reduction: a comparative review. J. Mach. Learn. Res. **10**, 66–71 (2009)
21. Štrumbelj, E., Kononenko, I.: Explaining prediction models and individual predictions with feature contributions. Knowl. Inf. Syst. **41**(3), 647–665 (2014). https://doi.org/10.1007/s10115-013-0679-x

To Select or Not to Select? The Role of Meta-features Selection in Meta-learning Tasks with Tabular Data

Irina Deeva$^{(\boxtimes)}$ and Alena Kropacheva

ITMO University, St. Petersburg 197101, Russia
iriny.deeva@gmail.com

Abstract. In meta-learning tasks with tabular data, the choice of meta-features significantly impacts model performance and interpretability. This study investigates the necessity and methods of meta-feature selection in the context of meta-learning, particularly for tabular datasets. We address the fundamental question: Is it better to select a subset of meta-features or use the entire feature set? We examine various selection techniques, including filter, wrapper, and embedded methods, as well as a novel causal-based approach utilizing counterfactual reasoning. Our experiments demonstrate that feature selection generally enhances performance, with causal-based methods, especially those leveraging counterfactual generation, showing superior efficiency and generalizability. Furthermore, we explore how these methods fare under shifts in data, particularly when non-informative features are added. The results reveal that the counterfactual method maintains high efficacy across different meta-learners and exhibits a favorable balance between model performance and interpretability. These findings underscore the importance of meta-feature selection in improving the adaptability and transparency of meta-learners for tabular data tasks.

Code and supplementary materials for this research are available on GitHub: https://github.com/ITMO-NSS-team/MetaSelect.

Keywords: Meta-learning · Meta-features selection · Causal-based methods · Counterfactual reasoning · Tabular data

1 Introduction

Motivation. In the world of machine learning (ML), there exist domains that are characterised by a substantial and varied array of models, such as machine learning on tabular data. This domain is replete with models of diverse classes, giving rise to a plethora of research endeavours aimed at ascertaining the relative merits of these models. A notable example of this is the investigation into whether deep models possess an advantage in the context of tabular data [5, 23, 24]. Recent studies have indicated that a more effective strategy might be to focus on optimizing the hyperparameters of a specific tabular model rather

© The Author(s), under exclusive license to Springer Nature Switzerland AG 2025
M. H. Lees et al. (Eds.): ICCS 2025, LNCS 15904, pp. 294–308, 2025.
https://doi.org/10.1007/978-3-031-97629-2_21

than on selecting the most optimal model [16]. This is understandable given the complexity and time-consuming nature of selecting the best model. However, there exist methods and approaches, such as meta-learning, that aim to simplify this process. The fundamental components of meta-learning encompass the meta-characteristics of the data and the meta-learner itself, whose function is to facilitate the adaptation of performance information from disparate models to novel data without the necessity of directly training and evaluating those models [10]. The creation of an initial space of meta-characteristics (meta-features) is pivotal to the success of the meta-learning task [20]. Nevertheless, there is a paucity of research on the advisability of reducing the initial meta-characteristics space [9,18], despite the plethora of studies on meta-characteristics selection [7,8,19].

Contribution. Whilst the reduction of the dimensionality of the meta-features space through selection appears reasonable in terms of efficiency and interpretability, a fundamental investigation of this problem was deemed necessary in order to answer the main research question: *when it comes to meta-features, is it better to pick and choose, or is it always better to train on the whole set of meta-features?* In contrast to previous studies, our analysis encompasses not only the impact of selection on meta-learner performance, but also the impact on interpretability. To this end, we have suggested metric for the latter and have analysed a large number of meta-features selection approaches. Our analysis enables us to draw conclusions about the effectiveness and generalisability of these approaches to different meta-learners. In addition, an approach for meta-features selection based on causal analysis and counterfactual reasoning is proposed. The experimental studies demonstrate the high efficiency and generalisability of this approach.

2 Related Works

2.1 Meta-learning for Tabular Models

Given that the focus of this paper is on tabular data, it would be prudent to analyse studies that address the problem of selecting algorithms for such data. In essence, meta-learning based on meta-characteristics bears resemblance to a conventional machine learning task, with meta-features serving as predictors and the outcomes of machine learning models (e.g., performance) designated as the target variable. In [23], the authors conducted experiments on a benchmark of 111 datasets using 20 machine learning models for tabular data classification. The meta-model employed is a logistic classifier, the purpose of which is to minimise the error of predicting the win of a particular model based on the characteristics of the data. The authors [25] propose an alternative approach, which involves the initial reduction of the dimensionality of the original meta-features space. This is followed by the training of a meta-learner SVM in this space. In a separate publication [22], the authors conducted an investigation into a variety of models, encompassing both classical models (logistic regression, decision tree models) and deep models (FT-transformer, MLP) as meta-learners. The authors observe that while deep models demonstrate optimal efficiency, they exhibit a concomitant loss in interpretability. Also the authors [29] propose an alternative predictive problem, namely the prediction of changes in the performance curve based on initial points. The proposed meta-model involves the identification of a mapping between meta-characteristics and initial values of dynamics, as well as validation statistics.

It is evident that the domain of meta-learning for tabular data is undergoing continuous development. Currently, a prevalent practice is the selection of either classical or deep models for tabular data, as the efficacy of these methods remains a subject of ongoing research. Meta-learning has emerged as a potential solution to automate this decision-making process. It is also noteworthy that in the vast majority of papers, authors select meta-features in one way or another, yet rarely elucidate the rationale behind their selection and the underlying principles that guided their decision-making. This underscores the necessity and pertinence of undertaking a more comprehensive investigation into the matter of reducing the set of meta-features as a part of meta-learning pipeline.

2.2   Meta-features Selection Methods

With regard to the selection of meta-features, it is first necessary to recognise that the task of meta-features selection is a particular instance of the more general task of feature selection in machine learning. Consequently, all the established approaches to feature selection are applicable to the selection of meta-features. The classification of feature selection methods can be divided into three distinct groups: *filter*, *embedded* and *wrapper*. Each of these groups possesses its own advantages and disadvantages. Filter methods comprise a range of approaches, including correlation-based selection, selection based on statistical tests (e.g., the Chi-square test and the ANOVA F-test), and selection based on information criteria (e.g., mutual information) [11]. These methods are simple and computationally efficient, and can also be termed model-free. Wrapper methods address the feature selection problem through the lens of an optimisation problem. The quality of the machine learning model for which the features are selected is frequently employed as a metric to gauge the efficacy of the selection process. The optimisation algorithms employed in this context often encompass evolutionary algorithms [28], greedy algorithms [21], and more complex algorithms such as recursive feature elimination [27] or forward-backward selection with early dropping [4]. Wrapper methods have been shown to be efficient, but they are sensitive to the dimensionality of the source space and therefore not always computationally efficient. Finally, a group of embedded methods involves the direct selection of features during the training of machine learning (ML) models. The most popular approach here is lasso regression, and there are also its modifications, for example, LassoNet [12] and Deep Lasso [6]. The primary disadvantage of embedded methods is that they are model-free and thus contingent on the properties of the model itself.

In this paper, we propose the identification of an additional group of methods, which we designate as *causal − based*. These methods have application in both the filter and wrapper groups; however, we consider them as a discrete group, given that they are predicated on entirely distinct principles of causal machine learning. Current methods for the selection of features do not invariably elucidate the features that engender a result in the data. In such instances, the principles of causal machine learning can prove beneficial. The majority of causal-based methods are predicated on the identification of the Markov Blanket (MB) of the target variable, since it is the attributes included in the MB that contain the main information about the target variable [31]. MB search-based approaches are informative, but they are also sensitive to a large number of features. Another group of methods is based on causal inference, for example, feature selection is based on calculated Average Treatment Effect values [15] or on counterfactual reasoning [13,30]. To the best of our knowledge, causal approaches have not yet been investigated by authors for the task of meta-features selection, so in this study we are going to fill this gap by comparing them with other classical approaches to meta-features selection.

3   Backgrounds

The goal of meta-feature selection is to identify an optimal subset of features $S \subseteq F$ from a candidate set $F = \{X_1, X_2, \ldots, X_n\}$ that maximizes the predictive performance of a meta-learner M while minimizing subset size. This is formalized as a regularized optimization problem:

$$S^* = \arg\min_{S \subseteq F} \left[\mathcal{L}\big(M(S)\big) + \lambda |S| \right], \tag{1}$$

where $\mathcal{L}$ represents the meta-learner's loss function and λ controls the trade-off between model performance and meta-feature set cardinality. The mathematical formulations of the methods to be investigated in this paper are described below.

3.1   Filter Methods

Filter methods rank features using statistical measures of relevance:

– **Spearman's Rank Correlation:**

$$\rho(X_i, Y) = \left| \frac{\mathrm{Cov}(\mathrm{rank}(X_i), \mathrm{rank}(Y))}{\sigma_{\mathrm{rank}(X_i)} \sigma_{\mathrm{rank}(Y)}} \right| \tag{2}$$

– **Mutual Information (MI):**

$$\mathrm{MI}(X_i, Y) = \sum_{x \in X_i} \sum_{y \in Y} P(x, y) \log \left(\frac{P(x, y)}{P(x)P(y)} \right) \tag{3}$$

– **ANOVA F-Test (continuous vs categorical):**

$$F(X_i, Y) = \frac{\frac{1}{K-1} \sum_{k=1}^{K} n_k (\bar{X}_i^{(k)} - \bar{X}_i)^2}{\frac{1}{N-K} \sum_{k=1}^{K} \sum_{j=1}^{n_k} (X_{ij}^{(k)} - \bar{X}_i^{(k)})^2} \tag{4}$$

3.2 Embedded Methods

Feature selection is integrated into the learning algorithm:

– **Lasso Regression (L1 Regularization):**

$$\beta^* = \arg\min_{\beta} \left\{ \frac{1}{2N} \|Y - X\beta\|_2^2 + \lambda\|\beta\|_1 \right\} \tag{5}$$

Non-zero coefficients ($\beta_j \neq 0$) indicate selected features.
– **XGBoost Feature Importance:**

$$\text{Importance}(X_j) = \sum_{t=1}^{T} \sum_{s \in \mathcal{S}_j(t)} \text{gain}(s), \tag{6}$$

where $\mathcal{S}_j(t)$ denotes splits on feature X_j in tree t. Non-zero $Importance(X_j)$ indicates sekected features.

3.3 Wrapper Methods

Recursive Feature Elimination (RFE). RFE iteratively removes the least important features:

1. Initialize with full feature set $S_0 = F$
2. For each iteration t:
    – Train model M_t on current features S_t:

$$\theta_t = \arg\min_{\theta} \mathcal{L}(M_t(S_t; \theta)) \tag{7}$$

    – Compute feature importances $\{w_i^{(t)}\}$
    – Remove lowest-ranked features:

$$S_{t+1} = S_t \setminus \left\{ X_j \mid w_j^{(t)} = \min_{X_i \in S_t} w_i^{(t)} \right\} \tag{8}$$

3. Terminate when $|S_t| = k$ (desired subset size). Here desired subset size is equal to half of the original number of meta-features.

3.4 Causal-Based Meta-feature Selection

Average Treatment Effect Based Selection. For robust estimation of heterogeneous treatment effects in meta-feature selection, we employ the Causal-ForestDML estimator [3], which combines double machine learning (DML) with causal forests [2]. For a meta-feature X_j acting as treatment variable T, and outcome Y representing algorithm performance, we model:

$$Y = \theta(X_j)T + g(X_{-j}) + \epsilon, \quad T = f(X_{-j}) + \eta, \tag{9}$$

where $X_{-j} = F \setminus \{X_j\}$ are confounders, $g(\cdot)$ and $f(\cdot)$ are nuisance functions, $\epsilon \perp \eta \perp X$.

The CausalForestDML estimator implements the following steps:

1. **Orthogonalization:** Fit regularized models to estimate

$$\tilde{Y} = Y - \hat{g}(X_{-j}) \tag{10}$$

$$\tilde{T} = T - \hat{f}(X_{-j}) \tag{11}$$

2. **Causal Forest:** Estimate conditional average treatment effect (CATE) via

$$\hat{\theta}(x_j) = arg\min_{\theta} \sum_{i=1}^{N} \alpha_i(x_j)(\tilde{Y}_i - \theta\tilde{T}_i)^2, \tag{12}$$

where weights $\alpha_i(x_j)$ are determined by forest similarity kernel.

The feature importance for X_j is computed as:

$$\text{ATE-Importance}(X_j) = \mathbb{E}\left[\hat{\theta}(X_j)\right] \cdot \sqrt{\text{Var}(\hat{\theta}(X_j))}, \tag{13}$$

prioritizing features with both large average effects and effect heterogeneity.

Counterfactual Based Selection. For causal interpretation through intervention analysis, we propose a counterfactual feature selection mechanism. Existing counterfactual generation-based feature selection approaches typically utilise these counterfactuals as supplementary data [13]. In contrast, the present study proposes a direct utilisation of the counterfactual generation efficiency as a feature selection criterion. The rationale underlying this approach is that if this generation is successful in a given feature space, it can be assumed that these are the features that influence the target variable.

Given a classifier f and meta-feature $s_j \in S$, we formulate counterfactual generation as the optimization problem:

$$\delta^* = arg\min_{\delta} \underbrace{\mathcal{L}_{\text{pred}}\left(f(x+\delta), y^{\text{target}}\right)}_{\text{Prediction Loss}} + \lambda \underbrace{\frac{1}{N}\sum_{i=1}^{N} \frac{|x_i^{\text{cf}} - x_i^{\text{orig}}|}{\text{MAD}}}_{\text{Robust Distance Loss}} \tag{14}$$

where δ is the perturbation, y^{target} is the opposite to $f(x)$ label, λ controls minimal intervention strength, $\text{MAD} = \text{median}(|x_i - \text{median}(x)|)$ scales perturbations by feature robustness. It is asserted that a counterfactual generation which compelled the classifier to reverse the label, whilst concomitantly effecting only negligible alterations to the data, is to be regarded as effective. Then the feature selection criterion evaluates:

$$\text{Efficiency}(s_j) = \frac{1}{N}\sum_{i=1}^{N} \mathbb{I}\left[f(x_i) \neq f(x_i + \delta_i^*)\right] \tag{15}$$

A meta-feature s_j is selected if $\text{Efficiency}(s_j) > \epsilon$, where ϵ is a tolerance threshold. In this study, the linear classifier was utilised as the classification model (f), while the cross-entropy function was employed to calculate the prediction loss. Additionally, the Adam optimiser, with a learning rate set to 0.01, was employed for optimisation purposes.

4   Design of Experiments

4.1   Dataset and Meta-features Description

We use experiment results provided in paper [16]. Authors present metrics for 19 algorithms (including GBDTs, neural networks and baselines) on 176 classification datasets from OpenML [26]. The results table is organized as follows: datasets are split into 10 folds, with each fold corresponding to a row containing metrics for every model with different hyperparameter sets. The evaluated metrics include logarithmic loss, AUC, accuracy, F1 score, and runtime (in seconds), calculated for three data splits: train, validation, and test. In our study, we focus on the F1 score from the test splits as the target metric for the following models: Logistic Regression, Random Forest, XGBoost, MLP, ResNet, and FT-Transformer.

Experiment results also include meta-features dataset. For each dataset fold meta-features are calculated with Python package PyMFE [1]. They include general features (such as number of attributes, number of distinct classes, etc.), statistical features (such as skewness or kurtosis), information-theoretic features (such as noisiness of attributes or joint entropy), landmarking features (such as performance of the Naive Bayes classifier) and model-based features (such as number of leaf nodes in the Decision Tree model). Some meta-features are represented as vectors (e.g., maximum value from each attribute). These are aggregated using various functions, including average, maximum, minimum, standard deviation, kurtosis, skewness, and interquantile range.

The meta-dataset used in our research consists of meta-features and target metrics for each dataset. The data preprocessing steps include:

- Aggregating meta-features by folds using median values,
- Removing meta-features with more than half of the values undefined,
- Excluding meta-features with large absolute values,
- Eliminating constant meta-features,
- Filtering meta-features based on the aggregation function,
- Removing duplicate columns and rows,
- Excluding datasets with undefined target metrics,
- Scaling data using Yeo-Johnson transform,
- Filtering meta-features with large VIF [17].

The final meta-dataset consists of 134 datasets, 123 meta-features.

Meta-learning Tasks and Meta-learners. In the present study, the following meta-learning task was considered: predicting the F1 score of a ML model. The task was formulated as binary classification task. The F1 score was predicted as follows: if the F1 score exceeded the median of the target scores then 1 was assigned; otherwise, 0 was assigned. The models employed as meta-learning models were KNN Classifier (6 neighbors, uniform weight function, leaf size: 40), XGBoost Classifier (maximal depth: 7, learning rate: 0.1, 50 estimators, evaluation metric: accuracy) and MLP Classifier (hidden layer size: 25, logistic activation function, L-BFGS solver, strength of the L2 regularization: 0.05, adaptive learning rate with initial value 0.05).

5   Results

The following research questions are formulated in this study:

- RQ1. Should we make a selection of meta-features, and how do the different ways of selecting features compare?
- RQ2. How do methods behave under shifts in the data when non-informative features are added?
- RQ3. How do the methods relate to each other in terms of the interpretability of the selected features?

5.1   Study of the Performance of Feature Selection Methods

Initially, a baseline study was conducted in order to ascertain the performance of the selection methods. These were then compared to their profit performance on all meta-features. The results are presented in Fig. 1. The initial conclusion that can be drawn is that feature selection almost invariably improves the quality of the result, thereby demonstrating its necessity. With regard to the performance of feature selection methods, causal-based approaches, and in particular the proposed algorithm based on **counterfactual reasoning**, demonstrate favourable results. Indeed, the algorithm is among the top three methods in more than half of the results. In the context of a group of methods, filter-based and causal-based methods demonstrate optimal performance, with wrapper approaches and embedded methods exhibiting average performance.

A thorough analysis of the performance of the methods with respect to the target type reveals that the method based on counterfactual generations exhibited equivalent proficiency in predicting the productivity of classical models and the performance of deep models. With regard to meta-learning models, the counterfactual method demonstrated notable efficacy in the MLP and XGBoost models.

A study was also conducted on the computational complexity of feature selection methods. As illustrated in the Fig. 2, the outcomes of measuring the execution time of selection methods in relation to the dimensionality of the initial feature space are presented. As would be anticipated, causal-based methods demonstrate the greatest longevity and exhibit increased sensitivity to the augmentation in the number of features.

5.2   Feature Selection Under Data Shifting Conditions

In order to undertake a comprehensive investigation into the performance of feature selection methods, experiments were conducted with the objective of understanding how the methods respond to shifts in the data, namely the addition of uninformative features. In order to achieve this, the logic of the experiments presented in the paper [6] was utilised. Three categories of uninformative features were generated based on the said study:

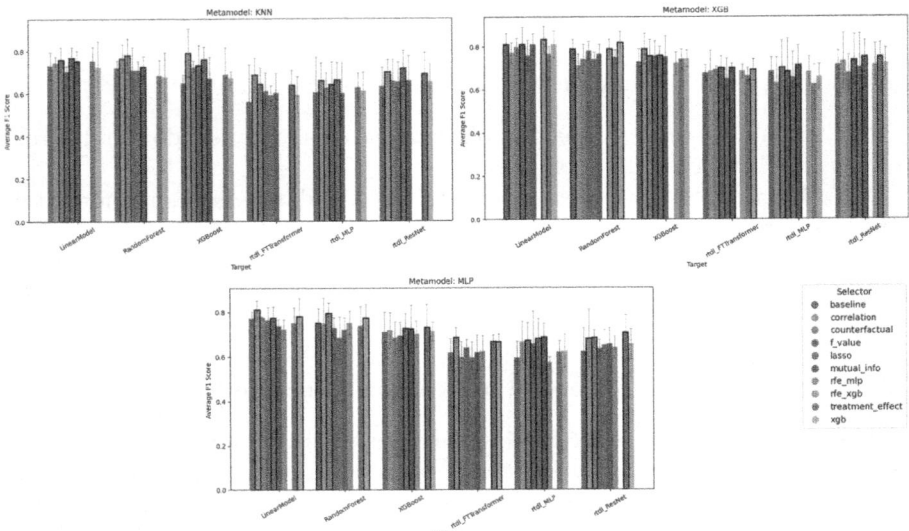

Fig. 1. Results comparing feature selection methods, here *base* is a run on all meta-features. In each category, the top 3 methods are highlighted with a black box.

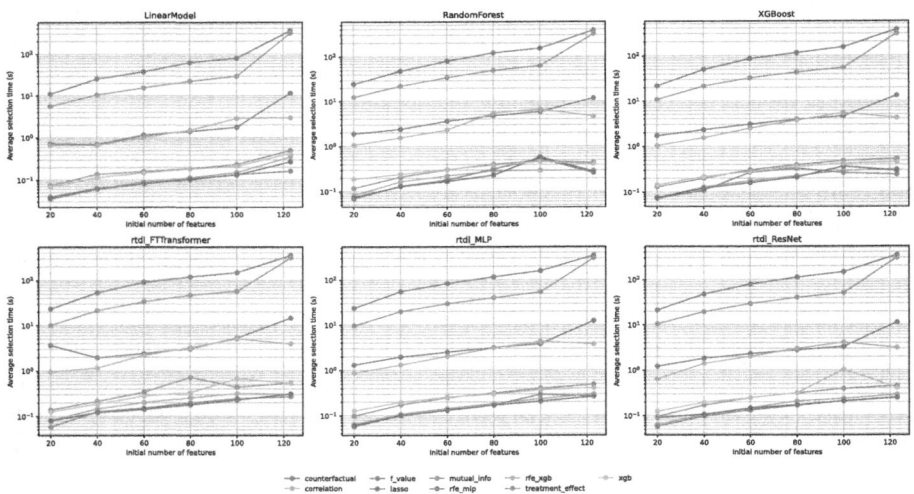

Fig. 2. Results comparing the running time of feature selection methods as a function of the initial number of features.

- Random features - generated from a random Gaussian distribution.
- Corrupted features - the original features are randomly selected and Gaussian noise is added to them.

– Second-order features - initial features are randomly selected and new features generated from them by means of squaring, addition, multiplication, etc.

In the course of the experiment, 50% of uninformative features were appended to the dataset, with the objective of investigating the performance of meta-features selection methods in terms of classification performance. The results of the experiments are presented in Fig. 3. In order to facilitate the perception of results, the mean value was calculated for each target type. This enabled the prediction of the performance of both the classical and deep models. It is evident that the quality remains largely consistent with the calculated efficiency outlined in the preceding section (Sect. 5.1). The f-value statistical test emerged as the most robust approach when confronted with variations in uninformed features. Nonetheless, the counterfactual generation method consistently yielded commendable results, consistently ranking within the top three in over half of the cases examined. The performance of the embedded methods (lasso and xgb) was found to be suboptimal in the addition of uninformative features.

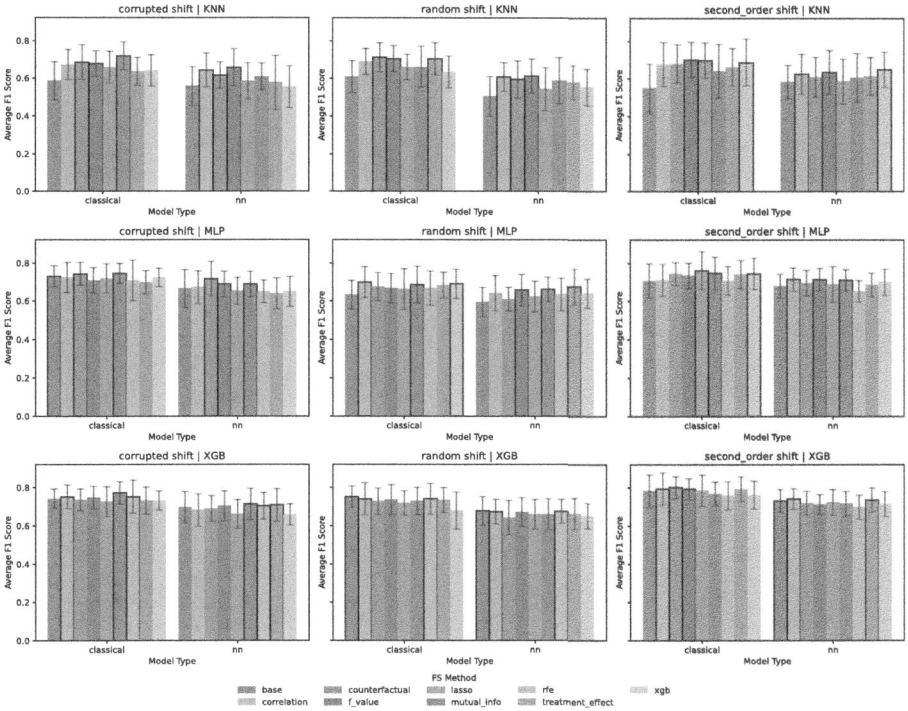

Fig. 3. Results comparing feature selection methods with different types of uninformative features, here *base* is a run on all meta-features. In each category, the top 3 methods are highlighted with a black box.

5.3   Feature Selection and Interpretability

In addition, an investigation was conducted into feature selection methods with regard to the interpretability of the results obtained. It is intuitive to assume that if features are selected, their influence can be more easily analysed at a later stage. In order to evaluate the interpretability of the results obtained after feature selection, a metric **importance fraction score** is introduced (Eq. 16). The prevailing logic in this context posited that an elevated concentration of importance among a reduced number of meta-features would result in enhanced interpretability. Consequently, the significance of each feature (I) was calculated using the SHAP method [14]. The ratio of the sum of the top five features significances to the sum of the significances of all the selected features was then determined. Consequently, if the significance of the features was "distributed" over all the selected features, this indicated a low interpretability of the selected features and resulted in low importance fraction scoring.

$$Importance\ fraction\ score = \frac{\sum_{i=1}^{k} I(s_i)}{\sum_{i=1}^{n} I(s_i)} \tag{16}$$

The results of the feature importance score comparison are displayed in the Fig. 4. It is evident that the lasso and xgboost methods are the most prominent. However, this can be readily explained by the fact that these methods typically select a minimal number of meta-features (no more than 10), resulting in a high importance fraction score. It is noteworthy that the counterfactual and correlation methods typically select a comparable number of features (Fig. 5), yet the importance fraction score of the counterfactual method is marginally higher on average. This finding suggests that the counterfactual method does indeed select causal features. The absence of feature selection naturally engenders low interpretability values of the results, which once again confirms the need to select meta-features.

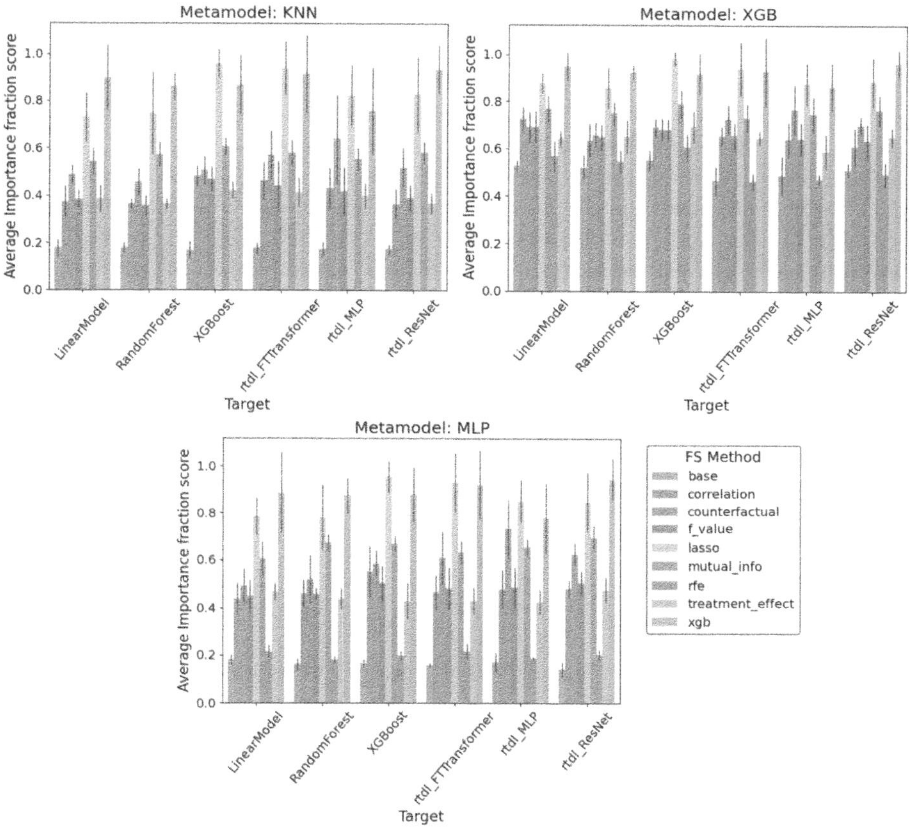

Fig. 4. Results of comparison of importance fraction score for different feature selection methods, here *base* is a run on all meta-features.

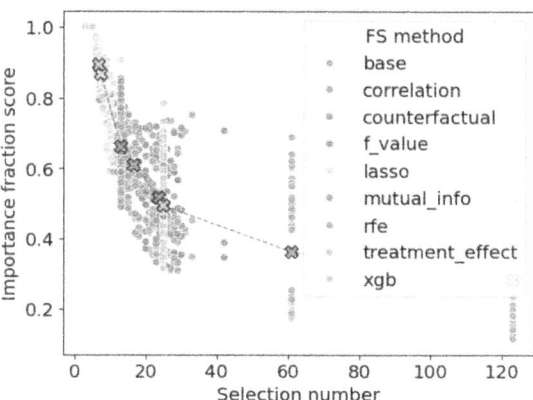

Fig. 5. Dependence of average importance fraction score on the number of selected meta-features, here the cross - feature selection method's cluster centroid.

6   Conclusion and Discussion

In this study, we examined the impact of meta-feature selection on the performance and interpretability of meta-learning models for tabular data. Our investigation addressed the fundamental question of whether it is more advantageous to select a subset of meta-features rather than relying on the entire feature set. Through extensive experimentation across various selection methodologies—including filter, wrapper, embedded, and our proposed causal-based counterfactual approach—we demonstrated that judicious feature selection can substantially enhance both predictive performance and the clarity of model interpretations.

Our experimental results consistently indicated that meta-feature selection improves model outcomes. Notably, the causal-based methods, particularly the counterfactual generation approach, emerged as a robust solution, often ranking among the top performers across multiple meta-learners. This method not only maintained high efficiency under standard conditions but also exhibited strong resilience when datasets were augmented with uninformative features. Such robustness underscores its potential for real-world applications where data distributions may shift or noise is prevalent.

From an interpretability standpoint, our findings reveal that selecting a smaller, more informative set of meta-features allows for a concentrated distribution of feature importance. Methods such as Lasso and XGBoost tended to select a minimal number of features, thereby yielding high importance fraction scores. However, the counterfactual method, while selecting a comparable number of features to correlation-based approaches, provided marginally higher interpretability. This suggests that causal-based selection not only filters out redundant information but also preserves the intrinsic causal relationships that are critical for understanding model behavior.

While the advantages of meta-feature selection are evident, our study also highlights several challenges. The performance of embedded methods was notably affected by the introduction of non-informative features, indicating a potential sensitivity to data quality. Additionally, the computational complexity of some selection methods, particularly those based on causal inference and counterfactual reasoning, warrants further investigation to ensure scalability and efficiency in larger datasets.

In conclusion, our work reinforces the importance of incorporating meta-feature selection into meta-learning pipelines, particularly for tasks involving tabular data. By carefully selecting meta-features, practitioners can achieve a more efficient and interpretable modeling process without sacrificing predictive accuracy. Future research may focus on refining causal-based selection techniques, exploring their applicability across diverse data modalities, and developing more computationally efficient algorithms. Ultimately, these efforts will contribute to the broader goal of automating and enhancing the decision-making process in machine learning model selection.

Acknowledgments. This research is financially supported by The Russian Scientific Foundation, Agreement No 24-71-10093, https://rscf.ru/en/project/24-71-10093/.

References

1. Alcobaça, E., Siqueira, F., Garcia, L.P., Rivolli, A., Oliva, J., de Carvalho, A.: MFE: towards reproducible meta-feature extraction. J. Mach. Learn. Res. **21**, 1–5 (2020)
2. Athey, S., Tibshirani, J., Wager, S.: Generalized random forests. Ann. Stat. (2019)
3. Battocchi, K., et al.: EconML: a python package for ml-based heterogeneous treatment effects estimation (2021). https://github.com/microsoft/EconML
4. Borboudakis, G., Tsamardinos, I.: Forward-backward selection with early dropping. J. Mach. Learn. Res. **20**(8), 1–39 (2019). http://jmlr.org/papers/v20/17-334.html
5. Borisov, V., Leemann, T., Seßler, K., Haug, J., Pawelczyk, M., Kasneci, G.: Deep neural networks and tabular data: a survey. IEEE Trans. Neural Netw. Learn. Syst. (2022)
6. Cherepanova, V., et al.: A performance-driven benchmark for feature selection in tabular deep learning. In: Advances in Neural Information Processing Systems, vol. 36 (2024)
7. Filchenkov, A., Pendryak, A.: Datasets meta-feature description for recommending feature selection algorithm. In: 2015 Artificial Intelligence and Natural Language and Information Extraction, Social Media and Web Search FRUCT Conference (AINL-ISMW FRUCT), pp. 11–18. IEEE (2015)
8. Garouani, M., Ahmad, A., Bouneffa, M.M.: Explaining meta-features importance in meta-learning through shapley values. In: 25th International Conference on Enterprise Information Systems (ICEIS 2023), vol. 1, pp. 591–598. SCITEPRESS-Science and Technology Publications (2023)
9. Kalousis, A., Hilario, M.: Feature selection for meta-learning. In: Pacific-Asia Conference on Knowledge Discovery and Data Mining, pp. 222–233. Springer (2001)
10. Khan, I., Zhang, X., Rehman, M., Ali, R.: A literature survey and empirical study of meta-learning for classifier selection. IEEE Access **8**, 10262–10281 (2020)
11. Lazar, C., et al.: A survey on filter techniques for feature selection in gene expression microarray analysis. IEEE/ACM Trans. Comput. Biol. Bioinf. **9**(4), 1106–1119 (2012)
12. Lemhadri, I., Ruan, F., Abraham, L., Tibshirani, R.: LassoNet: a neural network with feature sparsity. J. Mach. Learn. Res. **22**(127), 1–29 (2021)
13. Liu, D., Yue, X.: Efficient feature selection algorithm based on counterfactuals. In: 2024 6th International Conference on Communications, Information System and Computer Engineering (CISCE), pp. 541–544. IEEE (2024)
14. Lundberg, S.M., Lee, S.I.: A unified approach to interpreting model predictions. In: Guyon, I., et al. (eds.) Advances in Neural Information Processing Systems, vol. 30, pp. 4765–4774. Curran Associates, Inc. (2017). http://papers.nips.cc/paper/7062-a-unified-approach-to-interpreting-model-predictions.pdf
15. Mangal, A., Rathore, S.S.: ATE-FS: an average treatment effect-based feature selection technique for software fault prediction. ACM Trans. Intell. Syst. Technol. (2025)
16. McElfresh, D., et al.: When do neural nets outperform boosted trees on tabular data? In: Advances in Neural Information Processing Systems, vol. 36 (2024)

17. O'brien, R.M.: A caution regarding rules of thumb for variance inflation factors. Qual. Quant. **41**, 673–690 (2007)

18. Pereira, G.T., Santos, M.R.D., de Carvalho, A.C.P.D.L.F.: Evaluating meta-feature selection for the algorithm recommendation problem. arXiv preprint arXiv:2106.03954 (2021)

19. Pinto, F., Soares, C., Mendes-Moreira, J.: Towards automatic generation of metafeatures. In: Bailey, J., Khan, L., Washio, T., Dobbie, G., Huang, J.Z., Wang, R. (eds.) PAKDD 2016. LNCS (LNAI), vol. 9651, pp. 215–226. Springer, Cham (2016). https://doi.org/10.1007/978-3-319-31753-3_18

20. Rivolli, A., Garcia, L.P., Soares, C., Vanschoren, J., de Carvalho, A.C.: Meta-features for meta-learning. Knowl.-Based Syst. **240**, 108101 (2022)

21. Sağbaş, E.A.: A novel two-stage wrapper feature selection approach based on greedy search for text sentiment classification. Neurocomputing **590**, 127729 (2024)

22. Shao, X., Wang, H., Zhu, X., Xiong, F., Mu, T., Zhang, Y.: EFFECT: explainable framework for meta-learning in automatic classification algorithm selection. Inf. Sci. **622**, 211–234 (2023)

23. Shmuel, A., Glickman, O., Lazebnik, T.: A comprehensive benchmark of machine and deep learning across diverse tabular datasets. arXiv preprint arXiv:2408.14817 (2024)

24. Shwartz-Ziv, R., Armon, A.: Tabular data: deep learning is not all you need. Inf. Fusion **81**, 84–90 (2022)

25. Smith-Miles, K., Muñoz, M.A.: Instance space analysis for algorithm testing: methodology and software tools. ACM Comput. Surv. **55**(12), 1–31 (2023)

26. Vanschoren, J., van Rijn, J.N., Bischl, B., Torgo, L.: OpenML: networked science in machine learning. ACM SIGKDD Explor. Newsl. **15**(2), 49–60 (2014)

27. Xia, S., Yang, Y.: A model-free feature selection technique of feature screening and random forest-based recursive feature elimination. Int. J. Intell. Syst. **2023**(1), 2400194 (2023)

28. Xue, B., Zhang, M., Browne, W.N., Yao, X.: A survey on evolutionary computation approaches to feature selection. IEEE Trans. Evol. Comput. **20**(4), 606–626 (2015)

29. Ye, H.J., Liu, S.Y., Cai, H.R., Zhou, Q.L., Zhan, D.C.: A closer look at deep learning on tabular data. arXiv preprint arXiv:2407.00956 (2024)

30. You, D., et al.: Counterfactual explanation generation with minimal feature boundary. Inf. Sci. **625**, 342–366 (2023)

31. Yu, K., et al.: Causality-based feature selection: methods and evaluations. ACM Comput. Surv. (CSUR) **53**(5), 1–36 (2020)

Simple Error Estimation for PIES in 2D Elasticity Problems

Agnieszka Bołtuć[(✉)] [iD] and Eugeniusz Zieniuk [iD]

Faculty of Computer Science, University of Bialystok, Bialystok, Poland
{a.boltuc,e.zieniuk}@uwb.edu.pl

Abstract. The paper presents a posteriori error estimation strategy in the parametric integral equation system (PIES). It uses collocation point differences between solutions obtained numerically by PIES and another solutions interpolated based on the initial PIES analysis. Various techniques for interpolation are proposed: by repeating the interpolation omitting one collocation point, by interpolating using only values from adjacent collocation points and by the one degree higher polynomial obtained using the least squares approximation. This allows the calculation of local and global percentage error using the integral over the mentioned differences. Finally, it can be applied to ensure convergence of the solutions using the PIES method or to adaptive refinement of distribution or number of collocation points by identifying boundary regions where the error is relatively high.

Keywords: Error estimation · Parametric integral equation system (PIES) · Collocation points · Distribution refinement

1 Introduction

Parametric integral equation system (PIES) is the method for solving boundary value problems, developed by the authors as an alternative to well-known numerical approaches like the finite element method (FEM) [1–3], the boundary element method (BEM) [4–6] or so-called meshless methods [7–9]. A distinctive aspect of PIES lies in its approach to the geometry representation. Instead of relying on meshing or discretization (into various kinds of elements), the method uses a limited set of points: corner points of the polygonal geometry or key boundary points reflecting the curvilinear shape of the considered body. It is possible, because PIES is an analytical modification of the boundary integral equation (BIE) [4–6], consisting of the analytical incorporation of the shape into the formalism of the equation. For this reason, the shape can be represented in various ways, but the authors chose very effective and flexible parametric curves (e.g. Bezier), which are well-known in computer graphics [10, 11]. They allow for simple modeling using dedicated control points. The analytically incorporated shape, described using formulas representing curves, means that each shape modification is automatically reflected in the PIES formalism.

Moreover, the proposed approach enables a clear separation between shape modeling and solution approximation - two fundamental stages in solving boundary value

© The Author(s), under exclusive license to Springer Nature Switzerland AG 2026
M. H. Lees et al. (Eds.): ICCS 2025, LNCS 15904, pp. 309–322, 2025.
https://doi.org/10.1007/978-3-031-97629-2_22

problems. In FEM and BEM, they are often dependent on each other, which means more elements equal more accurate solutions. The PIES solution is approximated by a series with arbitrary basis functions (e.g. Lagrange polynomials in this paper) and forces the PIES equation to be satisfied at selected points (collocation points). Such an approach allows for the accuracy of the solutions to be influenced by changing only one parameter of the approximating series without the need to perform cumbersome re-discretization, as is the case with the methods mentioned above. Meshless methods also do not rely on traditional elements like PIES. However, despite their variations - such as the boundary node method (BNM) [12], where input data includes only boundary nodes - they still require domain partitioning (division into cells) for integration.

PIES has been successfully applied to solving various boundary value problems, starting with potential problems [13], through acoustic [14], elastic [15], and transient heat conduction [16] to elastoplastic [17]. The effectiveness of the described above way of shape modeling was then examined, and the accuracy of the obtained solutions was compared to the analytical results. The conclusions of this analysis were very satisfactory. However, an analytical solution only sometimes exists, and there may also be situations where there is no even numerical solution to compare. Therefore, developing a dedicated error estimation strategy for PIES is essential. It refers to a set of techniques designed to evaluate the accuracy of numerical solutions to boundary value problems without comparing it to any other exact results. They are crucial to reliably assess solution accuracy, guarantee convergence, and enable adaptive refinement, ensuring the method's robustness and practical applicability.

The error estimation schemes may differ between methods for solving boundary value problems [18–20]. Focusing on BEM, as it is a predecessor of PIES, they can be classified into several types: the residual type, the interpolation type, the integral equation type, the node sensitivity type and the solution difference type [21]. Since PIES aims to improve efficiency (by e.g. simplifying modeling), from available approaches was selected the one which is computationally cheaper than the others – interpolation error estimation. It has been widely used in other numerical methods [19, 20], even though the accuracy of the predicted solutions is not guaranteed. This approach generally compares the original numerical solution obtained by the applied method with the interpolated solution (from now on referred to as the predicted solution). The interpolation is made based on the initial numerical solution mentioned above. The difference between predicted and numerical solutions is estimated as the error.

This paper presents three approaches to interpolation of the predicted solution in PIES. The first consists of omitting one of the collocation points (the one where a new solution is calculated) and re-interpolating by the Lagrange polynomial based on the existing PIES solutions in the remaining collocation nodes. In the second variant, the Lagrange interpolation is performed only based on the values from the neighboring collocation nodes (one from the left and two from the right of the estimated collocation point). The last approach uses the least square approximation with a polynomial of arbitrary degree based on the PIES values in the collocation points, skipping the currently predicted one. A sequence of monomials is assumed as the basis function. The relative percentage errors, local and global, are calculated as integrals over the boundary from the obtained differences between the initial numerical solution and the predicted solution. Some examples have been used to demonstrate the behavior of the error estimators.

2 Parametric Integral Equation System (PIES) for 2D Elasticity

A parametric integral equation system (PIES) for 2D elastic problems without body forces can be presented in the following form [15]

$$
0.5u_l(\bar{s}) = \sum_{j=1}^{n} \int_{s_{j-1}}^{s_j} \left\{ \overline{U}_{lj}^*(\bar{s}, s)p_j(s) - \overline{P}_{lj}^*(\bar{s}, s)u_j(s) \right\} J_j(s)ds, \tag{1}
$$

where $s_{l-1} \leq \bar{s} \leq s_l$, $s_{j-1} \leq s \leq s_j$, $l = 1, 2, 3, \ldots, n$, $J_j(s)$ is the Jacobian of transformation to a 1D parametric reference system in which the boundary in PIES is defined and n is the number of boundary segments.

The first integrand, $\overline{U}_{lj}^*(\bar{s}, s)$, is a modified fundamental solution and, as mentioned in the introduction, takes into account in its mathematical formalism the shape of the boundary defined in a general way. For the plane strain state, it can be represented by [15]

$$
\overline{U}_{lj}^*(\bar{s}, s) = -\frac{1}{8\pi(1-v)\mu} \begin{bmatrix} (3-4v)\ln(\eta) - \frac{\eta_1^2}{\eta^2} & -\frac{\eta_1\eta_2}{\eta^2} \\ -\frac{\eta_1\eta_2}{\eta^2} & (3-4v)\ln(\eta) - \frac{\eta_2^2}{\eta^2} \end{bmatrix}, \tag{2}
$$

where $\eta = \left[\eta_1^2 + \eta_2^2\right]^{0.5}$, $\eta_1 = \Gamma_j^{(1)}(s) - \Gamma_l^{(1)}(\bar{s})$, $\eta_2 = \Gamma_j^{(2)}(s) - \Gamma_l^{(2)}(\bar{s})$, and v is Poisson's ratio and μ is a shear modulus. The parametric function $\Gamma(s)$ can be represented by various curves known from computer graphics [10, 11].

The following expression $\overline{P}_{lj}^*(\bar{s}, s)$ represents the second kernel [15]

$$
\overline{P}_{lj}^*(\bar{s}, s) = -\frac{1}{4\pi(1-v)\eta} \begin{bmatrix} P_{11} & P_{12} \\ P_{21} & P_{22} \end{bmatrix}, \quad l, j = 1, 2, \ldots n, \tag{3}
$$

where

$$
P_{11} = \left\{ (1-2v) + 2\frac{\eta_1^2}{\eta^2} \right\} \frac{\partial\eta}{\partial n}, \quad P_{22} = \left\{ (1-2v) + 2\frac{\eta_2^2}{\eta^2} \right\} \frac{\partial\eta}{\partial n},
$$

$$
P_{21} = P_{12} = \left\{ 2\frac{\eta_1\eta_2}{\eta^2} \frac{\partial\eta}{\partial n} - (1-2v)\left[\frac{\eta_1}{\eta}n_2(s) + \frac{\eta_2}{\eta}n_1(s) \right] \right\},
$$

$$
\frac{\partial\eta}{\partial n} = \frac{\partial\eta_1}{\partial\eta}n_1(s) + \frac{\partial\eta_2}{\partial\eta}n_2(s),
$$

and $n_1(s)$ and $n_2(s)$ are direction cosines of the external normal to jth segment of the boundary.

Functions $u_j(s)$ and $p_j(s)$ are parametric boundary functions. They are known or searched depending on individual segments' boundary conditions. They can be approximated using series with arbitrary basis functions

$$
u_j(s) = \sum_{k=0}^{M-1} u_j^{(k)} L_j^{(k)}(s), \quad p_j(s) = \sum_{k=0}^{M-1} p_j^{(k)} L_j^{(k)}(s), \quad j = 1, \ldots, n, \tag{4}
$$

where $\boldsymbol{u}_j^{(k)}$, $\boldsymbol{p}_j^{(k)}$ are unknown coefficients, M is the number of coefficients on segment j and $L_j^{(k)}(s)$ is the Lagrange polynomial on segment j. Various polynomials can be used as basis functions, like Chebyshev, Legendre, etc., but the Lagrange polynomials are applied in this paper.

Equation (1) is then written for all collocation points and takes the form of the equation system with $2 \times n \times M$ equations. After solving it, the unknown coefficients from (4) are obtained. In the case of Lagrange polynomials, they are solutions at collocation points at the same time.

The accuracy of PIES solutions depends on two main factors: the number M in (4) and the arrangement of collocation points on each boundary segment. Various approaches have been tested over time, but two are mostly used: uniform distribution and at places corresponding to the roots of Chebyshev polynomials of the first kind (degree M). Since the recursive formula by which the roots of the Chebyshev polynomial are generated is known [22], the arrangement of collocation points at the locations of these roots is automatic. The examination of the convergence technically comes down to the choice of just two parameters: the number M and the way of arranging the collocation points.

3   Boundary Modeling by Curves

The main advantage of PIES is that the approximation of the solutions and the shape are separated. It comes from the fact that the boundary is modeled using parametric curves known from computer graphics [10, 11]. Instead of classical discretization used in BEM or FEM, the whole segments are created by a single curve. Moreover, they are incorporated into the PIES formula analytically, so each change in the shape of the curve involves an automatic modification of the PIES formalism. The curves are easily modified by changing only their control points. This process is incomparably more effective than re-discretization in the so-called element methods. Figure 1 presents how the boundary is modeled depending on whether the segment is straight or curved.

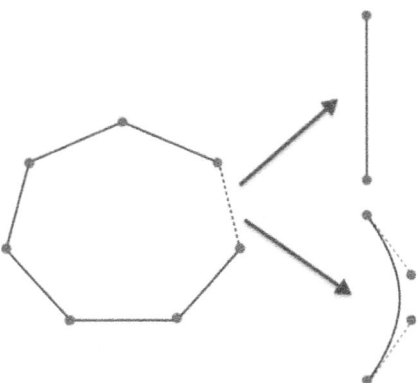

Fig. 1. Modeling of the boundary in PIES.

As shown in Fig. 1, the boundary is modeled using curves. This paper uses Bezier curves of various degrees depending on the required shape. Thus, the polygonal shapes are modeled by curves of the first degree, defined by only two control points, while curved shapes by the curves of the third degree using four control points. Higher degrees are unnecessary because cubic curves can model all the necessary shapes and are not too computationally expensive.

Modification of such defined geometry is also straightforward. It is enough to change the position of small number of control points to modify the shape significantly. An example is shown in Fig. 2, where three control points of the curved geometry change its shape.

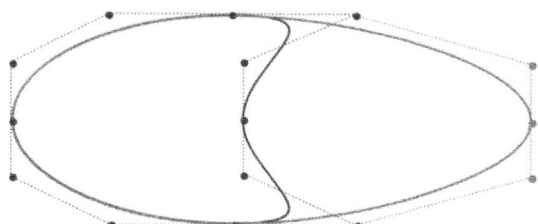

Fig. 2. Modification of the boundary in PIES.

The above-described incorporation of curves into the PIES formalism allowed for separating the shape approximation from the solution approximation, which in PIES bases on changing the parameters of the approximating series (4).

4 Error Estimation

The exact error of the boundary solutions is calculated as

$$e_u = u - u_n, \quad e_p = p - p_n, \tag{5}$$

where u, p are the exact solutions and u_n, p_n are the numerical displacements and tractions obtained by PIES. It is known that analytical solutions only exist for some engineering problems. Therefore, its prediction should be used instead. For this reason, formula (5) takes the following form

$$\breve{e}_u = \breve{u} - u_n, \quad \breve{e}_p = \breve{p} - p_n, \tag{6}$$

where $\breve{u}, \breve{p}$ are the predicted approximations of displacements and tractions.

The predicted approximations $\breve{u}, \breve{p}$ can be obtained by interpolating the PIES original numerical solution using various-degree polynomials. The original PIES solution is just a solution of the system of equations at all collocation points. Therefore, the first approach (in the following chapters called approach 1) determines the new prediction at the particular collocation point by re-interpolating based on the original values, omitting the considered node. As shown in Fig. 3, if the new value is predicted for node 2, the

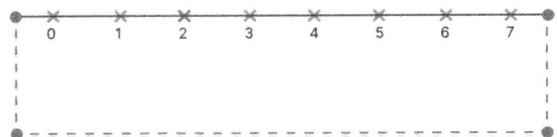

Fig. 3. The boundary segment with collocation nodes - diagram for approach 1.

interpolation is performed using the remaining nodes (without 2). The same interpolation approach (Lagrange polynomials) is applied as in the PIES method.

The second approach (approach 2) concerns interpolation using only the nearest neighbors. To have the second-degree polynomial, three neighbors are considered - one to the left of the considered node and two to the right. For example, in Fig. 4, to predict the value of node 2, values from nodes 1, 3, 4 should be interpolated. There are some extreme cases: for the right-most node and the penultimate node, where an appropriate number of nodes from the left are taken into account to preserve the degree of the polynomial. The left-most node takes into account only nodes from its right.

Fig. 4. The boundary segment with collocation nodes - diagram for approach 2.

Both presented above ideas consider the interpolation of the numerical solution by a lower degree of interpolation function than in the initial PIES analysis. Therefore, the third approach (approach 3) is also proposed. It uses least squares approximation [22]. To obtain the interpolation polynomial

$$Q_m(x) = a_0 + a_1 x + \cdots + a_m x^m, \tag{7}$$

one must choose its degree (m) and then find the coefficients ($a_0, a_1, \ldots, a_m$). They are chosen so that the value of the squared deviation between the polynomial values ($Q_m(x_i)$) on the set of collocation points and the PIES initial results ($f(x_i)$) is as small as possible

$$S = \sum_{i=1}^{n} [Q_m(x_i) - f(x_i)]. \tag{8}$$

By finding partial derivatives (8) concerning all coefficients of the polynomial (7) and equating them to zero, a system of m + 1 equations is obtained, based on which the required coefficients of the polynomial (7) are determined. Then (7) can be used to predict new values at collocation points. This time, like in the first approach, values from all collocation nodes are used except the considered one (Fig. 3), and the polynomial degree for predicting is one bigger than in the initial PIES analysis.

Errors (6) are estimated collocation points errors. To calculate the error for the boundary segment i, the following equations can be used

$$\|e_u\|_i = \int_{\Gamma_i} (\tilde{u} - u_n) d\Gamma, \quad \|e_p\|_i = \int_{\Gamma_i} (\tilde{p} - p_n) d\Gamma. \tag{9}$$

The global error for the whole boundary is expressed by

$$\|e_u\| = \int_\Gamma (\breve{u} - u_n)d\Gamma, \ \|e_p\| = \int_\Gamma (\breve{p} - p_n)d\Gamma. \tag{10}$$

The global relative percentage error η is then written as

$$\eta_u = \frac{\|e_u\|}{\|u_n\|} \times 100\%, \ \eta_p = \frac{\|e_p\|}{\|p_n\|} \times 100\%. \tag{11}$$

5 Numerical Examples

5.1 Initial Verification of Proposed Approaches for Global Error Estimation in Comparison to Analytical Solutions

The first example concerns a cylinder subjected to an internal pressure $p = 22.5\,\text{MPa}$ under plane strain conditions. Young's modulus $E = 21000\,\text{MPa}$ and Poisson's ratio $v = 0.3$ are the elastic material properties. Figure 5 presents the quarter of the cylinder because of the symmetry.

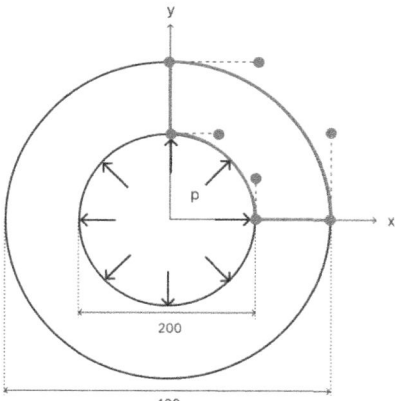

Fig. 5. The quarter of the cylinder under internal pressure.

As can be seen, the geometry is modeled using four curves: two of the first degree and two of the third degree. Only eight nodes are defined (●).

As mentioned earlier, the accuracy of solutions depends on the arrangement and the number of collocation points. The two most tested variants are: uniform and at places corresponding to the roots of Chebyshev polynomials. As the interpolation is performed on collocation nodes, the uniform variant is excluded from the tests, because at high degrees of the interpolation polynomial we can expect distortions known as Runge's phenomenon [23].

Global relative percentage errors for the whole boundary (11) for solutions u_1, p_1 are calculated. Figure 6 presents values obtained for 4–20 collocation nodes at each boundary segment, and three approaches are used to predict new values at existing nodes (Sect. 4).

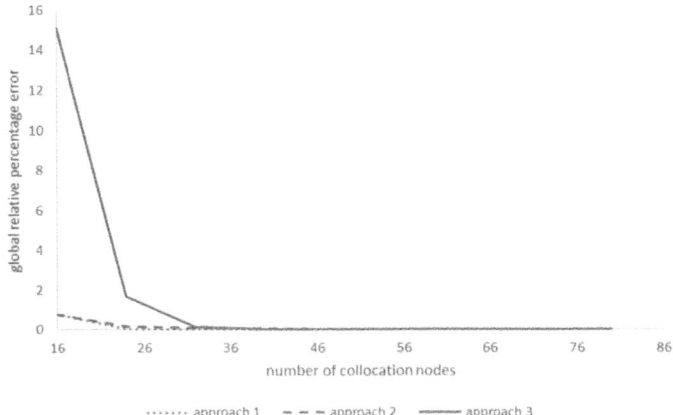

Fig. 6. Global relative percentage error η_{u_1}.

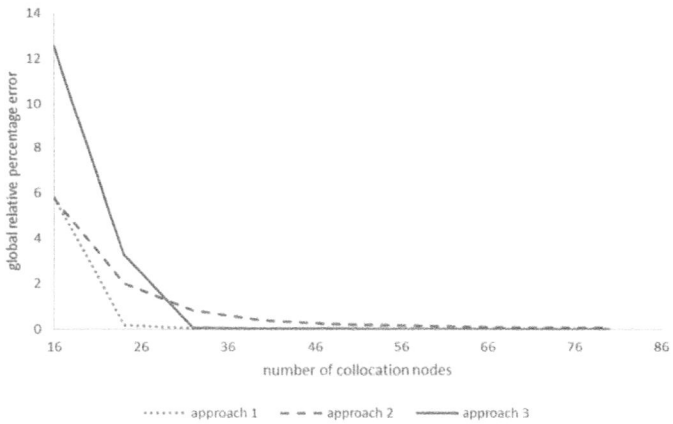

Fig. 7. Global relative percentage error η_{p_1}.

As shown in Fig. 6 and 7, global relative percentage errors calculated by approaches 1,2 and 3 have similar trends for both u_1 and p_1. Starting from higher values, they stabilize when using 8–10 collocation points per segment (32–40 in total). With 20 (80 in total) collocation points used, of the three considered estimation approaches, the second method (interpolation using only nearest neighbors, always of degree 2) had the most significant error. The other two are characterized by the similar level of error.

The analytical results [24] confirm the trend in Fig. 6 and 7. Displacements u_1 were calculated at the lower linear boundary of the quarter. The average relative error decreased from 0.1035% (16 collocation nodes in total) to 0.0277% (80 collocation nodes in total), but as shown in Fig. 6 the almost final stable level is reached at 32 collocation points (0.0278%).

5.2   Proposed Approaches in Applications for Global and Local Error Estimating

In the second example, the elastic plate with the circular hole subjected to the tensile load p at its ends is considered (Fig. 8). Plane stress conditions are assumed with the following material properties $E = 1 \times 10^6$ MPa and $v = 0.3$. Due to the symmetry, only the upper right square quadrant is analyzed.

The boundary is modeled in PIES using five curves: four of the first and one of the third degrees. It requires posing seven nodes.

Once again, the error estimation is performed using various numbers of collocation points and multiple approaches for solution interpolation.

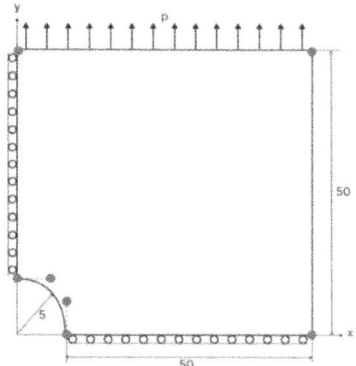

Fig. 8. The quarter of the plate with the circular hole.

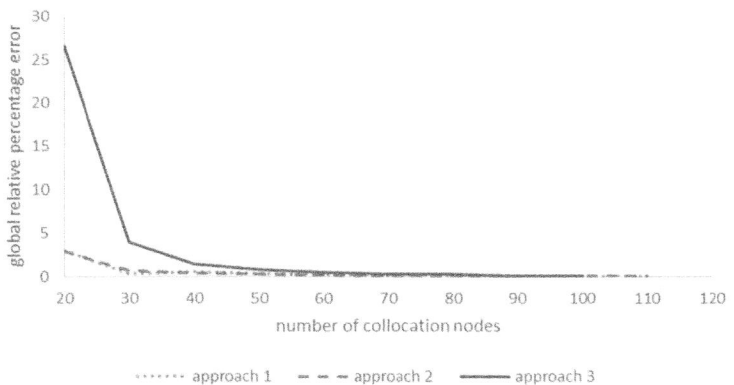

Fig. 9. Global relative percentage error η_{u_1}.

As shown in Fig. 9, error η_{u_1} for all approaches has the same trend and similar values. Once again, stabilization is visible with about 50 colocation points, where all error values are under 1%. The most significant value of η_{u_1} for 110 points is obtained by approach 2, which confirms the results from the previous example.

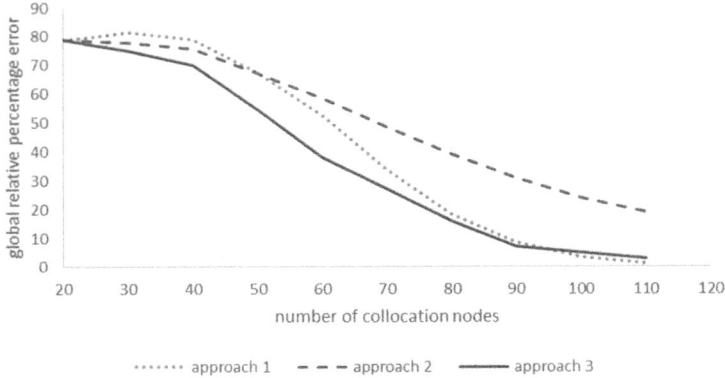

Fig. 10. Global relative percentage error η_{p_1}.

Analyzing η_{p_1} (Fig. 10) shows that all approaches start with a similar high error (~79%). With the highest number of collocation points, approaches 1 and 3 have their error reduced to 1–3%, while approach 2 remained at a high level of 19% error. It is still the approach with the most significant estimated error value.

Finally, analytical results [24] are also analyzed. Solution p_2 at the bottom linear boundary of the plate is calculated. The comparison of the exact with the numerical results obtained by PIES with 20 and 110 collocation nodes is presented in Fig. 11. As can be seen, there is a very significant improvement in the accuracy of the solutions with the increased number of collocation nodes. It should be emphasized that the numerical solutions for 70 collocation points already have an error close to that achieved with 110 points.

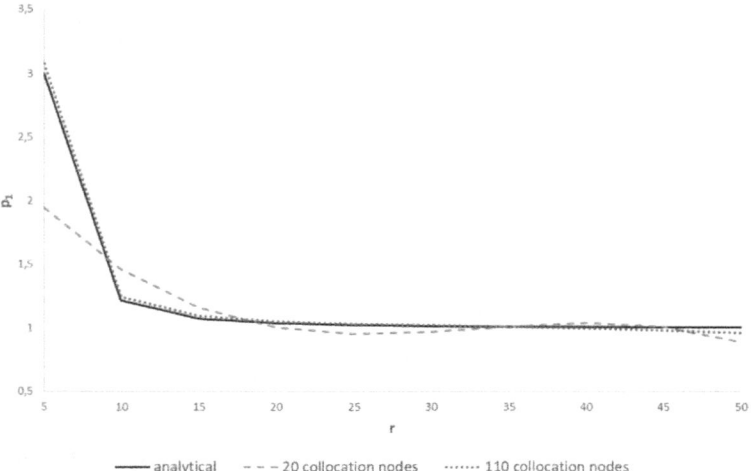

Fig. 11. Comparison of analytical and PIES results for various number of collocation nodes.

Analyzing errors more locally, i.e., on individual segments, it can be seen that the most significant relative percentage error was obtained for p_2 on the left vertical side and p_1 on the bottom side of the plate. It amounted to about 3% and 1% respectively. The highest error at the collocation point on the vertical segment is at the first point from the hole. The difference expressed by formula (6) is 0.012906. At the bottom side it is again the closest collocation node to the hole, with a difference equal to 0.0468041. The calculations described in this paragraph were performed using approach 3. Having the sides and points with the highest errors, it is possible to densify the collocation points to improve the accuracy of the obtained results.

5.3 Global and Local Error Estimating on Example with Numerical Solutions Only

The last example concerns the polygonal L-shaped plate under uniform tension and plane stress conditions. The boundary conditions and dimensions are presented in Fig. 12. The assumed material properties are $E = 1 \times 10^5 \, \text{N}/\text{cm}^2$ and $v = 0.3$.

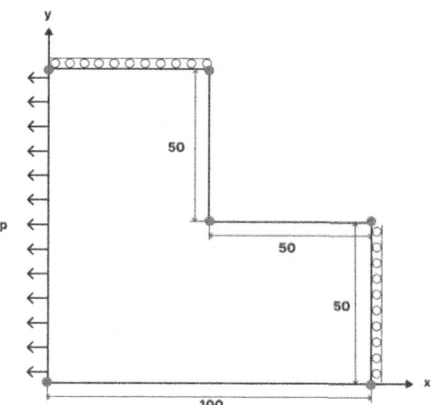

Fig. 12. The L-shaped plate under the uniform tension.

The considered polygonal boundary is modeled using six curves of the first degree by six corner points.

Global relative percentage errors for u_1 and p_1 are calculated by taking various numbers of collocation nodes and applying multiple approaches proposed in the paper. The results are presented in Tables 1 and 2.

As can be seen from Tables 1 and 2, each approach generates a similar trend. For both η_{u_1} and η_{p_1}, it can be observed that since using 48 points, estimated errors are less than 1% for all proposed approaches Values obtained at the maximum number of collocation points are close to zero and similar in each tested approach.

These conclusions are confirmed by analyzing numerical solutions obtained for different numbers of collocation points. Such a test was carried out due to the lack of

Table 1. Global relative percentage error η_{u_1}.

no of collocation points	η_{u_1}		
	approach 1	approach 2	approach 3
24	4.4496	4.4496	27.304
36	0.8971	1.0572	3.9844
48	0.2671	0.5383	0.3969
60	0.1201	0.3124	0.6612
72	0.0718	0.1955	0.1376
84	0.0481	0.1301	0.1687
96	0.0353	0.0902	0.0915
108	0.0269	0.066	0.0433
120	0.0211	0.0496	0.0469

Table 2. Global relative percentage error η_{p_1}.

no of collocation points	η_{p_1}		
	approach 1	approach 2	approach 3
24	2.5624	2.5624	38.2614
36	0.7162	0.2973	7.9506
48	0.0486	0.1059	0.1371
60	0.0768	0.0546	0.1236
72	0.0297	0.0313	0.0188
84	0.0262	0.0201	0.0219
96	0.0194	0.0139	0.0197
108	0.0156	0.0101	0.0104
120	0.0125	0.0074	0.0093

analytical solutions for this example. Tractions p_1 at the right boundary of the L-shaped plate are compared for 24, 48, 120, and 180 collocation points and presented in Table 3.

As can be seen, solutions obtained for 48 collocation nodes are very similar to those for 120 and, finally, 180 points. Therefore, if we assume that the solution for the most significant number of collocation points is exact, the obtained global relative percentage errors reflect the results shown in Table 3.

The boundary relative percentage errors are calculated to use the proposed estimation approach to solutions' adaptive refinement. The approach 3 with 96 collocation points is used. The most significant errors were obtained for the sides at the concave corner and were 0.1783% (u_1) and 0.3086% (u_2) for the horizontal and vertical edges, respectively.

Table 3. Values of p_1 at the right edge of the plate.

y	no of collocation points			
	24	48	120	180
5	13.5781	13.1634	13.2307	13.2247
10	58.4243	60.4602	60.4068	60.4028
15	104.528	107.376	107.288	107.285
20	151.543	153.99	153.977	153.976
25	199.125	200.539	200.575	200.576
30	246.927	247.102	247.118	247.119
35	294.603	293.545	293.541	293.544
40	341.809	339.659	339.719	339.726
45	388.198	385.421	385.549	385.558

This may mean that the density of collocation points on these sides improves the final accuracy. Such a process can be done automatically.

6 Conclusions

The paper presents the approach to error estimating in PIES for 2D elastic problems using various techniques for obtaining predicted solutions. The first performs interpolation at all collocation nodes with received numerical solutions besides the one where the expected value is searched. The second one carries out interpolation more locally, based on three neighboring collocation nodes. The last one uses the approximation polynomial of a degree one greater than the number of collocation points (without the node at which the value is predicted) obtained by the least square approximation.

The proposed approaches are applied to three elasticity examples. Both the global and local (boundary) relative percentage errors are calculated for different numbers of collocation points using three techniques to obtain predicted solutions. In general, all proposed approaches gave similar error trends. Approach 3 always starts with the most significant error value, while approach 2 usually ends with it. The errors decreased with the increasing number of collocation points, so in most cases they fell to 1–3% with a total number of points of 80–120, depending on the example. Moreover, the local analysis is also reliable because the analyzed approaches returned the most significant errors on those boundaries that are actually characterized by the most considerable variability of solutions and thus require an increase in the number and change in the arrangement of collocation points.

The next step in research carried out by the authors should be the implementation of adaptive refinement of the number and distribution of collocation points based on the error estimation in PIES. Combined with optimization algorithms, it can be an excellent tool to improve the accuracy of the obtained PIES solutions.

Disclosure of Interests. The authors have no competing interests to declare that are relevant to the content of this article.

References

1. Ameen, M.: Computational elasticity. Alpha Science International Ltd., Harrow (2005)
2. Zienkiewicz, O.C.: The Finite Element Methods. McGraw-Hill, London (1977)
3. Liu, G.R., Quek, S.S.: The Finite Element Method: A Practical Course. Butterworth Heinemann, Oxford (2003)
4. Gao, X.W., Davies, T.G.: Boundary Element Programming in Mechanics. Cambridge University Press, Cambridge (2002)
5. Becker, A.A.: The Boundary Element Method in Engineering. A Complete Course. McGraw-Hill, Cambridge (1992)
6. Aliabadi, M.H.: The Boundary Element Method, vol. 2. Applications in Solids and Structures. Wiley, Chichester (2002)
7. Liu, G.R., Gu, Y.T.: Meshfree Methods: Moving Beyond the Finite Element Method. CRC Press, Boca Raton (2009)
8. Chen, Y., Lee, J.D., Eskandarian, A.: Meshless Methods in Solid Mechanics. Springer, Cham (2006)
9. Jaworska, I.: Multipoint meshless FD schemes applied to nonlinear and multiscale analysis. In: Groen, D., de Mulatier, C., Paszynski, M., Krzhizhanovskaya, V.V., Dongarra, J.J., Sloot, P.M.A. (eds.) Computational Science, ICCS 2022. Lecture Notes in Computer Science, vol. 13353. Springer, Cham (2022)
10. Farin, G.: Curves and Surfaces for CAGD: A Practical Guide. Morgan Kaufmann Publishers, San Francisco (2002)
11. Salomon, D.: Curves and Surfaces for Computer Graphics. Springer, Cham (2006)
12. Chati, M.K., Mukherjee, S.: The boundary node method for three-dimensional problems in potential theory. Int. J. Numer. Meth. Eng. **47**, 1523–1547 (2000)
13. Zieniuk, E.: Bézier curves in the modification of boundary integral equations (BIE) for potential boundary-values problems. Int. J. Solids Struct. **40**(9), 2301–2320 (2003)
14. Zieniuk, E., Boltuc, A.: Bézier curves in the modeling of boundary geometry for 2D boundary problems defined by Helmholtz equation. J. Comput. Acoust. **14**(3), 353–367 (2006)
15. Zieniuk, E., Boltuc, A.: Non-element method of solving 2D boundary problems defined on polygonal domains modeled by Navier equation. Int. J. Solids Struct. **43**(25–26), 7939–7958 (2006)
16. Zieniuk, E., Sawicki, D., Bołtuć, A.: Parametric integral equations systems in 2D transient heat conduction analysis. Int. Commun. Heat Mass Transfer **78**, 571–587 (2014)
17. Bołtuć, A.: Parametric integral equation system (PIES) for 2D elastoplastic analysis. Eng. Anal. Boundary Elem. **69**, 21–31 (2016)
18. Zienkiewicz, O.C., Zhu, J.Z.: A simple error estimator and adaptive procedure for practical engineering analysis. Int. J. Numer. Meth. Eng. **24**(2), 337–357 (1987)
19. Zhao, Z.: Error estimation in adaptive BEM by postprocessing interpolation. Commun. Numer. Methods Eng. **14**, 633–645 (1998)
20. Ebrahimnejad, M., Fallah, N., Khoei, A.R.: Adaptive refinement in the meshless finite volume method for elasticity problems. Comput. Math. Appl. **69**, 1420–1443 (2015)
21. Kita, E., Kamiya, N.: Error estimation and adaptive mesh refinement in boundary element method, an overview. Eng. Anal. Boundary Elem. **25**, 479–495 (2001)
22. Ralston, A., Rabinowitz, P.: A First Course in Numerical Analysis, 2nd edn. Dover Publications, New York (2001)
23. Atkinson, K.E.: An Introduction to Numerical Analysis. Wiley, Hoboken (1988)
24. Timoshenko, S.P., Goodier, J.N.: Theory of Elasticity. McGraw-Hill, Tokyo (1970)

Author Index

M. H. Lees et al. (Eds.): ICCS 2025, LNCS 15904, pp. 323–324, 2025.
https://doi.org/10.1007/978-3-031-97629-2

The manufacturer's authorised representative in the EU is Springer
Nature Customer Service Centre GmbH, Europaplatz 3, 69115 Heidelberg,
Germany. If you have any concerns regarding our products, please
contact ProductSafety@springernature.com

Printed and bound by CPI Group (UK) Ltd, Croydon, CR0 4YY

28/04/2026

02098522-0003